Hydroclimatology: Principles and Applications

Hydroclimatology: Principles and Applications

Editor: Adrian Wolf

R CALLISTO REFERENCE

www.callistoreference.com

Callisto Reference,
118-35 Queens Blvd., Suite 400,
Forest Hills, NY 11375, USA

Visit us on the World Wide Web at:
www.callistoreference.com

ISBN: 978-1-64116-057-5 (Hardback)

Trademark Notice: Registered trademark of products or corporate names are used only for explanation and identification without intent to infringe.

Cataloging-in-Publication Data

Hydroclimatology : principles and applications / edited by Adrian Wolf.
    p. cm.
Includes bibliographical references and index.
ISBN 978-1-64116-057-5
1. Hydrometeorology. 2. Climatology. 3. Hydrologic cycle. I. Wolf, Adrian.
GB2805 .H93 2019
551.57--dc23

Table of Contents

Permissions

List of Contributors

Index

Preface

The world is advancing at a fast pace like never before. Therefore, the need is to keep up with the latest developments. This book was an idea that came to fruition when the specialists in the area realized the need to coordinate together and document essential themes in the subject. That's when I was requested to be the editor. Editing this book has been an honour as it brings together diverse authors researching on different streams of the field. The book collates essential materials contributed by veterans in the area which can be utilized by students and researchers alike.

The field of study concerned with the changes in the hydrological cycle, due to climate systems is known as hydroclimatology. It focuses on developing a systematic structure in order to interpret such changes with respect to space and time on a global as well as local scale. Hydroclimatology integrates the principles of climate science and hydrology in order to better understand the natural phenomena of floods, global warming and droughts. This book brings forth some of the most innovative concepts and unexplored aspects of this field. It studies, analyses and upholds the pillars of hydroclimatology and its utmost significance in modern times. It aims to serve as a resource guide for students, experts, researchers, academicians, climatologists and hydrologists.

Each chapter is a sole-standing publication that reflects each author's interpretation. Thus, the book displays a multi-facetted picture of our current understanding of application, resources and aspects of the field. I would like to thank the contributors of this book and my family for their endless support.

Editor

Evaluation of TRMM 3B42 V7 Rainfall Product over the Oum Er Rbia Watershed in Morocco

Hamza Ouatiki [1,*], Abdelghani Boudhar [1,*], Yves Tramblay [2], Lionel Jarlan [3], Tarik Benabdelouhab [4], Lahoucine Hanich [5], M. Rachid El Meslouhi [6] and Abdelghani Chehbouni [3]

[1]   Faculté des Sciences et Techniques, Université Sultan Moulay Slimane, B.P. 523, Béni-Mellal 23030, Maroc
[2]   IRD-HydroSciences, Montpellier 34090, France; yves.tramblay@ird.fr
[3]   CESBIO (Université de Toulouse, CNRS, CNES, IRD), 18 Av. Edouard Belin BPI 280,
      Toulouse 31401 CEDEX 9, France; lionel.jarlan@cesbio.cnes.fr (L.J.); Ghani.Chehbouni@ird.fr (A.C.)
[4]   Institut National de la Recherche Agronomique, B.P 415 R.P, Rabat 10000, Maroc;
      tarik.benabdelouahab@gmail.com
[5]   Faculté des Sciences et Techniques, Université Cadi ayyad, B.P 549, Marrakech 40000, Maroc;
      l.hanich@uca.ma
[6]   Agence du bassin Hydraulique d'Oum Er Rabia, B.P 511, Béni Mellal 23000, Maroc; meslouhi@gmail.com
*   Correspondence: hamza.ouatiki@gmail.com (H.O.); ab.boudhar@usms.ma (A.B.)

Academic Editor: Zhong Liu

Abstract: In arid and semi-arid areas, rainfall is often characterized by a strong spatial and temporal variability. These environmental factors, combined with the sparsity of the measurement networks in developing countries, constitute real constraints for water resources management. In recent years, several spatial rainfall measurement sources have become available, such as TRMM data (Tropical Rainfall Measurement Mission). In this study, the TRMM 3B42 Version 7 product was evaluated using rain gauges measurements from 19 stations in the Oum-Er-Bia (OER) basin located in the center of Morocco. The relevance of the TRMM product was tested by direct comparison with observations at different time scales (daily, monthly, and annual) between 1998 and 2010. Results show that the satellite product provides poor estimations of rainfall at the daily time scale giving an average Pearson correlation coefficient (r) of 0.2 and average Root Mean Square Error (RMSE) of 10 mm. However, the accuracy of TRMM rainfall is improved when temporally averaged to monthly time scale (r of 0.8 and RMSE of 28 mm) or annual time scale (r of 0.71 and RMSE of 157 mm). Moreover, improved correlation with observed data was obtained for data spatially averaged at the watershed scale. Therefore, at the monthly and annual time scales, TRMM data can be a useful source of rainfall data for water resources monitoring and management in ungauged basins in semi-arid regions.

Keywords: evaluation; rainfall; remote sensing; TRMM; semi-arid; Oum Er Rbia; Morocco

1. Introduction

The quantification of water supply is a major concern worldwide, and particularly in regions characterized by a semi-arid to arid climate. These regions frequently face problems of water resources management in an environment of scarcity and droughts [1,2]. Certainly, the availability of water in these areas is one of the main factors influencing socio-economic development, especially when these regions are based largely on revenues of the agriculture [3–5]. However, the constraints in such context marked by precipitations with a high spatial heterogeneity and a strong irregularity in time [6,7] pose great challenges for any attempt to accurately measure this resource [8–10].

The rain gauges are the main instruments to provide basic measures of precipitation in a watershed. Each individual rain gauge provides a continuous measurement of rainfall over time, but these measures can only be representative of relatively small areas. The WMO (World Meteorological Organization) in 1994 indicated that in mountainous regions a rain gauge is required every 250 km^2 in order to have an optimal coverage [11]. Nevertheless, the sparsity of the networks in developing countries, particularly in mountainous areas, may strongly limit the estimation of these intakes on the watershed.

In recent years, remote sensing data have been widely used in the climate field, particularly for estimating precipitations [12–15]. Several precipitation data sets derived from various remote sensing data have been released. They provide estimation of rainfall and allow the spatiotemporal monitoring of rainfall variability at a global scale [16]. On the one hand, there are models that provide rainfall estimates using infrared satellite imagery, such as PERSIANN (Precipitation Estimation from Remote Sensing Information using Artificial Neural Network, [17]), algorithms, and techniques like MIRAA (Microwave/Infrared Rain Rate Algorithm, [18]) and CMORPH (CPC MORPHing technique, [19]) which estimates rainfall based on passive microwave and infrared satellite data. On the other hand, there is TRMM (Tropical Rainfall Measurement Mission), a space mission that provides several products of rain estimates from a combination of passive microwave, visible/infrared, and rainfall radar data. Furthermore, these products can be an important source of rainfall data at the regional and local scales due to their global availability and high measurement frequencies. Combining rainfall data derived from these missions with punctual data measured by rain gauges can greatly improve the accuracy of rainfall estimations at the catchment scale, overcoming the challenges related to data availability [20,21].

Over the years, TRMM, which was primarily designed to monitor and study tropical rainfall, has proved to be an important source of information in many application fields, such as monitoring the global hydrological cycle [22–24], floods [25–27], and drought [28–30]. Indeed, the evaluation of TRMM data against rain gauges has been conducted in various previous studies around the world. Some studies have evaluated the rainfall estimates from TRMM in parallel with other satellite products [31–34]. Al Mazroui [35] and Mantas et al. [36] have focused their work on the TRMM product by evaluating its accuracy on different time steps and in different geographical and topographical contexts. In addition to the direct comparison against rain gauges, Collischonn et al. [37] provided an assessment based on hydrological modeling at a daily time step. Islam and Uyeda [38] compared daily rainfall from TRMM 3B42 to rain-gauge measurements over Bangladesh. Huang et al. [39] evaluated the TMPA V7 products with a relatively dense rain gauge network in Beijing and adjacent regions for an extreme precipitation event. In Morocco, only a few studies have been undertaken with the TRMM products. Milewski et al. [40] have evaluated four TMPA 3B42 products (V6, V7 temporary, V7, reel time V7) against rain gauges data depending on climatic zones and topography. The study covered a large part of the Moroccan territory, but the analysis was focused only for the average annual precipitation. The work done by Tramblay [41] over the Oued El Makhazine basin in northern Morocco shows the potential of these data in hydrological modeling.

The objective of this study is to evaluate the robustness and the accuracy of 3B42 V7 TRMM (Tropical Rainfall Measuring Mission) product to estimate rainfall over the Oum Er-Rbia (OER) basin in the center of Morocco. We evaluate the 3B42 V7 rainfall estimates at three different time scales (daily, monthly, and annually). Furthermore, we compared the product and rain gauges data based on a zonal scale, considering the average rainfall from rain gauges and TRMM pixels within the same sub watershed.

2. Materials and Methods

2.1. Study Area

The OER watershed (Figure 1), one of the biggest watersheds of the Kingdom of Morocco, has an area of nearly 35,000 km^2, covering 7% of the total area of Morocco. It lies between 31°15′N and 33°22′N for latitudes and between 5°00′W and 9°20′W for longitudes. It is characterized by a marked topography with elevations that vary between 0 m above sea level (a.s.l.) at its outflow and 3890 m a.s.l. in the Atlas Mountains located in the eastern part of the watershed. The Atlas chain ensures the main water supply to compensate the demand for potable water and economic activities, particularly the agricultural sector in the plain. There are four main sub-watershed located in the upstream part of Oum Er-Rbia which are Tassaout-Lakhdar, Oued El Abid, Central Oum Er-Rbia, and the Upper Oum Er-Rbia.

In this study, we have focused on the upstream part of the OER watershed where most rain gauges are available and where several dams and reservoirs are located, including the Bin El Ouidane dam that is the main hydropower unit in Morocco, which produces about two thirds of the hydraulic energy of the country.

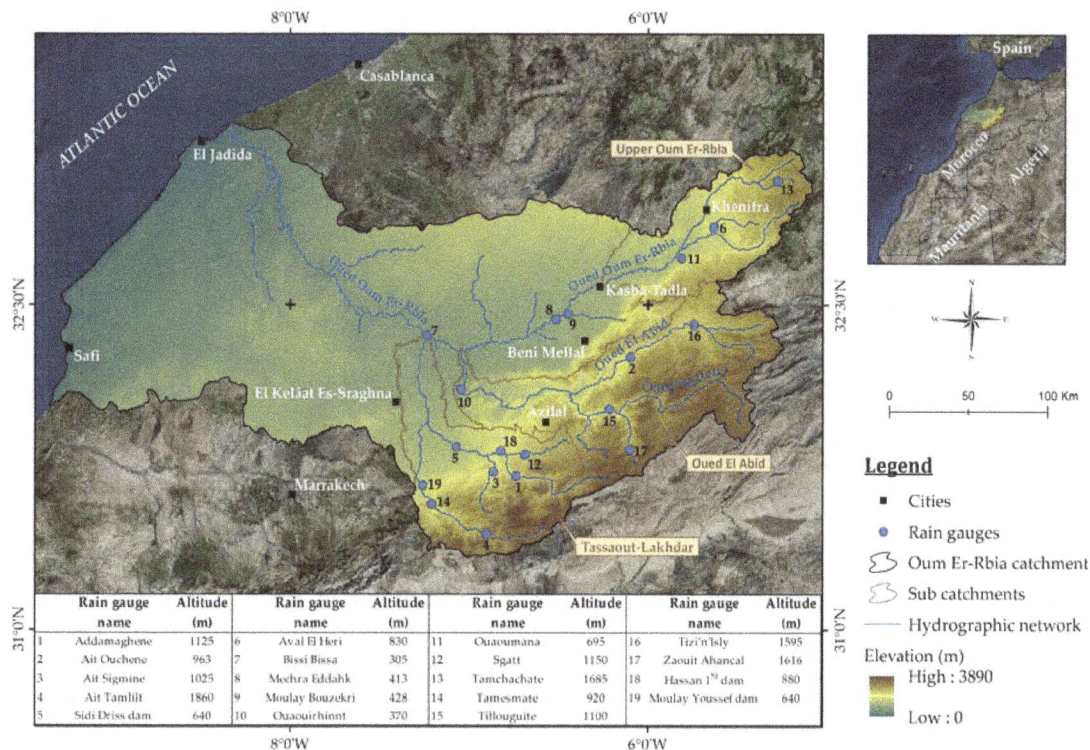

Rain gauge name	Altitude (m)		Rain gauge name	Altitude (m)		Rain gauge name	Altitude (m)		Rain gauge name	Altitude (m)
1 Addamaghene	1125	6	Aval El Heri	830	11	Ouaoumana	695	16	Tizi'n'Isly	1595
2 Ait Ouchene	963	7	Bissi Bissa	305	12	Sgatt	1150	17	Zaouit Ahancal	1616
3 Ait Sigmine	1025	8	Mechra Eddahk	413	13	Tamchachate	1685	18	Hassan 1st dam	880
4 Ait Tamlilt	1860	9	Moulay Bouzekri	428	14	Tamesmate	920	19	Moulay Youssef dam	640
5 Sidi Driss dam	640	10	Ouaouirhinnt	370	15	Tillouguite	1100			

Figure 1. Geographical location of the study area and locations of used rain gauges.

The climate of the OER watershed ranges from semi-arid in the plain to sub-humid at the Atlas chain [42], with a dominance of semi-arid. The study area is characterized by two distinct seasons: a wet season that extends between October and April, and a dry season that begins in May and ends in September. It experiences an oceanic influence manifested by the fact that the moisture wind comes frequently from the west [43]. Based on the data (rainfall and temperature) obtained from the ABHOER (Oum Er-Rbia hydraulic Basin Agency), the rainy season in the study area usually begins in October. A peak is often marked during December, where the mean monthly rainfall could exceed 100 mm in some regions. During July (the driest), the mean monthly rainfall rarely surpasses 10 mm over the whole study area (Figure 2a). Likewise, the snowfall generally occurs between November and May in the high mountains [44,45]. The influence of the continental climate is stronger when moving away from the coastal areas. It induces high temperatures in summer and very low temperature during winter [6]. The lowest temperatures (under $-3\,°C$ in the mountains area and around $0\,°C$ in the foothill) are recorded during the period from December to March where January is the coldest (Figure 2d,e). The highest temperature appears during the period from June to August with a peak generally in July (around $41\,°C$ in the mountains and $46\,°C$ in the foothills) (Figure 2b,c).

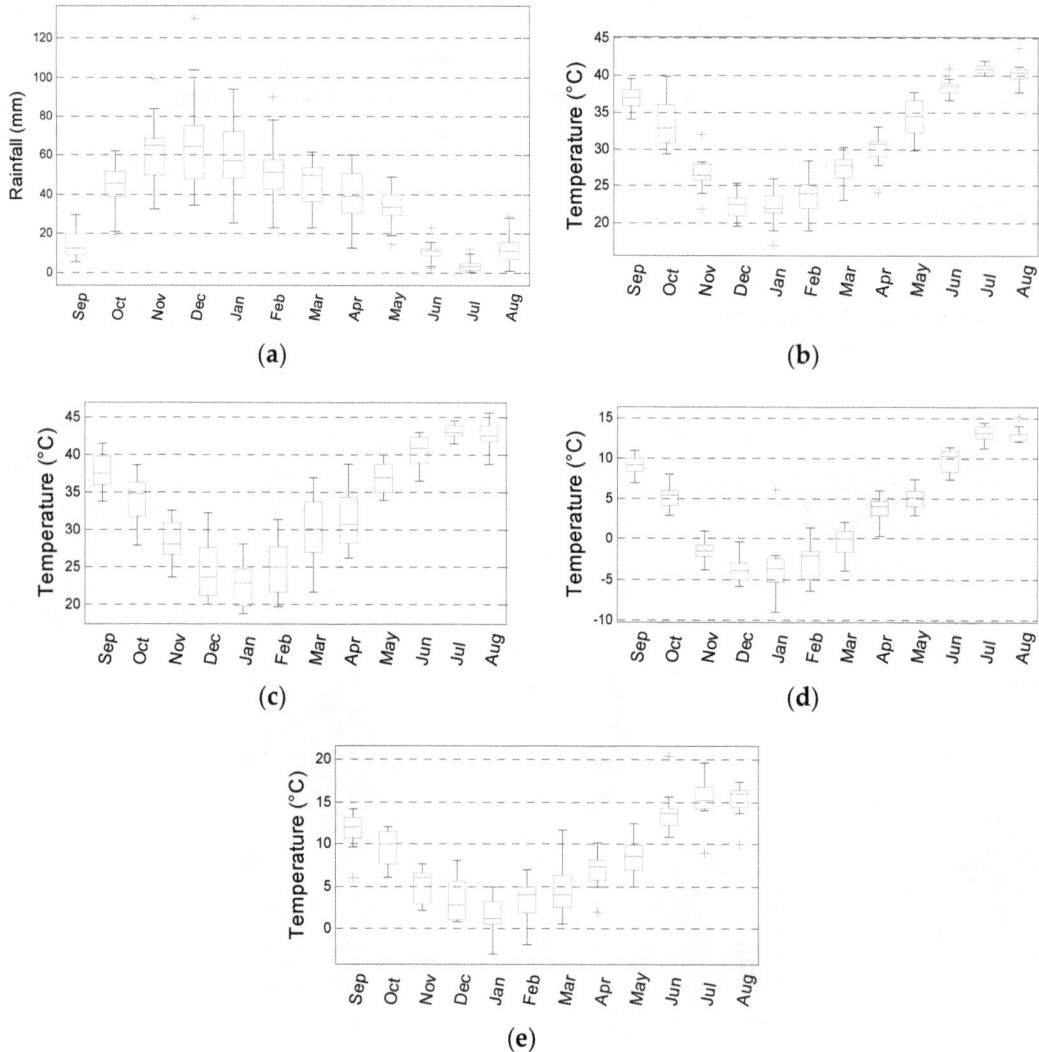

Figure 2. (a) Overall inter-annual mean monthly rainfall of the study area (1998–2010); (b) and (c) Maximal monthly temperature at Tillouguite (1100 m a.s.l.) and Sidi Driss dam's rain gauge (640 m a.s.l.), respectively (1998–2010); (d) and (e) Minimal monthly temperature at Tillouguite and Sidi Driss dam's rain gauge, respectively (1998–2010).

2.2. Data

In order to evaluate the TRMM product for estimating rainfall we have used a daily data set of 19 rain gauges located in the study area. The time series covers a period between September 1998 and August 2011 concomitant with the availability of TRMM products (13 hydrological seasons, from 1 September to 31 August). These data sets were delivered by the Oum Er-Rbia hydraulic Basin Agency (ABHOER). All rain gauges managed by the ABHOER provide daily-accumulated rainfall data. They are distributed in the plain and the mountainous areas, following the development of the hydrographic network of Oum Er-Rbia River and its tributaries. The location of the rain gauges are presented in Figure 1.

The original ABHOER time series were explored to identify and fill the missing values. The data for the month of February for the 2007–2008 season were completely missing in the time series of Sidi Driss dam's rain gauge. These gaps, which represent less than 2%, were filled by linear regression using a neighbor rain gauge with a height correlation coefficient, in considering the recommendations of WMO [11].

TRMM (Tropical Rainfall Measuring Mission) is the first space mission dedicated to the quantitative measurement of rainfall. It was launched from NASDA (National Space Development Agency of Japan) on 28 November 1997 [46]. The mission is a collaboration between NASA (National Aeronautics and Space Administration) and JAXA (Japan Aerospace eXploration Agency). The satellite launched as a part of the TRMM mission covers a band between 50°N and 50°S. It operates in an orbit with 400 km of altitude, an inclination of 35° and a period of 92.5 min allowing it to rotate around the Earth 16 times a day, although the satellite revisits the same scene of the earth's surface twice a day. It embeds five measuring instruments, the Precipitation Radar (PR), the TRMM Microwave Imager (TMI), Visible and Infrared Scanner (VIRS) Clouds and the Earth's Radiant Energy System (CERES), and Lightning Imaging Sensor (LIS) [47,48]. Three of these instruments (PR, TMI, and VIRS) are dedicated for the precipitation measurement system.

The TRMM program provides a variety of algorithms, including, the TMPA (TRMM Multi-satellite Precipitation Analysis) 3B42 Version 7 [49,50] that provides rainfall estimation products through the combination of a set of data from TRMM and various satellite sensors that provide similar data as those on board of TRMM [51]. The seventh version of the 3B42 algorithm (3B42 V7), available since 22 May 2012, incorporated several important changes compared to its predecessor (3B42 V6). In addition to the data used in the previous version, 3B42 V7 algorithm uses new data sources in order to enhance the rainfall estimations [52]. In this study, we have decided to use the 3B42 V7 product based on results from previous studies ([40,53,54], which demonstrate the improvement of the 3B42 V7 product rainfall estimation.

The data were extracted as NetCDF images with spatial resolution of 0.2° and daily temporal resolution. These data were acquired from NASA's official website of Goddard Earth Sciences Data and Information Services Center [55] for the period between 1 January 1998 and 30 August 2011. The monthly and yearly time series were prepared by accumulating the daily data.

2.3. Methodology

Several studies, in various areas around the world, have evaluated the rainfall estimations derived from the 3B42 V7 product. Moazami et al. [56] evaluated daily rain rates derived from three high-resolution satellite precipitation products including TMPA 3B42 V7 over the entire country of Iran. Cai et al. [57] assessed the accuracy of TMPA 3B42 V7 data over the mid-high latitudes region of China. Nastos et al. [58] evaluated the space borne TMPA 3B43 V7 research products with the respective interpolated monthly rain gauge data over Greece. In this study, the use of the interpolation approach can be inefficient to produce a representative distribution of precipitations due to the low spatial density of rain gauges in a rough topography context. Using a moderate density measurement network can bring several mistakes, which will affect the evaluation process [59]. As a matter of fact, the TRMM data were evaluated by direct comparison between punctual rainfall data obtained from rain gauges and the amount of rainfall under the corresponding pixel for TRMM.

The rain gauges and the 3B42 V7 product time series were compared to analyze their variation and relationship at different timescales using various indices and statistical parameters (daily, monthly, and yearly). First of all, we used the verification indices: POD (Probability Of Detection), FAR (False Alarm Ratio), and FBI (Frequency Bias Index) [60]. These indices, which are based on contingency tables, emphasize the product's ability to identify precipitation events above a given threshold [32]. Different thresholds have been adopted in various studies throughout the world to differentiate between rainy or dry days, including 0.1 mm to define a rainy day in Saudi Arabia [35], 0.5 mm in the Korean peninsula [33], and 1 mm in the Peruvian Andes [36]. In this study, we have chosen to apply 0.5 mm as a precipitation threshold based on an analysis where different precipitation thresholds were tested, which were 0.1, 0.3, 0.5, 0.7, and 1 mm.

The POD (the Probability Of Detection), which represents the fraction of observed events that were correctly estimated, is also referred to as the success rate. It is defined as follows (Equation (1)):

$$POD = \frac{a}{a + c},$$
(1)

The FAR (the False Alarm Ratio) is the estimated proportion of events that tend to be falsely detected. It is calculated by the following formula (Equation (2)):

$$FAR = \frac{b}{a + b},$$
(2)

The FBI (the Frequency Bias Index) is the ratio of the number of estimated and observed rainfall events. It is calculated in Equation (3):

$$FBI = \frac{a + b}{a + c},$$
(3)

where, a, b and c represents the number of rain events that fulfilled the conditions in Table 1.

Table 1. Contingency table showing the meaning of parameters used in Equations (1)–(3) (rain ≥ threshold → rainy day; rain < threshold → day without rain).

		Rain Gauge	
		Rain ≥ Threshold	Rain < Threshold
Satellite	Rain ≥ Threshold	a	b
	Rain < Threshold	c	d

The comparison of the satellite product data against observed data from rain gauges was also conducted through a statistical analysis based on standard parameters. These include the Pearson correlation coefficient (Equation (4)) to evaluate the linear relationship between the two data sets. There is also the RMSE (Root Mean Square Error) in mm (Equation (5)) and the bias (Equation (6)) that are used to quantify the difference between the TRMM and the observed data.

$$r = \frac{\sum_{i=1}^{n}(E_i - \overline{E}_i)(O_i - \overline{O}_i)}{\sqrt{\sum_{i=1}^{n}(E_i - \overline{E}_i)(O_i - \overline{O}_i)^2}},$$
(4)

$$RMSE = \sqrt{\frac{\sum_{i=1}^{n}(E_i - O_i)}{n}},$$
(5)

$$Bias = \frac{\sum TRMM}{\sum Station},$$
(6)

E_i and O_i are the 3B42 V7 product and rain gauge data, respectively, at a time i. $\overline{E}_i$ and $\overline{O}_i$ are the average values of the product and rain gauge data, respectively. n is the total number of the rain-gauge records.

3. Results

3.1. Evaluation at Daily Time Step

The 3B42 V7 product's performance varies from one area to another. It has been found that at a daily time step, there is a remarkable difference between 3B42 V7 and rain gauges. Over the entire study area, the calculated statistics are generally low (Table 2). According to the POD, about 36%–60% of the rain gauge records have been properly detected by TRMM. Nevertheless, at 75% of the rain gauges, the correctly detected events are usually below 50%. On one hand, the FAR shows that at half of the studied rain gauges, TRMM gave false alarms for 38 to 55% of the measurements. On the other hand, for the rain gauges the proportion of false alarms can reach degrees higher than 70%, particularly at those located in mountainous areas receiving snowfall. Furthermore, the FBI varies between 0.59 and 2.63. At 50 % of the studied rain gauges, TRMM overestimates the number of rainy days with FBI that are greater than one.

At a daily time scale, the two data sets exhibit a poor linear relationship across the study area. This is well demonstrated by the Pearson correlation coefficient values that are lower than 0.38 at all rain gauges (Table 2). This low efficiency of TRMM can be explained by the rapid cloud dynamics and the stormy nature of precipitation events in the study region that influence the spatial and temporal distribution of rainfall. Baik and Choi [33] have pointed out that rapid temporal evolution of convective rainfall may not be captured by TRMM due to differences in overpass time. Moreover, there is a mismatch of scale since the satellite product is extracted for a 25 km area, unlike the rain gauges that provide punctual measurements.

Table 2. Obtained statistics (POD, FAR, FBI, and Pearson correlation coefficient) for daily time scale. POD, Probability Of Detection; FAR, False Alarm Ratio; FBI, Frequency Bias Index.

Station	POD	FAR	FBI	Pearson
Addammaghene	0.60	0.77	2.59	0.08
Ait Ouchene	0.48	0.43	0.84	0.29
Ait Sigmine	0.50	0.81	2.63	0.10
Ait Tamlilt	0.54	0.77	2.31	0.14
Sidi Driss dam	0.47	0.51	0.97	0.22
Aval El Heri	0.49	0.51	1.00	0.31
Bissi Bissa	0.48	0.62	1.26	0.31
Mechra Eddahk	0.36	0.39	0.59	0.35
Moulay Bouzekri	0.44	0.38	0.71	0.38
Ouaouirhinnt	0.44	0.47	0.83	0.21
Ouaoumana	0.43	0.45	0.77	0.31
Sgatt	0.44	0.54	0.97	0.13
Tamchachate	0.42	0.72	1.51	0.15
Tamesmate	0.40	0.72	1.44	0.00
Tillouguite	0.52	0.62	1.37	0.18
Tizi N lsly	0.46	0.53	0.97	0.16
Zaouit Ahancal	0.60	0.68	1.88	0.16
Hassan 1st dam	0.45	0.53	0.95	0.14
Moulay Youssef dam	0.42	0.50	0.85	0.20

3.2. Evaluation at Monthly Time Step

At the monthly time scale, the linear relationship between the rain gauges and TRMM data shows a considerable improvement, compared to the daily one (Figures 3 and 4). The two data sets demonstrate a good relationship with correlation coefficients varying between 0.63 and 0.91 with an overall average of 0.80. Furthermore, 17 rain gauges have correlation coefficients greater than 0.7, and two gauges have correlation coefficients of 0.63 and 0.68. As for the RMSE, which shows sensitivity to the accumulated rainfall amounts, we have found that it is less than 25 mm for over 50% of the studied tiles. However, most of the rain gauges located at the mountainous area present the highest RMSE, especially those with the highest monthly rainfall averages. This is maybe due to the topographic effect that reduces the accuracy of the measurements or the presence of snow in high-elevation areas.

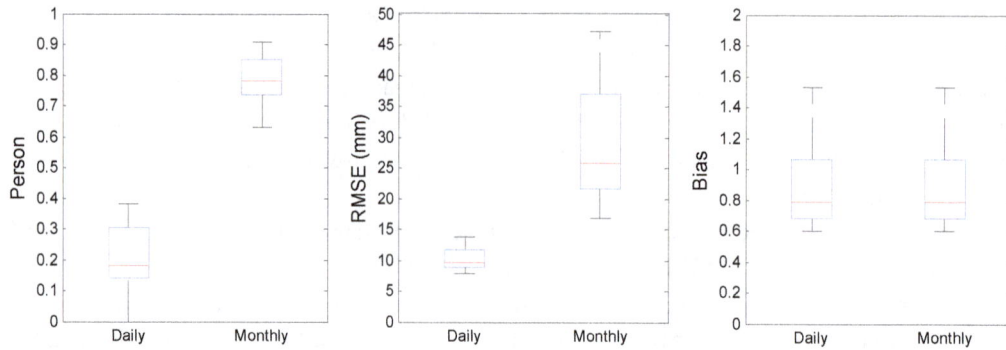

Figure 3. Obtained statistics (Pearson correlation coefficient, RMSE, and Bias) for daily and monthly time scale.

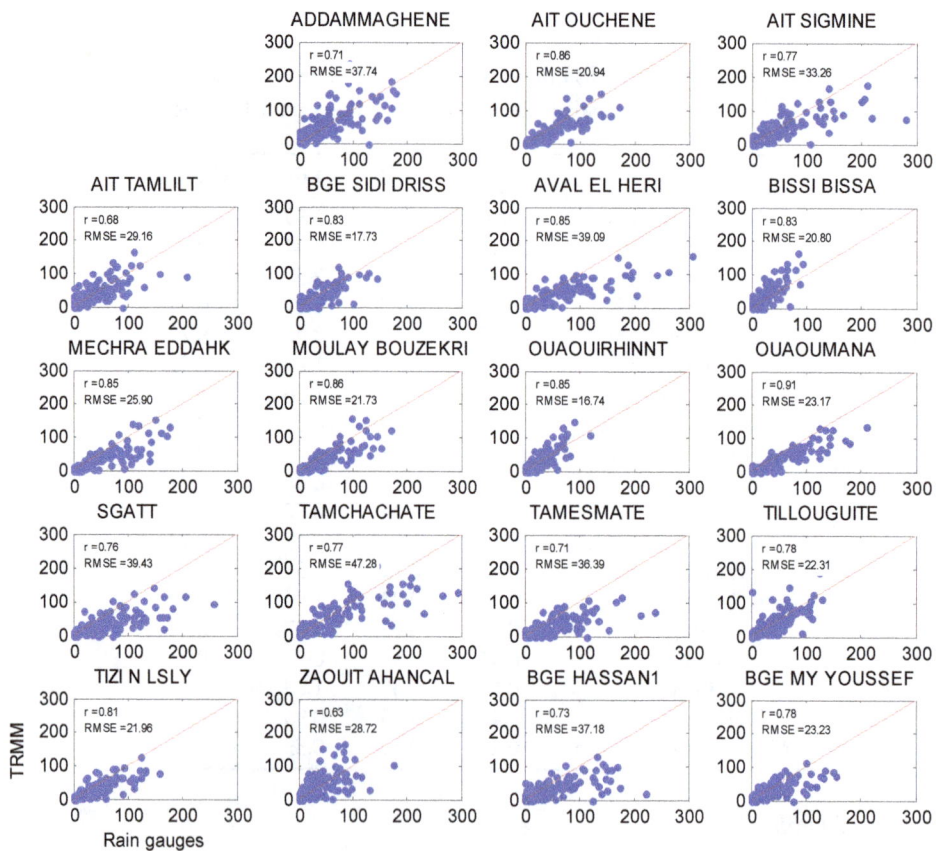

Figure 4. Scatterplot between the 3B42 V7 product and the rain gauges at a monthly time scale over the period 1998–2010.

To explore further the relationship between the two data sets, an analysis was conducted which considered the obtained statistical indices in relation to the following different factors: the measurement time, the topographic dependence, and the recorded rainfall amounts. Actually, the relevance of the statistical indices changes over the hydrological season. In a semi-arid climate, heavy rainfall events typically occur during summer [9,61]. These events are generally characterized by low spatial extension and short time scales, which make their detection a real challenge for both the rain gauges and satellite. In the study area, these short-term events generally occur during July and August. The statistical indices month by month showed that July and August exhibit the lowest correlation coefficients (Figure 5).

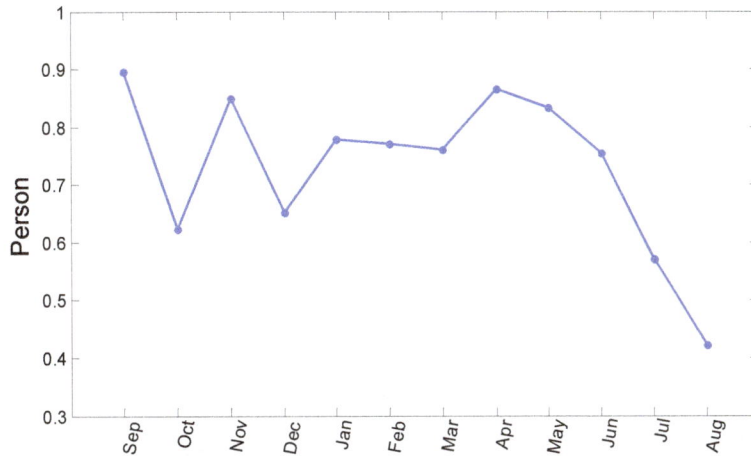

Figure 5. Average correlation coefficient of each month of the time series.

The second factor, topography, has a considerable impact on the relationship between rain gauges and TRMM data. Unlike those found for the plain, most of the TRMM pixels located at the mountainous area have lower agreement with the corresponding rain gauges (Figure 6). These pixels have correlation coefficients that vary between 0.7 and 0.8, except those corresponding to Ait Tamlilt and Zaouit Ahencal rain gauges, which present correlation coefficients less than 0.7. This is can be explained by the fact that the two pixels are located in an area experiencing snowfall.

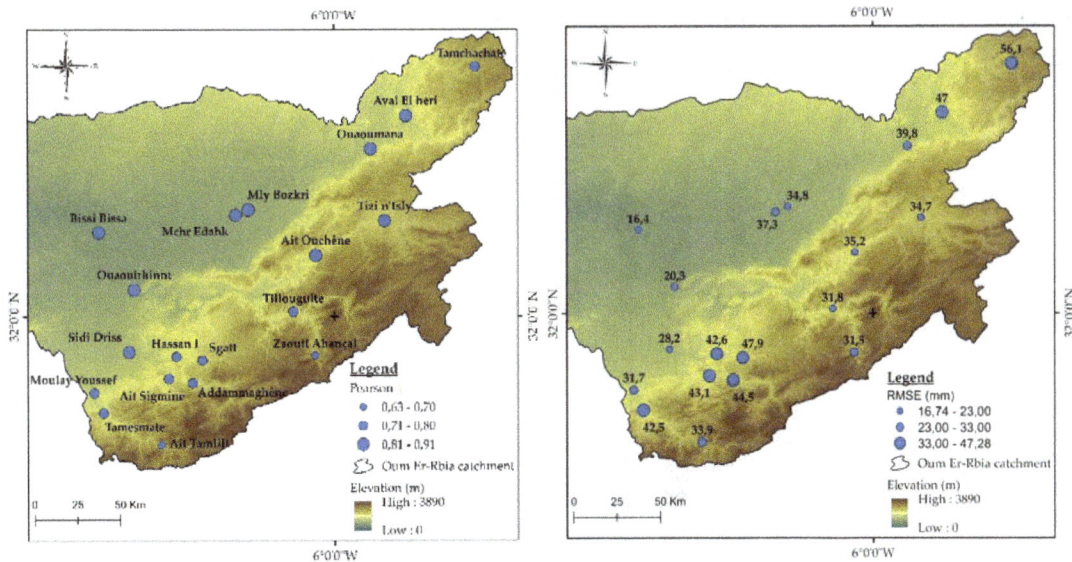

Figure 6. Pearson correlation coefficient based on topography and RMSE based on mean monthly rainfall.

Finally, an analysis of bias values sorted in ascending order according to the monthly average rainfall (Figure 7) reveals that in the most arid zones, TRMM tends to overestimate rainfall amounts as compared to rain gauges. Bissi Bissa's and Ouaouirinth's rain gauges present the lowest monthly average rainfall all over the study area. The bias computed for the pixels corresponding to these two rain gauges (which happen to be located at the most arid sites) shows that TRMM overestimates the amount of rainfall in such areas. On the other hand, for rain gauges with high monthly averages, TRMM underestimates rainfall. This is observed for most of the rain gauges, except for the ones that may receive snowfall, such as Zaouit Ahencal and Ait Tamlilt. At these sites, TRMM tends to provide overestimated rainfall amounts.

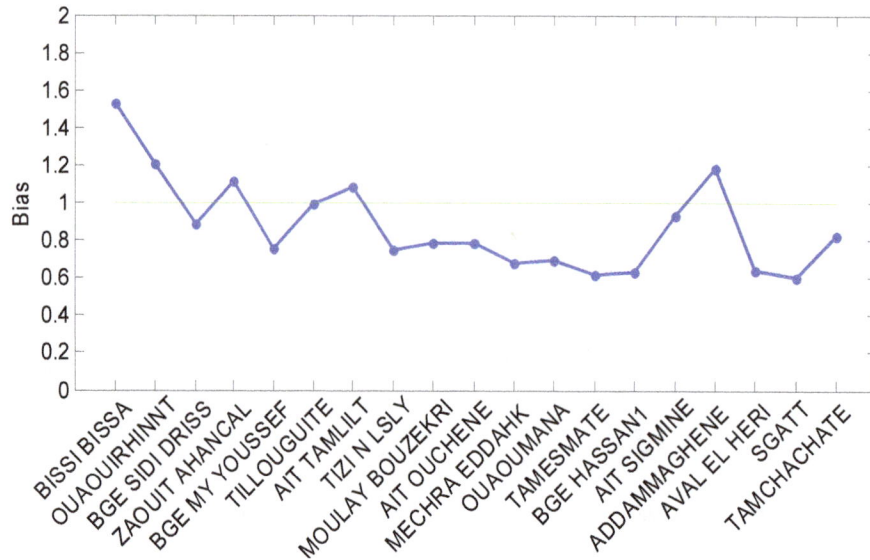

Figure 7. Bias sorted according to the monthly average rainfall.

In addition, the annual cycle of rainfall derived from TRMM data is fairly close to that of rain gauges, as is shown in Figure 8. The annual cycle of rainfall demonstrates a maximum during the rainy season around December and a minimum in the dry season from June to August. Nevertheless, the TRMM estimations are indeed influenced by a bias that is manifested very often by under-estimations of the rainfall amounts. Partially, this bias was expected considering the fact that the comparison is made between punctual measurements and spatial estimations.

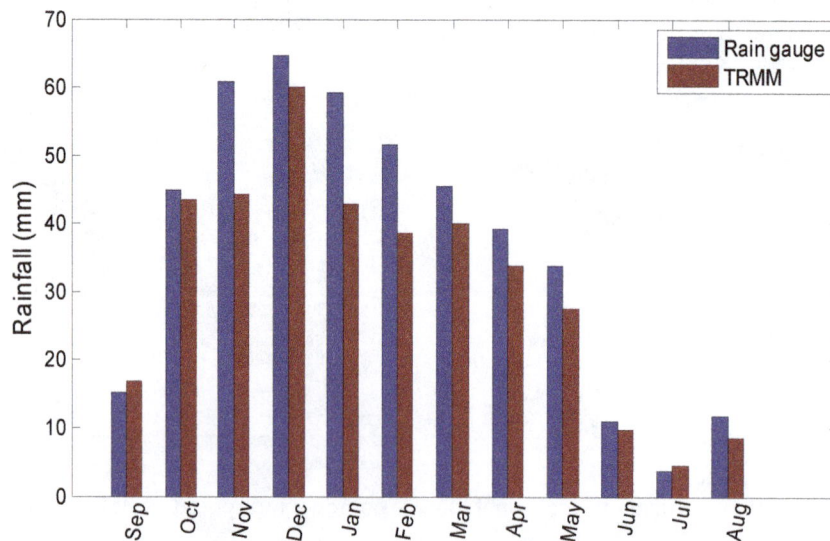

Figure 8. Annual cycle of rainfall obtained from Tropical Rainfall Measurement Mission (TRMM) data and rain gauges for the period of 1998–2010.

3.3. Evaluation at Annual Time Step

At the annual scale, there is a reasonable overall linear relationship between the TRMM and rain gauges data, according to the Figure 9. However, this relationship is irregular from one part to another of the study area. The correlation coefficients range between 0.43 and 0.88, where nearly 75% of the studied rain gauges have coefficients higher than 0.66. Compared to the monthly time scale, the correlation coefficients are slightly reduced for most rain gauges but much lower for Mechra Eddahk.

Both AitTamlilt and Zaouit Ahencal rain gauges still have one of the lowest correlation coefficients of the study area. This highlights the problematic effect of snowfall on the 3B42 V7 estimations. Some low coefficients are found for rain gauges situated in the plain, which maybe related to the impact fewer records of the considered time series. This is a limiting factor in the correlation analysis in judging the acceptability of the calculated coefficients.

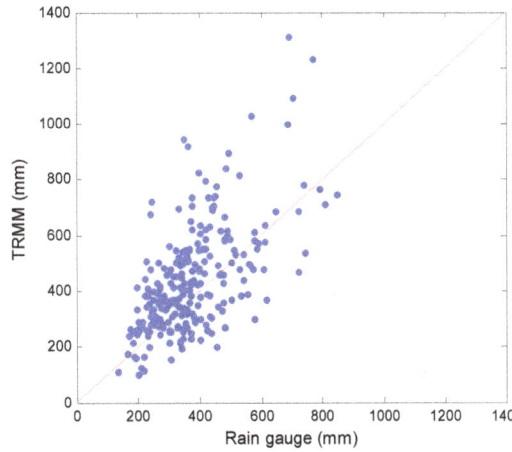

Figure 9. The overall scatterplot between TRMM and rain gauges at annual time scale.

The RMSE between observed and TRMM annual precipitation ranged from 70 to 261 mm with 50% of the pixels having values lower than 140 mm. For some other pixels the RMSE remains high generally due to the large difference between the two data sets that often occurs during the wetter seasons where TRMM tends to underestimate the rainfall amounts particularly in the seasons of 2003–2004, 2008–2009, and 2009–2010 (Figure 10).

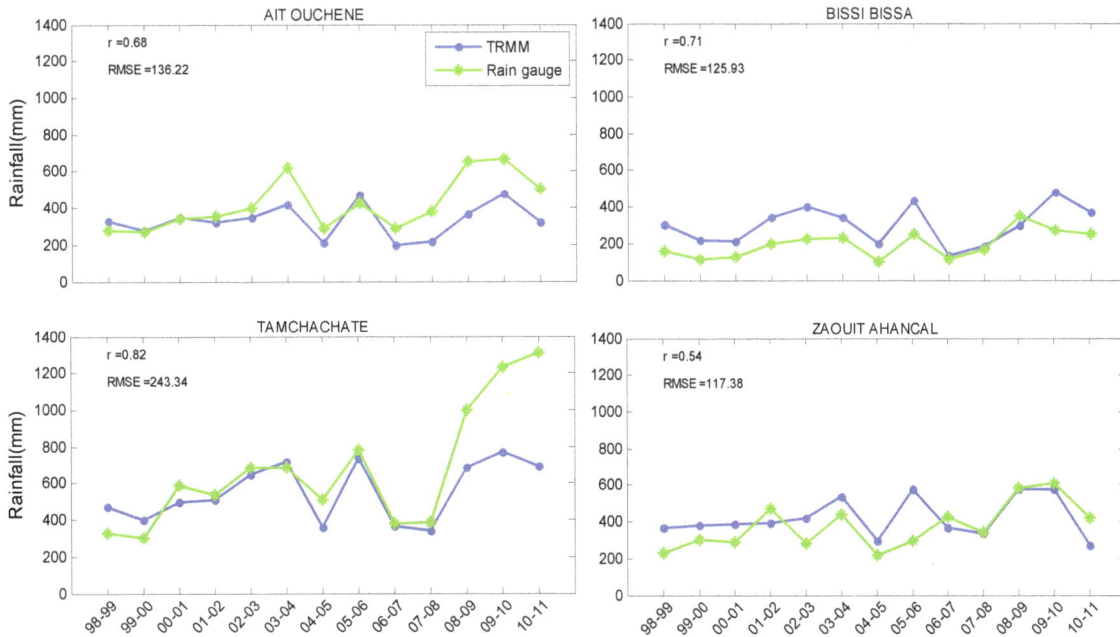

Figure 10. Annual rainfall obtained from representative TRMM pixels and rain gauges data for the period 1998–2010.

3.4. The Evaluation at The Catchment Scale

A comparison between TRMM and rain gauges data was conducted based on a zonal scale, considering the average rainfall collected in rain gauges and pixels within the same sub-watershed. Because of the very low statistics at the daily time scale, this comparison was carried out monthly (Figure 11) and annually (Figure 12).

In the three sub-watersheds (High Oum Er-Rbia Oueld El Abid and Tassaout Lekhdar) included in this analysis, the two datasets show a good agreement at both monthly and annual time scale. At the monthly scale, the Pearson correlation coefficients vary between 0.82 and 0.87, the RMSE between 17 mm and 35 mm, and the bias between 0.67 and 0.91. Among the three sub-watersheds, Ouad El Abid presents the best agreement with TRMM data followed by Tassaout-Lakhdar. However, at the annual scale, the correlation coefficients are slightly decreased with values ranging between 0.7 and 0.84, the RMSE values are more important between 82 mm and 278 mm, and the bias between 0.61 and 0.94. The two watersheds of Oued El Abid and Tassaout show good correlation coefficients (greater than 0.83) and acceptable RMSE (92 mm and 82 mm, respectively) with bias close to one. However, the upper watershed of Oum Er-Rbia shows relatively high correlation coefficients and low bias, associated with high RSME. This is again caused by the presence of snow during winter months, which is not detected by TRMM data.

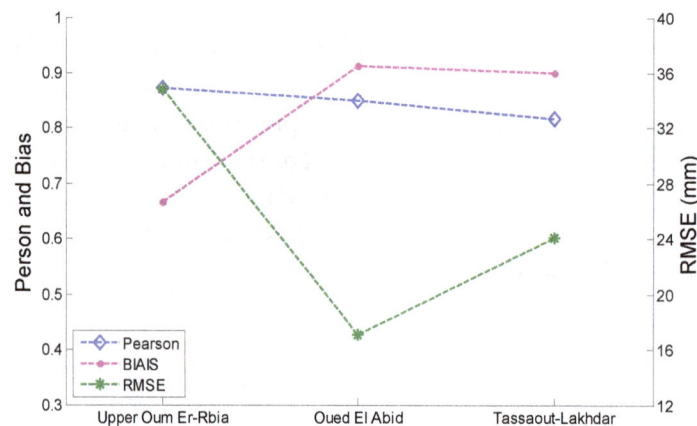

Figure 11. Evaluation statistics for the three studied sub watershed at monthly time scale.

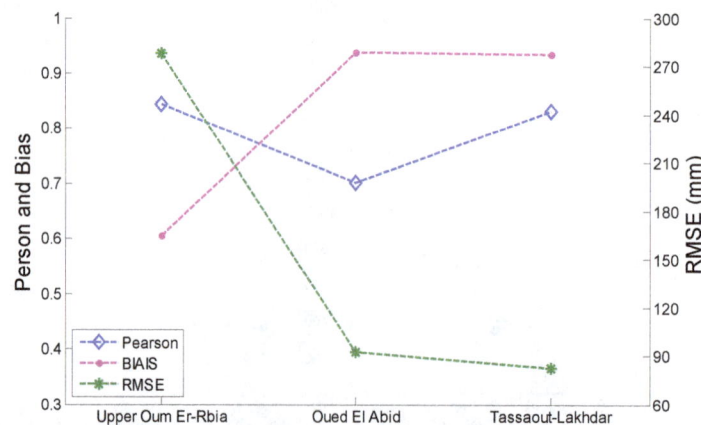

Figure 12. Evaluation statistics for the three studied sub-watershed at annual time scale.

4. Discussion and Conclusions

In the last few decades, water resources availability in the Oum Er Rbia watershed, which contains major river systems in Morocco, has faced alarming reductions of surface water and groundwater resources. According to future projections, the hydrological response in the upstream part of this watershed will be characterized by a reduction of total flow due to the increasing temperature and decreasing precipitation [62]. The rainfall, as a key parameter of the hydrological cycle, required a deep analysis to monitor the variability at spatial and temporal scale in order to ensure an operational water resources management. However, remote sensing data, such as TRMM, offers an opportunity to compensate the lack of rainfall measurements over a large region. This valuable information is not yet used in the Oum Er Rbia watershed, particularly as input data for hydrological modeling purpose. In this regards, we assessed the performance of the 3B42 V7 rainfall product at different time scales over the upstream part of the Oum Er Rbia watershed. Due to the varying topographic systems in the studied area, we conducted this assessment by a direct comparison between the rain gauges measurements and the corresponding pixels data obtained from TRMM. The working methodology has allowed us to evaluate the satellite product in different climatic and topographical contexts, thereby avoiding the uncertainties related to interpolation approaches based on a low-density measurement network.

The results, at a daily time scale, showed a remarkable difference between 3B42 V7 rainfall product and rain gauges, with correlation coefficients lower than 0.38 all over the study area. Mantas et al. [36] suggested that some of the poor agreement at smaller time resolution is due to intra-tile variability. For aggregated time scales (monthly and annual), the 3B42 V7 product and rain gauges show a stronger agreement compared to the daily time scale. These improvements for large time scales are probably because TRMM precipitation estimates are calibrated and rescaled using monthly rain gauge data [63]. Generally, TRMM and rain gauges data are well correlated for most of the study sites, with a mean correlation coefficient of 0.8 and 0.71 for monthly and annual time scales, respectively. However, the analysis showed that the TRMM product had difficulties in retrieving rainfall amounts over mountainous regions. Similarly, Almazroui [35], Mantas et al. [36], and Milweski et al. [40] emphasized the important impact of the accidental rugged topography on the accuracy of both satellite and rain gauges measurements. They all demonstrated the low agreement between TRMM and rain gauges over the highest elevations areas. Considering the fact that snowfall occurs usually in most of these areas during the wet season. This remains problematic for rainfall estimations [40]. In our case study, we have found that TRMM tends to overestimate rainfall amounts in areas receiving snowfall, contrary to what was reported by Anders et al. [64] and Krakauer et al. [65]. This observation may be explained by the unmeasured snow water equivalent by the rain gauges. Moreover, the study showed that the 3B42 V7 product is able to adequately reproduce the annual cycle of rainfall at large scale with bias compared to the rain gauges. Therefore, TRMM tends, in general, to overestimate rainfall in dry regions and underestimate rainfall in humid regions, which usually located at the highest elevations. During winter, TRMM mostly underestimates the rainfall amounts. Thus, the product faces a real constraint in capturing convective and short time events that occur often during July and August.

The rainfall data from the 3B42 V7 product, at monthly and annual time scale, can be a valuable source of information at a watershed scale, especially for hydrological modeling on one hand, and on the other hand, can also be valuable information for managers and decision makers. The use of these freely available data can represent an alternative in data-sparse regions and an important source of data for ungauged watersheds. These data, with their global spatial coverage, can be considered in further studies for hydrological and climatic research in the Oum Er Rbia watershed. Moreover, the use of this date will constitute a valuable source of information to assess the vulnerability of the various studied watersheds, and evaluate the sustainability of the anthropic activities under the actual and future hydrological and climatic contexts.

Acknowledgments: The authors are grateful to the Oum Er-Rbia hydraulic Basin Agency for providing the rain gauges data that we used in this study. We would like to thank NASA's and JAXA's teams, for creating and making their precipitation products freely available. We also would like to thank the reviewers of the manuscript for their helpful remarks.

Author Contributions: Hamza Ouatiki conducted the analysis, the interpretation of results, and the preparation of the manuscript under the supervision and the effective guidance and suggestions of Abdelghani Boudhar; Yves Tramblay, Lionel Jarlan, and Tarik Benabdelouhab contributed to the design and revisions of the manuscript, and provided important suggestions and reviews; Lahoucine Hanich and Abdelghani Chehbouni provided reviews and suggestions; Mohamed Rachid El Meslouhi processed the rain gauges data. All authors contributed to the writing of the manuscript.

Conflicts of Interest: The authors declare no conflict of interest.

References

1. Esper, J.; Frank, D.; Büntgen, U.; Verstege, A.; Luterbacher, J.; Xoplaki, E. Long-term drought severity variations in Morocco. *Geophys. Res. Lett.* **2007**, *34*, 1–5. [CrossRef]
2. Jarlan, L.; Khabba, S.; Er-Raki, S.; Le Page, M.; Hanich, L.; Fakir, Y.; Merlin, O.; Mangiarotti, S.; Gascoin, S.; Ezzahar, J.; et al. Remote sensing of water resources in semi-arid Mediterranean areas: The joint international laboratory TREMA. *Int. J. Remote Sens.* **2015**, *36*, 4879–4917. [CrossRef]
3. Jellali, M.M. Développement des ressources en eau au Maroc. In *Aspects économiques de la Gestion de l'eau dans le Bassin Méditerranéen*; CIHEAM: Bari, Italie, 1997; pp. 51–68.
4. Houdret, A. Les conflits autour de l'eau au Maroc: Origines Sociopolitiques et écologiques et Perspectives Pour une Transformation des Conflits. Ph.D. Thesis, University Duisburg-Essen, Duisburg/Essen, Allemagne, 2009.
5. Schilling, J.; Freier, K.P.; Hertig, E.; Scheffran, J. Climate change, vulnerability and adaptation in North Africa with focus on Morocco. *Agric. Ecosyst. Environ.* **2012**, *156*, 12–26. [CrossRef]
6. Knippertz, P.; Christoph, M.; Speth, P. Long-term precipitation variability in Morocco and the link to the large-scale circulation in recent and future climates. *Meteorol. Atmos. Phys.* **2003**, *83*, 67–88. [CrossRef]
7. Boudhar, A.; Duchemin, B.; Hanich, L.; Jarlan, L.; Chaponnière, A.; Maisongrande, P.; Boulet, G.; Chehbouni, A. Long-term analysis of snow-covered area in the Moroccan High-Atlas through remote sensing. *Int. J. Appl. Earth Obs. Geoinf.* **2010**, *12*, S109–S115. [CrossRef]
8. Yvan, C. Modélisation des écoulements d'Origine Pluvio-nivo-glaciaire en Contexte de Haute Montagne Tropicale: Application à la Haute Vallée du Zongo (Bolivie). Ph.D. Thesis, University Montpellier II, Montpellier, France, 2001.
9. Güntner, A. Large-Scale Hydrological Modelling in the Semi-Arid North-East of Brazil. Ph.D. Thesis, Departement Geoecology, University Potsdam, Potsdam, Germany, 2002.
10. Boudhar, A.; Hanich, L.; Boulet, G.; Duchemin, B.; Berjamy, B.; Chehbouni, A. Evaluation of the snowmelt runoff model in the Moroccan High Atlas Mountains using two snow-cover estimates. *Hydrol. Sci. J.* **2009**, *54*, 1094–1113. [CrossRef]
11. Organisation Météorologique Mondiale. *Guide Des Pratiques Hydrologiques: Acquisition et Traitement des Données, Analyses, Prévision et Autres Applications*; OMM: Genève, Suisse, 1994.
12. Adler, R.F.; Negri, A.J.; Keehn, P.R.; Hakkarinen, I.M. Estimation of monthly rainfall over Japan and surrounding waters from a combination of low-orbit microwave and geosynchronous IR data. *J. Appl. Meteorol.* **1993**, *32*, 335–356. [CrossRef]
13. Adler, R.F.; Huffman, G.J.; Keehn, P.R. Global tropical rain estimates from microwave-adjusted geosynchronous IR data. *Remote Sens. Rev.* **1994**, *11*, 125–152. [CrossRef]
14. Bennartz, R.; Thoss, A.; Dybbroe, A.; Michelson, D.B. Precipitation analysis using the advanced microwave sounding unit in support of nowcasting applications. *Meteorol. Appl.* **2002**, *9*, 177–189. [CrossRef]
15. Muñoz, E.A.; Di Paola, F.; Lanfri, M.A.; Arteaga, F.J. Observing the troposphere through the Advanced Microwave Technology Sensor (ATMS) to retrieve rain rate. *IEEE Lat. Am. Trans.* **2016**, *14*, 586–594. [CrossRef]
16. Huffman, G.J.; Adler, R.F.; Morrissey, M.M.; Bolvin, D.T.; Curtis, S.; Joyce, R.; McGavock, B.; Susskind, J. Global precipitation at one-degree daily resolution from multisatellite observations. *J. Hydrometeorol.* **2001**, *2*, 36–50. [CrossRef]
17. Hsu, K.; Gao, X.; Sorooshian, S.; Gupta, H.V. Precipitation estimation from remotely sensed information using artificial neural networks. *J. Appl. Meteorol.* **1997**, *36*, 1176–1190. [CrossRef]
18. Miller, S.W.; Arkin, P.A.; Joyce, R. A combined microwave/infrared rain rate algorithm. *Int. J. Remote Sens.* **2001**, *22*, 3285–3307. [CrossRef]

19. Joyce, R.J.; Janowiak, J.E.; Arkin, P.A.; Xie, P. CMORPH: A method that produces global precipitation estimates from passive microwave and infrared data at high spatial and temporal resolution. *J. Hydrometeorol.* **2004**, *5*, 487–503. [CrossRef]

20. Vila, D.A.; de Goncalves, L.G.G.; Toll, D.L.; Rozante, J.R. Statistical evaluation of combined daily gauge observations and rainfall satellite estimates over continental South America. *J. Hydrometeorol.* **2009**, *10*, 533–543. [CrossRef]

21. Rozante, J.R.; Moreira, D.S.; de Goncalves, L.G.G.; Vila, D.A. Combining TRMM and surface observations of precipitation: Technique and validation over South America. *Weather Forecast.* **2010**, *25*, 885–894. [CrossRef]

22. Arvor, D.; Dubreuil, V.; Ronchail, J.; Penello, S.; Paris, U. Apport des données TRMM 3B42 à l'étude des précipitations au Mato Grosso. *Climatologie* **2008**, *5*, 49–69. [CrossRef]

23. Su, F.; Hong, Y.; Lettenmaier, D.P. Evaluation of TRMM Multisatellite Precipitation Analysis (TMPA) and its utility in hydrologic prediction in the La Plata Basin. *J. Hydrometeorol.* **2008**, *9*, 622–640. [CrossRef]

24. Gu, H.; Yu, Z.; Yang, C.; Ju, Q.; Lu, B.; Liang, C. Hydrological assessment of TRMM rainfall data over Yangtze River Basin. *Water Sci. Eng.* **2010**, *3*, 418–430.

25. Di Paola, F.; Ricciardelli, E.; Cimini, D.; Romano, F.; Viggiano, M.; Cuomo, V. Analysis of catania flash flood case study by using combined microwave and infrared technique. *J. Hydrometeorol.* **2014**, *15*, 1989–1998. [CrossRef]

26. Li, L.; Hong, Y.; Wang, J.; Adler, R.F.; Policelli, F.S.; Habib, S.; Irwn, D.; Korme, T.; Okello, L. Evaluation of the real-time TRMM-based multi-satellite precipitation analysis for an operational flood prediction system in Nzoia Basin, Lake Victoria, Africa. *Nat. Hazards* **2009**, *50*, 109–123. [CrossRef]

27. Moffitt, C.B.; Hossain, F.; Adler, R.F.; Yilmaz, K.K.; Pierce, H.F. Validation of a TRMM-based global flood detection system in Bangladesh. *Int. J. Appl. Earth Obs. Geoinf.* **2011**, *13*, 165–177. [CrossRef]

28. Zhang, A.; Jia, G. Monitoring meteorological drought in semiarid regions using multi-sensor microwave remote sensing data. *Remote Sens. Environ.* **2013**, *134*, 12–23. [CrossRef]

29. Du, L.; Tian, Q.; Yu, T.; Meng, Q.; Jancso, T.; Udvardy, P.; Huang, Y. A comprehensive drought monitoring method integrating MODIS and TRMM data. *Int. J. Appl. Earth Obs. Geoinf.* **2013**, *23*, 245–253. [CrossRef]

30. Yaduvanshi, A.; Srivastava, P.K.; Pandey, A.C. Integrating TRMM and MODIS satellite with socio-economic vulnerability for monitoring drought risk over a tropical region of India. *Phys. Chem. Earth* **2015**, *83*, 14–27. [CrossRef]

31. Brown, J.E.M. An analysis of the performance of hybrid infrared and microwave satellite precipitation algorithms over India and adjacent regions. *Remote Sens. Environ.* **2006**, *101*, 63–81. [CrossRef]

32. Dinku, T.; Chidzambwa, S.; Ceccato, P.S.; Connor, J.; Ropelewski, C.F. Validation of high-resolution satellite rainfall products over complex terrain. *Int. J. Remote Sens.* **2008**, *29*, 4097–4110. [CrossRef]

33. Baik, J.; Choi, M. Spatio-temporal variability of remotely sensed precipitation data from COMS and TRMM: Case study of Korean peninsula in East Asia. *Adv. Space Res.* **2015**, *56*, 1125–1138. [CrossRef]

34. Blacutt, L.A.; Herdies, D.L.; de Gonçalves, L.G.G.; Vila, D.A.; Andrade, M. Precipitation comparison for the CFSR, MERRA, TRMM3B42 and combined scheme datasets in Bolivia. *Atmos. Res.* **2015**, *163*, 117–131. [CrossRef]

35. Almazroui, M. Calibration of TRMM rainfall climatology over Saudi Arabia during 1998–2009. *Atmos. Res.* **2011**, *99*, 400–414. [CrossRef]

36. Mantas, V.M.; Liu, Z.; Caro, C.; Pereira, A.J.S.C. Validation of TRMM multi-satellite precipitation analysis (TMPA) products in the Peruvian Andes. *Atmos. Res.* **2015**, *163*, 132–145. [CrossRef]

37. Collischonn, B.; Collischonn, W.; Tucci, C.E.M. Daily hydrological modeling in the Amazon Basin using TRMM rainfall estimates. *J. Hydrol.* **2008**, *360*, 207–216. [CrossRef]

38. Islam, M.N.; Uyeda, V. Use of TRMM in determining the climatic characteristics of rainfall over Bangladesh. *Remote Sens. Environ.* **2007**, *108*, 264–276. [CrossRef]

39. Huang, Y.; Chen, S.; Cao, Q.; Hong, Y.; Wu, B.; Huang, M.; Qiao, L.; Zhang, Z.; Li, Z.; Li, W.; et al. Evaluation of version-7 TRMM multi-satellite precipitation analysis product during the Beijing extreme heavy rainfall event of 21 July 2012. *Water* **2014**, *6*, 32–44. [CrossRef]

40. Milewski, A.; Elkadiri, R.; Durham, M. Assessment and comparison of TMPA satellite precipitation products in varying climatic and topographic regimes in Morocco. *Remote Sens.* **2015**, *7*, 5697–5717. [CrossRef]

41. Tramblay, Y.; Thiemig, V.; Dezetter, A.; Hanich, L. Evaluation of satellite-based rainfall products for hydrological modelling in Morocco. *Hydrol. Sci. J.* **2016**, *61*, 2509–2519. [CrossRef]

42. Mokhtari, N.; Mrabet, R.; Lebailly, P.; Bock, L. Spatialisation des bioclimats, de l'aridité et des étages de végétation du Maroc. *Rev. Marocaine Sci. Agron. Vétérinaires* **2013**, *2*, 50–66.

43. Moniod, F. Étude hydrologicpe de l'Oum Er-Rbia. *Cah. ORSTOM Sér. Hydrol.* **1973**, *10*, 153–170.

44. Boudhar, A.; Hanich, L.; Boulet, G.; Outaleb, K.; Arioua, A.; Ben, B.; Hakkani, B. Etude de la disponibilité des ressources en eau à l'aide de la télédétection et la modélisation: Cas du bassin versant d'Oum Er Rbia (Maroc). In Proceedings of the 3éme Colloque International Eau–Climat'2014: Ressources en Eau Changement Climatique en Région Méditerranéenne, Hammamet, Tunisie, 21–23 Octobre 2014.

45. Marchane, A.; Jarlan, L.; Hanich, L.; Boudhar, A.; Gascoin, S.; Tavernier, A.; Filali, N.; Le Page, M.; Hagolle, O.; Berjamy, B. Assessment of daily MODIS snow cover products to monitor snow cover dynamics over the Moroccan Atlas mountain range. *Remote Sens. Environ.* **2015**, *160*, 72–86. [CrossRef]

46. Earth Observation Center National Space Development Agency of Japan. *TRMM Users Handbook*; NASDA: Tanegashima, Japan, 2001.

47. Kummerow, C.; Barnes, W.; Kozu, T.; Shiue, J.; Simpson, J. The tropical rainfall measuring mission (TRMM) sensor package. *J. Atmos. Ocean. Technol.* **1998**, *15*, 809–817. [CrossRef]

48. Kummerow, C.; Simpson, J.; Thiele, O.; Barnes, W.; Chang, A.T.C.; Stocker, E.; Adler, R.F.; Hou, A.; Kakar, R.; Wentz, F.; et al. The status of the Tropical Rainfall Measuring Mission (TRMM) after two years in orbit. *J. Appl. Meteorol.* **2000**, *39*, 1965–1982. [CrossRef]

49. Huffman, G.; Adler, R.; Bolvin, D.; Nelkin, E. The TRMM multi-satellite precipitation analysis (TMPA). In *Satellite Rainfall Applications for Surface Hydrology*; Springer: Berlin, Germany, 2010; pp. 3–22.

50. Bolvin, D.T.; Huffman, G.J. Transition of 3B42/3B43 Research Product from Monthly to Climatological Calibration/Adjustment. Available online: http://pmm.nasa.gov/sites/default/files/document_files/3B42_3B43_TMPA_restart.pdf (accessed on 15 Decembre 2015).

51. Huffman, G.J.; Bolvin, D.T.; Nelkin, E.J.; Wolff, D.B.; Adler, R.F.; Gu, G.; Hong, Y.; Bowman, K.P.; Stocker, E.F. The TRMM Multisatellite Precipitation Analysis (TMPA): Quasi-global, multiyear, combined-sensor precipitation estimates at fine scales. *J. Hydrometeorol.* **2007**, *8*, 38–55. [CrossRef]

52. Huffman, G.J.; Bolvin, D.T. TRMM and Other Data Precipitation Data Set Documentation. Available online: ftp://meso-a.gsfc.nasa.gov/pub/trmmdocs/3B42_3B43_doc.pdf (accessed on 15 Decembre 2015).

53. Xue, X.; Hong, Y.; Limaye, A.S.; Gourley, J.J.; Huffman, G.J.; Khan, S.I.; Dorji, C.; Chen, S. Statistical and hydrological evaluation of TRMM-based Multi-satellite Precipitation Analysis over the Wangchu Basin of Bhutan: Are the latest satellite precipitation products 3B42V7 ready for use in ungauged basins? *J. Hydrol.* **2013**, *499*, 91–99. [CrossRef]

54. Chen, S.; Hong, Y.; Gourley, J.J.; Huffman, G.J.; Tian, Y.; Cao, Q.; Yong, B.; Kirstetter, P.E.; Hu, J.; Hardy, J.; et al. Similarity and difference of the two successive V6 and V7 TRMM multisatellite precipitation analysis performance over China. *J. Geophys. Res. Atmos.* **2013**, *118*, 13060–13074. [CrossRef]

55. Tropical Rainfall Measuring Mission (TRMM) (2011), TRMM (TMPA) Rainfall Estimate L3 3 hour 0.25 degree x 0.25 degree V7, Greenbelt, MD, Goddard Earth Sciences Data and Information Services Center (GES DISC). Available online: http://disc.gsfc.nasa.gov/datacollection/TRMM_3B42_7.html (accessed on 17 December 2015).

56. Moazami, S.; Golian, S.; Kavianpour, M.R.; Hong, Y. Comparison of PERSIANN and V7 TRMM Multi-satellite Precipitation Analysis (TMPA) products with rain gauge data over Iran. *Int. J. Remote Sens.* **2013**, *34*, 8156–8171. [CrossRef]

57. Cai, Y.; Jin, C.; Wang, A.; Guan, D.; Wu, J.; Yuan, F.; Xu, L. Spatio-temporal analysis of the accuracy of tropical multisatellite precipitation analysis 3B42 precipitation data in mid-high latitudes of China. *PLoS ONE* **2015**, *10*, 1–22. [CrossRef] [PubMed]

58. Nastos, P.T.; Kapsomenakis, J.; Philandras, K.M. Evaluation of the TRMM 3B43 gridded precipitation estimates over Greece. *Atmos. Res.* **2016**, *169*, 497–514. [CrossRef]

59. Porcù, F.; Milani, L.; Petracca, M. On the uncertainties in validating satellite instantaneous rainfall estimates with raingauge operational network. *Atmos. Res.* **2014**, *144*, 73–81. [CrossRef]

60. Wilks, D.S. Forecast verification. In *Statistical Methods in the Atmospheric Sciences*; Elsevier: Burlington, MA, USA, 2006; pp. 255–335.

61. Chaponniere, A. Fonctionnement Hydrologique d'un Bassin Versant Montagneux Semi-aride: Cas du Bassin Versant du Rehraya (Haut Atlas marocain). Ph.D. Thesis, Institut National Agronomique Paris-Grignon, Paris, France, 2005.

62. Jaw, T.; Li, J.; Hsu, K.L.; Sorooshian, S.; Driouech, F. Evaluation for Moroccan dynamically downscaled precipitation from GCM CHAM5 and its regional hydrologic response. *J. Hydrol. Reg. Stud.* **2015**, *3*, 359–378. [CrossRef]

63. Gao, Y.C.; Liu, M.F. Evaluation of high-resolution satellite precipitation products using rain gauge observations over the Tibetan Plateau. *Hydrol. Earth Syst. Sci.* **2013**, *17*, 837–849. [CrossRef]

64. Anders, A.M.; Roe, G.H.; Hallet, B.; Montgomery, D.R.; Finnegan, N.J.; Putkonen, J. Spatial patterns of precipitation and topography in the Himalaya. *Geol. Soc. Am.* **2006**, *398*, 39–53.

65. Krakauer, N.Y.; Pradhanang, S.M.; Lakhankar, T.; Jha, A.K. Evaluating satellite products for precipitation estimation in mountain regions: A case study for Nepal. *Remote Sens.* **2013**, *5*, 4107–4123. [CrossRef]

2

Long Term Spatiotemporal Variability in Rainfall Trends over the State of Jharkhand, India

Shonam Sharma * and Prasoon Kumar Singh

Department of Environmental Science and Engineering, Indian Institute of Technology
(Indian School of Mines) Dhanbad, Jharkhand Pin-826004, India; singh.prasoon910@gmail.com
* Correspondence: shonamsharma@ese.ism.ac.in

Academic Editor: Yang Zhang

Abstract: The current study was conducted to examine the impact of climate change on rainfall in Jharkhand state of India. It deals with the analysis of the historical spatiotemporal variability of rainfall on the annual, seasonal and monthly scale in 18 districts of the state Jharkhand over a period of 102 years (1901–2002). Mann-Kendall trend test and Sen's slope method were applied to detect trends and the magnitude of change over the time period of 102 years (1901–2002). Mann Whitney Pettit's method and Cumulative deviations test were applied for detection of shift point in the series. The results obtained year 1951 to be the most probable shift point in annual rainfall. The trend analysis along with the percent change for the data series before (1901–1951) and after the shift point (1952–2002) was also done. A significant downward rainfall trend was found in annual, monsoon and winter rainfall over the period of 102 years. The maximum decrease was found for the Godda (19.77%) and minimum at Purbi Singhbum station (1.95%). Trend analysis before shift point, i.e., during 1901–1951 showed an upward trend in annual rainfall and after shift point (1952–2002) a downward trend. The trend analysis for entire Jharkhand demonstrated a significant downward trend in annual and monsoon rainfall with a decrease of 14.11% and 15.65% respectively. A downward trend in seasonal rainfall will have a more pronounced effect on agricultural activities in the area as it may affect the growth phase of the kharif crops (May–October) in the region.

Keywords: Mann Kendall test; rainfall trends; Jharkhand; climate change

1. Introduction

The study of regional/local climate change has been a subject of extensive research for few decades. One of the major reasons for the increased emphasis on these studies could be due to the changing pattern of rainfall as observed in several parts of the world [1–3]. Dore [4] observed that rainfall rich areas have become more rich while, dry and arid areas have experienced increased dryness for the past few years. On a global scale, the average precipitation is expected to increase, but on regional scale it is predicted to show the pattern of increase and vice versa [5]. However, since the end of the 19th century, the global terrestrial precipitation has augmented by about 2% [6,7]. The Phenomenon of El Niño Southern Oscillation (ENSO) is the most important driver of global climate variability, which modifies rainfall distribution temporarily. Global scale variability in rainfall tends to be substantially higher in ENSO (El Niño) affected areas [8].

Atmospheric circulation patterns have a direct impact on monsoon and its weakening leads to changes in precipitation. In India, 40% of the population is dependent on monsoon for agriculture. About 54% (75.5 million ha) of the net sown area is still dependent on monsoon rainfall [9]. The anomaly of sea surface temperature over the Indian Ocean influences the variability in the monsoon rainfall [10]. In the context of climate change, it is relevant to find out how the characteristics of Indian summer monsoon are changing. Hydrologists have put the emphasis on trend analysis of Indian summer monsoon rainfall [11].

Fluctuations in rainfall events are a result of the changes of the hydrological cycle due to global warming. An understanding of these changing pattern of rainfall are required for the sustainable agriculture and water resource management. Rotstayn and Lohmann [12] observed in their modelling studies, a shift in tropical rainfall trends over land (tropics) for the period 1900–1998 and found it as a result of indirect effects of sulphate aerosols. Krishnakumar et al. [13] found a significant decline in rainfall during the southwest monsoon and an increase during the post-monsoon season over Kerala. Subash et al. [14] investigated rainfall trends at four stations namely Madhepura, Sabour, Samastipur and Patna and found an upward annual rainfall trend over all the stations except Samastipur. Patra et al. [15] reported long term (1871–2006) insignificant downward trend of annual as well as monsoon rainfall, whereas an upward trend in post-monsoon season over Orissa. Basistha et al. [16] observed a downward rainfall trend as a sudden shift rather than gradual trend over Indian Himalayas. Kumar and Jain [17] also found downward trend in the annual rainfall as well as rainy days in 15 out of 22 basins in India. Studies over states of Chhattisgarh [18] and Madhya Pradesh also reported downward rainfall trends [19].

For the management and planning at regional or local scale it has been found that continental or global scale studies of climate variables are not very much useful [20,21]. Therefore, the regional and local level climatic variables studies are required for the same. The rainfall trend analysis is important to assess the impact of climate change; therefore, in this study, an effort has been made to determine the rainfall climatology at the district level over Jharkhand. The foremost aim of the present study is to analyze the changes in yearly and seasonal rainfall over each station over a period of 102 years (1901–2002).

2. Materials and Methodology

2.1. Description of Study Area

Jharkhand is a tribal dominant state and is often called as the "Land of Forest", extending over a geographical area of 74,677 km^2. The state is bordered by five other states- Bihar in the north, Uttar Pradesh, and Chhattisgarh in the west, West Bengal in the east and Orissa in the south. It lies between 23°37′3″ N and 24°4′ N latitude and between 86°6′30″ E and 86°50′ E longitude with an altitude ranging from 3 to 1359 m (Figure 1). The state has a cultivation area of about 1.8 million ha comprising 22% of the geographical area. There are three agro climatic zones in Jharkhand i.e., central, north eastern and south eastern plateau sub zones. It is rich in mineral deposits like iron ore, coal, mica, copper, bauxite and uranium. The state has high potential for higher production of horticulture and forest products. Here the soil is newly formed and has good capacity of humus formation. Kharif (June–September) is the main cropping season in the state that is heavily dependent on the monsoon. In Rabi season (October–November) cropping is not possible because of undependable rains during this season. In non-monsoon season, irrigation sources is very limited and mostly unavailable, while during monsoon season cultivation is possible due to the availability of rainfall and during non-monsoon season farmers cannot depend on rainfall due to its less occurrence frequency (undependable rainfall). The yield of Kharif crops, chiefly paddy (dominant crop) has been adversely affected due to decline in rainfall in recent years [22]. Hydro power production has been also declined in recent years due to the unavailability of adequate amount of water in river dams that requires heavy rainfall to feed the reservoir for its operation.

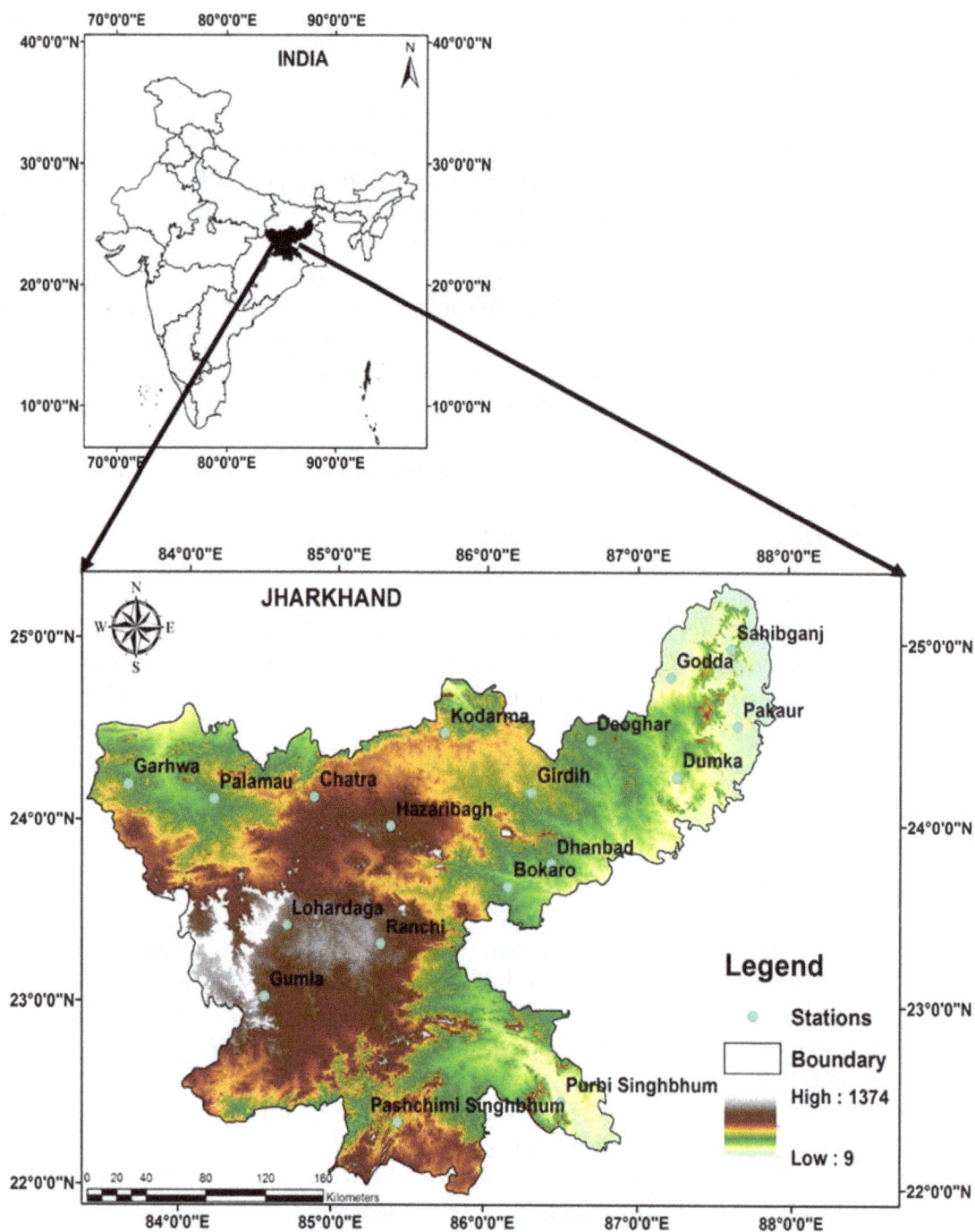

Figure 1. Jharkhand study area showing rain gauge stations along with elevation (in meters).

2.2. Data Collection

The monthly Rainfall data was downloaded from the website of India water portal for Jharkhand state for the period 1901–2002 [23]. India Metrological Department (IMD) is a centrally funded agency that is responsible for the collection of meteorological observation data, weather prediction and seismology. IMD has defined four seasons, namely winter (January–February), pre-monsoon (March–May), monsoon (June–September) and post-monsoon (October–December) so using monthly rainfall data, seasonal and annual rainfall series were prepared. For the homogeneity analysis, the cumulative deviations test was applied at 1% significance level and the results showed that, the series to be reliable and homogeneous. It is also used for shift point detection in the data series.

2.3. Methodology

As a first step of analysis, basic statistical parameters like mean, standard deviation (SD), skewness, kurtosis and coefficient of variation were estimated from the data for each station. Initially the Autocorrelation test was applied to check serial dependence in the dataset. Strong autocorrelations affect the significance assessment of trend estimates by inflating the distribution of the test statistics. These much larger critical values need to be employed as significance thresholds than in case of uncorrelated data. Apart from this, Loess regression curve was used to plot and check general patterns in data over the period of 1901–2002 for annual and seasonal series.

In general, both parametric and non-parametric tests are used for trend analysis but non parametric test is preferred as it does not require data to be normally distributed. Hydrometeorological data are generally non normal, hence non-parametric statistical tools like Mann-Kendall test and Sen's slope method were used to detect the direction and magnitude of a trend respectively in the present study. For the shift point detection the non-parametric tests like Mann-Whitney-Pettitt (PWM) test [24] and Cumulative Deviations (CD) [25] were also applied. Apart from this, for spatial and temporal trend analysis the linear regression slopes were interpolated using kriging in the present study. The detailed methodology is given below.

2.3.1. Autocorrelation

Lag-1 autocorrelation is used to check serial dependence between the data [26]. The lag-1 autocorrelation coefficient is the simple correlation coefficient of the first observations $N - 1$, X_t, $t = 1, 2, 3, \ldots, N - 1$ and the next observations, X_{t+1}, $t = 2, 3, \ldots, N$. The correlation between X_t and X_{t+1} is given by

$$r_1 = \frac{\sum_{t=1}^{N-1}(X_t - X)(X_{t+1} - \overline{X})}{\sum_{i=1}^{N}(X_t - \overline{X})^2} \tag{1}$$

where $\overline{X} = \sum_{t=1}^{N} X_t$ is the overall mean.

The lag-1 autocorrelation coefficient r_1 is tested for its significance. The probability limits on the correlogram of an independent series of the two tailed test is given below [18].

$$r_1(95\%) = \frac{-1 \pm 1.96\sqrt{N - k - 1}}{N - k} \tag{2}$$

where N is the sample size and k is the lag.

If the value of r_1 lie outside the confidence interval given above, the data are assumed to be serially correlated otherwise the sample data are considered to be serially independent.

2.3.2. Mann-Kendall trend test

Mann-Kendall test is a non-parametric test. It is frequently used for the detection of significant trend in hydrologic data series [15]. The MK test, S statistics for a series $X_1, X_2, \ldots, X_n$ can be given by:

$$S = \sum_{i=1}^{n-1} \sum_{j=i+1}^{n} sgn(X_j - X_i) \tag{3}$$

where X_i is ranked from $i = 1, 2, \ldots, n - 1$ and X_j is ranked from $j = i + 1$ and n is the length of the data set.

$$Sgn(\vartheta) = \begin{cases} 1 \ldots if \ \vartheta > 0 \\ 0 \ldots if \ \vartheta = 0 \\ -1 \ldots if \ \vartheta < 0 \end{cases} \tag{4}$$

Positive (negative) signs of the test statistics S indicate upward (downward) trend in the data. For the sample size $n \geq 8$, variance of the Mann-Kendall statistics is given by:

$$Var(S) = \frac{[n(n-1)(2n+1) - \sum_t t(t-1)(2t+5)]}{18} \tag{5}$$

where t_i is the number of ties present upto sample i.

The standardized MK test statistics (Z_{mk}) can be estimated by the following given formula:

$$Z_{mk} = \begin{cases} \frac{S-1}{\sqrt{V(S)}} & if\ S > 0 \\ 0 & if\ S = 0 \\ \frac{S+1}{\sqrt{V(S)}} & if\ S < 0 \end{cases} \tag{6}$$

The Z_{mk} follows a standard normal distribution and if its value is positive, it signifies an upward trend and if its value is negative it signifies a downward trend. If the value of Z_{mk} is greater than $Z_{\alpha/2}$ then it is considered as significant trend (where α is significance level) and the null hypothesis is rejected.

2.3.3. Sen's Method

This is a nonparametric method [27] that assumes a linear trend in the time series data and an uncorrelated data. This method is robust to missing data and outliers in the data series. It quantifies the median (50th percentile) concentration changes linearly with time and is used to determine the magnitude of the trend line. The slope of trend line in the sample of N pairs of data can be estimated by:

$$Q = \frac{X_j - X_i}{i - i'} \tag{7}$$

where X_j and X'_i are the data values at times i and i' ($i > i'$), respectively.

The median of these N values of Q is Sen's estimator of slope which is calculated as

$$\beta = Q\left(\frac{N+1}{2}\right) \text{ if } N \text{ is odd} \tag{8}$$

$$\beta = (Q(N/2) + Q(N+2)/2))/2 \text{ if } N \text{ is even} \tag{9}$$

A positive value of β indicates an upward (increasing) trend and a negative value indicates a downward (decreasing) trend in the time series.

The β sign reflects data trend direction, while its value indicates the steepness of the trend. To determine whether the median slope is statistically different than zero, one should obtain the confidence interval of β at specific probability. The confidence interval about the median slope can be computed as follows.

$$C_\alpha = Z_{1-\alpha/2} \times \sqrt{Var(S)} \tag{10}$$

where Variance (S) is

$$Var(S) = \frac{1}{18}\left[n(n-1)(2n+5) - \sum_{p-1}^{q} tp(tp-1)(2tp+5)\right] \tag{11}$$

$Z_{1-\alpha/2}$ is obtained from the standard normal distribution table.

The lower and upper limits of confidence interval, M_1 and M_2 are computed as

$$M_1 = \frac{N - C_\alpha}{2} \tag{12}$$

$$M_2 = \frac{N + C_\alpha}{2} \tag{13}$$

2.3.4. Mann-Whitney-Pettit Test

Let t be the most likely change point year of a time series $(X_1, X_2, \dots, X_n)$ with a length of n. Two partial time series, $\{X_1, X_2, \dots, X_t\}$ and $\{X_{t+1}, X_{t+2}, \dots, X_n\}$, can then be derived by dividing the time series at time t. The U_t statistics for the series can be given by

$$U_{tT} = \sum_{i=1}^{t} \sum_{j=t+1}^{n} sgn(X_i - X_j) \tag{14}$$

A continuous increase in the value of U_{tT} when plotted with t will indicate absence of change point but when the value of $|U_{tT}|$ increases initially and then decreases after a point, it will indicate presence of change point. The point where maximum value of U_t is obtained is considered as the most probable change point.

$$K_T = \max_{1 \le t \le T} |U_{tT}| \tag{15}$$

2.3.5. Cumulative Deviations Test

Here the null hypothesis says that the time series variable is independently and identically distributed and the alternate hypothesis says, that there is a shift in the mean value after a certain time.

The test can still be applied, however, when there are slight departures from normality. The departure from homogeneity is tested using the statistics, which is defined as

$$Q = \max_{1 \le k \le n} |S_k^*| \tag{16}$$

In which the rescaled adjusted partial sums are obtained by dividing the S_k by the sample standard deviation (D_X).

$$S_k^* = \frac{S_k}{D_X} \tag{17}$$

where S_k can be computed by

$$S_k = \sum_{i=1}^{k} (X_i - \overline{X}) \tag{18}$$

This test is based on the adjusted partial sums or cumulative deviations from the mean of the time series data. If the magnitude of $Q/\sqrt{n}$ exceeds the value at the considered critical level then the time series is heterogeneous (Yu et al., 2006). Critical values of $Q/\sqrt{n}$ for the 99% confidence limits are used in this study [25].

3. Results

3.1. Preliminary Investigation

Monthly and seasonal characteristics of rainfall over each station were calculated for the period 1901–2002 (Table 1). Annual precipitation varied between 1211 mm in the northwestern part (Chatra station) and 1383 mm in eastern part (Pakaur station) of Jharkhand. The standard deviation varied between 184 mm and 256 mm (Figure 2a). The skewness, which is a measure of the asymmetry in frequency distribution around the mean, varied between −0.04 and 0.81 indicating that annual rainfall during the period is asymmetric and it lies to the right of the mean over all the stations. Kurtosis varied from −0.34 to 1.73 which describes the peakedness of a symmetrical frequency distribution.

Table 1. Statistical Analysis of rainfall (mm) over the study area.

Stations	Mean (mm)	SD (mm)	Skewness	Kurtosis	CV (%)
Bokaro	1358	230	0.41	1.26	17.0
Chatra	1211	257	0.11	0.15	21.2
Deogarh	1312	234	0.61	0.55	17.8
Dhanbad	1341	217	0.60	1.51	16.2
Dumka	1346	236	0.60	0.57	17.5
Garhwa	1222	247	−0.04	−0.29	20.2
Giridih	1300	232	0.48	0.84	17.9
Godda	1285	228	0.38	−0.34	17.8
Gumla	1370	241	0.39	0.52	17.6
Hazaribagh	1305	242	0.29	0.72	18.5
Kodarma	1256	240	0.27	0.54	19.1
Lohardaga	1295	247	0.30	0.27	19.1
Pakaur	1383	246	0.41	0.05	17.8
Palamau	1225	257	0.05	−0.12	21.0
Paschimi Singhbhum	1280	187	0.81	1.73	14.6
Purbi Singhbhum	1310	184	0.65	1.40	14.1
Ranchi	1300	215	0.49	0.93	16.6
Sahibganj	1348	227	0.31	-0.24	16.8

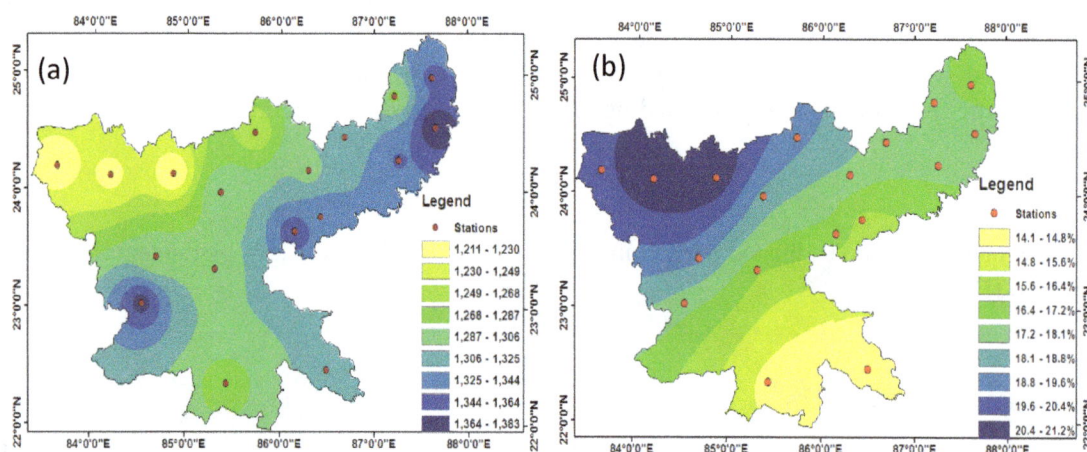

Figure 2. The spatial distribution of (**a**) rainfall and (**b**) Coefficient of variation in annual rainfall over the period of 1901–2002.

For the computation of rainfall variability, the formula of coefficient of variation was used. Coefficient of variation (CV) is a statistical measure of the dispersion of data points in a data series around the mean. The value of Coefficient of variation ranged between 14.1% (Purbi Singhbhum) and 21.2% (Chatra) with an average coefficient of variation 17.8% for the entire state (Figure 2b).

3.2. General Pattern Analysis in Precipitation

For the assessment of the long term pattern of precipitation, standardized series were prepared using 102 year period data for both annual and seasonal time series. The average of all the stations data was calculated for annual and seasonal scale and further standardization was performed using these averaged series. The moving average is not resistant (robust) to local fluctuations therefore, to reduce the local fluctuations, the standardized data series were fitted with LOESS [28–30] regression curves to identify patterns over time. The Loess curve of annual precipitation (Figure 3a) displayed a gradual rise in precipitation up to year 1941. It reached the highest value in 1941. From 1941 onwards, it showed a downward trend up to 2002 and reached the lowest value in 2002. This graph indicated a decrease in rainfall in the 2nd half of the century. Figure 3b presents the Loess regression curves for seasonal rainfall series. From Figure 3b it can be seen that monsoon rainfall series has followed a similar trend to annual rainfall series. The winter rainfall displayed a minor rise up to year 1939. From 1940 onwards, it indicated a decreasing trend up to 1963 and then slightly increased. The post-monsoon rainfall showed very slight increase up to 1962. From 1963 onwards, it decreased up to 1978 and then slightly increased. Pre-monsoon rainfall showed a decline in rainfall from 1924 onward up to 1960 and afterwards an increase was observed.

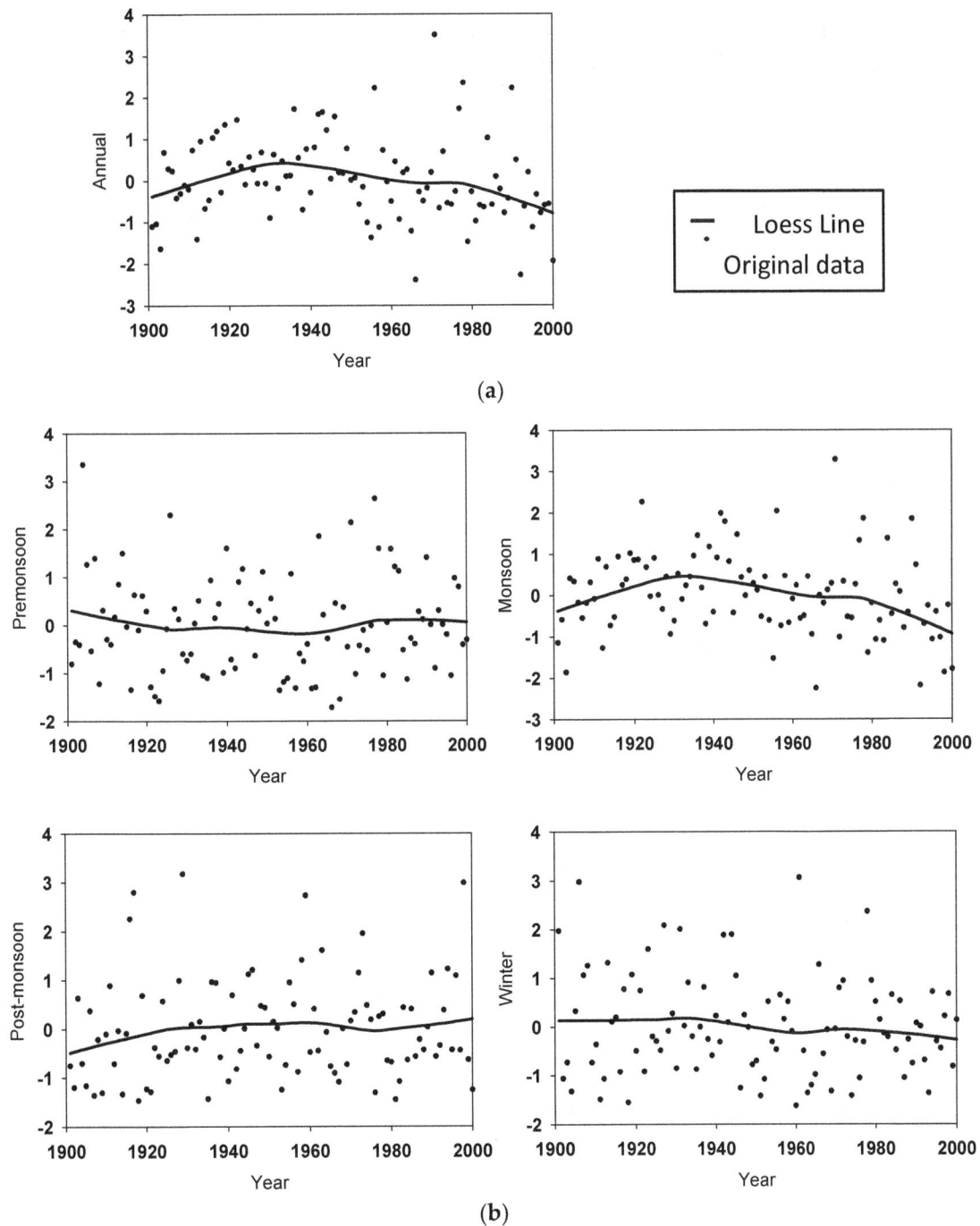

Figure 3. Loess regression curve of (**a**) Annual and (**b**) Seasonal rainfall.

3.3. Rainfall Trend and Percent Change Analysis

Lag-1 autocorrelation was used to detect the serial correlation in the dataset. None of the data series was found serially correlated so, the Mann-Kendall (MK) test was applied to compute the trend in the data. The results of MK test and Sen's slope are demonstrated in Figure 4 and Table 2 respectively. The percent change in annual and seasonal rainfall is shown in Figure 4. Significant downward trends in annual rainfall were observed over all the stations except Paschimi Singhbhum and Purbi Singhbhum which showed an insignificant downward trend. The slope of the downward trends in annual precipitation ranged between 0.25 (at Purbi Singhbhum station of Northeastern zone) to 2.49 mm per year (at Godda station of southeastern zone).

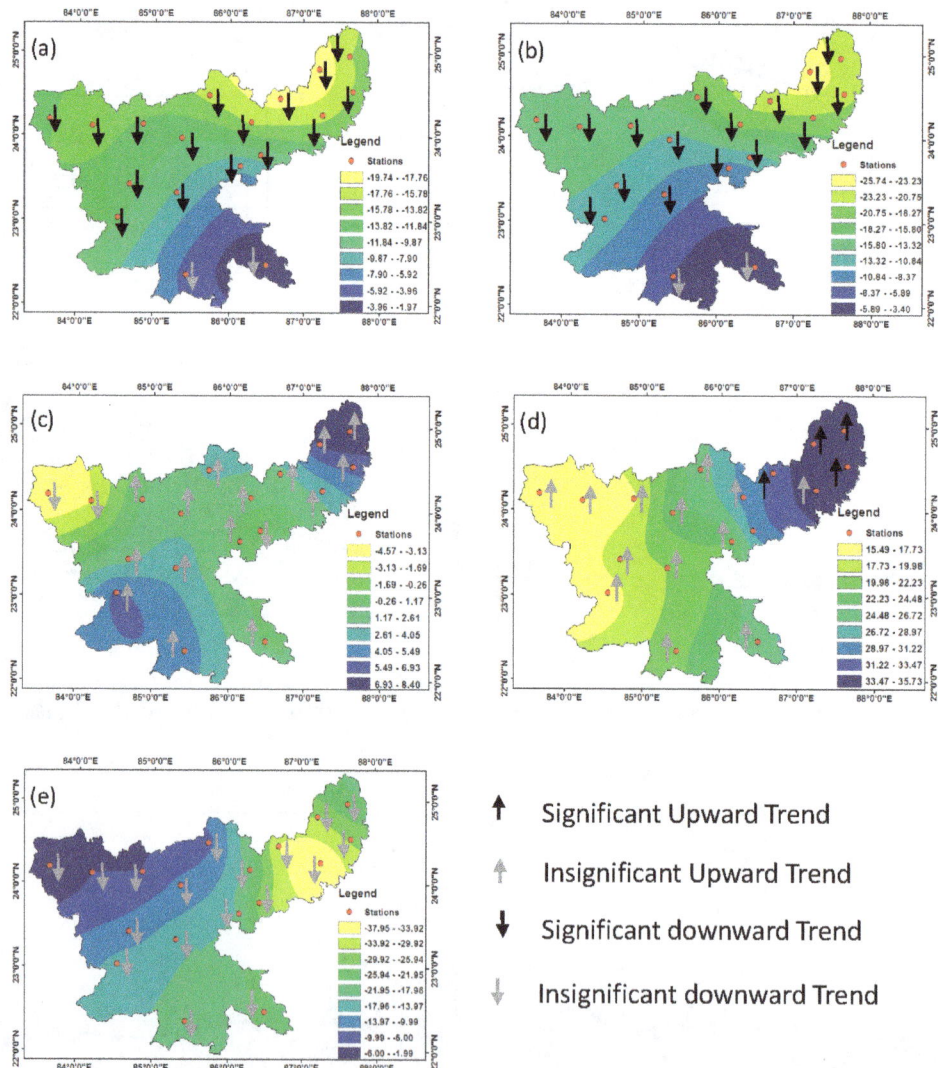

Figure 4. Spatial distribution of percent change over mean and trend analysis (Mann Kendall statistics) for period 1901-2002 of (**a**) Annual; (**b**) Monsoon; (**c**) Pre-monsoon; (**d**) Post-monsoon and (**e**) Winter season.

No significant results were found during Pre-monsoon rainfall. An insignificant upward trend was observed over all the stations except three stations namely, Palamau, Dhanbad and Garhwa that showed an insignificant downward trend. While monsoon rainfall showed the significant downward trends over all the stations except Purbi Singhbhum and Paschimi Singhum. The post monsoon rainfall showed significant upward rainfall trends at Dumka, Sahibganj, Deogarh, Pakaur and Godda.

During winter season, downward trends were prevalent in all the stations, but the significant trends were shown by Dumka and Deogarh only.

Table 2. Sen's slope (β) value of Annual and Seasonal rainfall (1901–2002).

Time (1901–2002) Series	Annual	Monsoon	Premonsoon	Postmonsoon	Winter
	β	β	β	β	β
Bokaro	−1.11	−1.08	0.00	0.20	−0.10
Chatra	−1.62	−1.52	0.01	0.13	−0.03
Deogarh	−2.40	−2.38	0.01	0.29	−0.14
Dhanbad	−1.41	−1.29	−0.01	0.27	−0.14
Dumka	−2.17	−2.20	0.03	0.35	−0.15
Garhwa	−1.64	−1.52	−0.02	0.11	−0.01
Giridih	−2.09	−2.05	0.02	0.26	−0.09
Godda	−2.49	−2.68	0.07	0.34	−0.10
Gumla	−1.66	−1.44	0.04	0.11	−0.07
Hazaribagh	−1.54	−1.34	0.02	0.20	−0.06
Kodarma	−2.01	−1.9	0.02	0.21	−0.04
Lohardaga	−1.60	−1.4	0.01	0.14	−0.06
Pakaur	−2.11	−2.4	0.07	0.38	−0.1
Palamau	−1.73	−1.6	0	0.11	−0.03
Pashchimi Singhbhum	−0.66	−0.6	0.05	0.18	−0.1
Purbi Singhbum	−0.25	−0.3	0.01	0.26	−0.11
Ranchi	−1.24	−1.1	0.03	0.16	−0.09
Sahibganj	−2.10	−2.3	0.1	0.36	−0.07

The downward trend of rainfall during monsoon (June-September) would adversely affect the yield of rice crops. The decrease in rainfall in almost all the seasons may drastically affect the agricultural activities and groundwater, surface water resources in the area. The Percent change more than 10% were shown by 14 annual series and 15 monsoon series. It was found least during pre-monsoon season and the maximum in winter season.

3.4. Shift Point Analysis

For shift point detection in annual series the Mann-Whitney-Pettitt (MWP) and the cumulative deviation test were used. The results were shown in Table 3. The most probable shift point was year 1951 as depicted from both the MWP and Cumulative deviation test.

Table 3. Results of MWP test and Cumulative deviation test.

Stations Name	MWP Test		Cumulative Deviation Test	
	Shift Point	p Value	Shift Point	Q√n
Bokaro	1951	0.033	1951	0.99
Chatra	1951	0.02	1951	1.18
Deogarh	1954	0.001	1957	1.46
Dhanbad	1952	0.015	1951	1.07
Dumka	1957	0.002	1957	1.40
Garhwa	1951	0.025	1951	1.31
Giridih	1952	0.002	1979	1.40
Godda	1960	0	1960	1.77
Gumla	1950	0.019	1950	1.16
Hazaribagh	1951	0.024	1979	1.11
Kodarma	1951	0.006	1979	1.37
Lohardaga	1950	0.02	1950	1.15
Pakaur	1960	0.001	1957	1.46
Sahibganj	1960	0	1957	1.60

Table 3. *Cont.*

Stations Name	MWP Test		Cumulative Deviation Test	
	Shift Point	*p* Value	Shift Point	Q√n
Palamau	1951	0.02	1951	1.27
Paschimi Singhbhum	1950	0.051	1950	0.83
Purbi Singhbhum	1950	0.307	1950	0.55
Ranchi	1950	0.029	1950	1.02
Sahibganj	1960	0	1957	1.60

3.5. Trend Analysis of Two Time Series (Annual) i.e., 1901–1951 and 1952–2002

The results of trend analysis for both partial time series were shown in Figures 5 and 6 for the periods 1901–1951 and 1952–2002 respectively. The trend analysis of two time series, i.e., before (1901–1951) and after the shift point (1952–2002) showed opposite results. Annually significant upward rainfall trends were found in all the stations except Purbi Singhbhum during the period 1901–1951. While, for the period 1901–2002 the downward trend was found in all the stations, but significant trends were shown by Godda and Deogarh. The results for monsoon season showed significant downward trends for 9 stations namely, Deogarh, Dhanbad, Garhwa, Giridih, Godda, Kodarma, Pakaur, Palamau and Sahibganj. Overall, it can be concluded that a decrease in rainfall has occurred over the entire time period. Percent change was computed for the period 1901–1951 and 1952–2002 and presented in Tables 4 and 5 respectively. Maximum increasing % change in annual rainfall was found at Chatra (20.55% during 1901–1951) and maximum decreasing % change in annual rainfall was found at Palamau (−18.15% during 1952–2002).

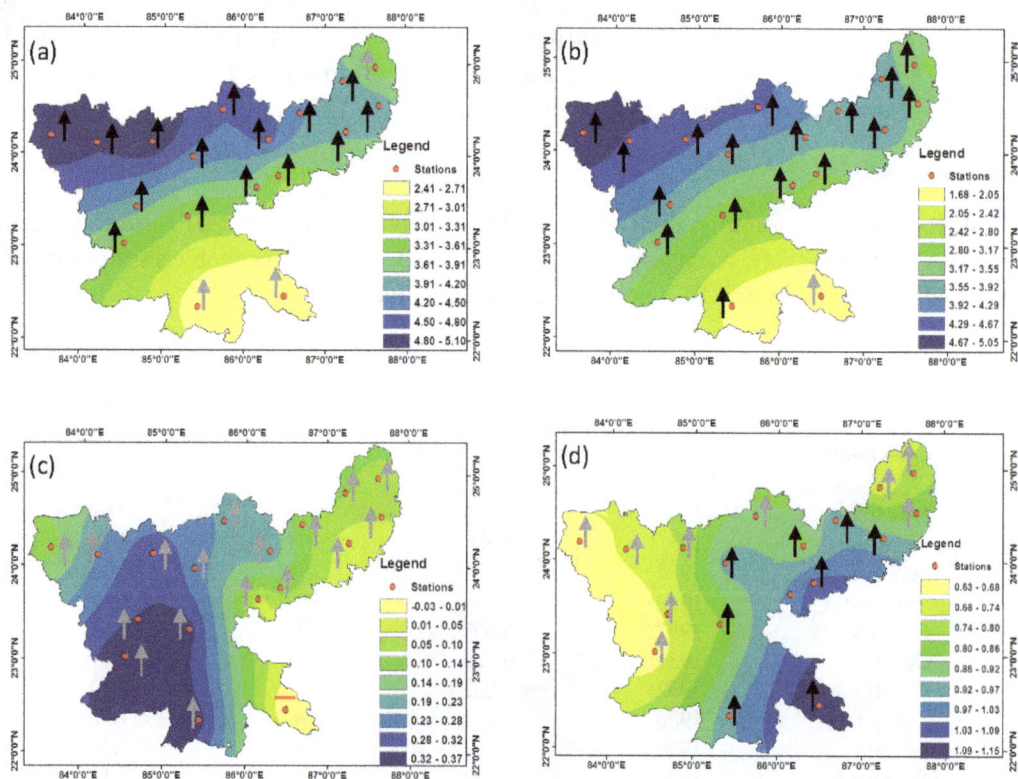

Figure 5. *Cont.*

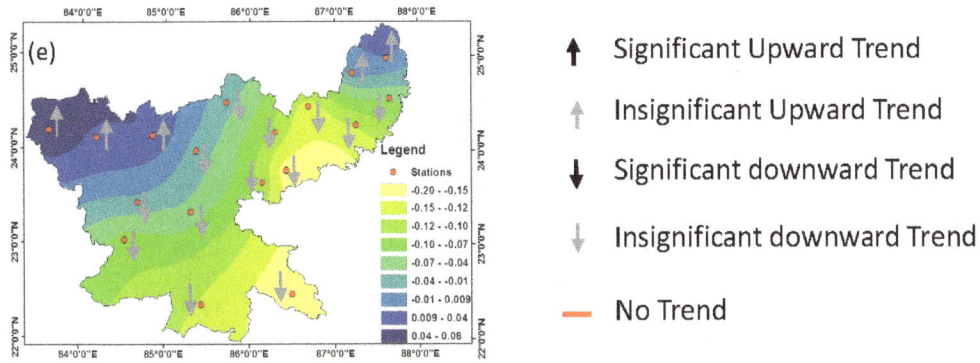

Figure 5. Spatial distribution of Sen's slope value and trend direction (Mann Kendall statistics) from 1901–1951of (**a**) Annual; (**b**) Monsoon; (**c**) Pre-monsoon; (**d**) Post-monsoon and (**e**) Winter season.

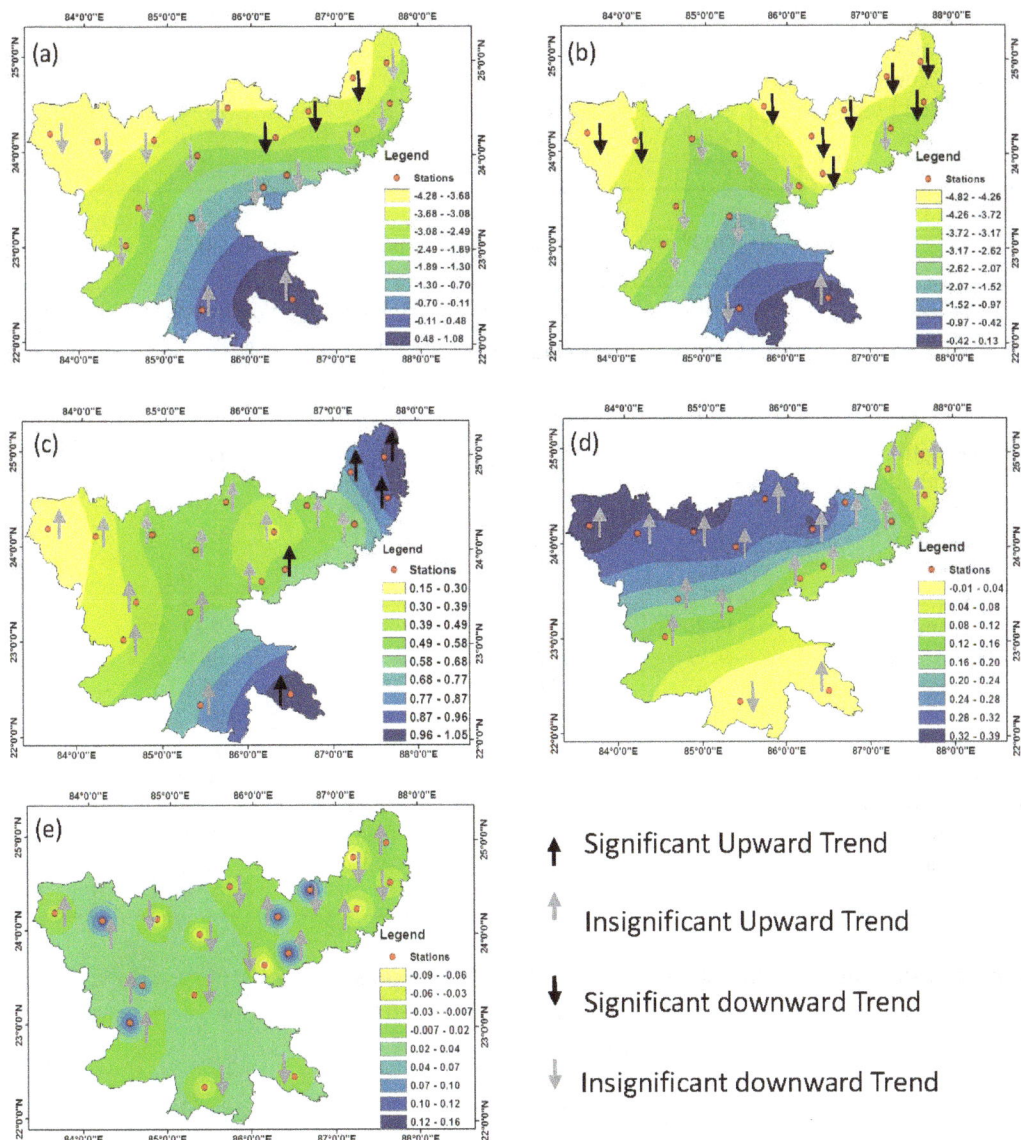

Figure 6. Spatial distribution of Sen's slope value and trend direction (Mann Kendall statistics) from 1952–2002 of (**a**) Annual; (**b**) Monsoon; (**c**) Pre-monsoon; (**d**) Post-monsoon and (**e**) Winter season.

Table 4. Percent change of Annual and Seasonal rainfall (1901–1951).

1901–1951	Annual	Pre-Monsoon	Monsoon	Post-Monsoon	Winter
Stations	% Change	% Change	% Change	% Change	% Change
Bokaro	13.65	2.09	14.48	56.53	−5.14
Chatra	20.55	24.99	21.14	49.71	3.31
Deogarh	15.76	7.15	16.80	55.67	−9.86
Dhanbad	12.28	6.49	14.26	60.75	−11.46
Dumka	14.34	1.43	16.66	51.11	−7.07
Garhwa	20.28	15.56	23.12	44.29	5.62
Giridih	17.44	14.40	16.75	50.10	−5.37
Godda	15.46	6.08	17.60	39.98	1.14
Gumla	12.62	29.74	13.79	49.34	−7.59
Hazaribagh	16.52	16.15	17.28	58.02	−2.44
Kodarma	17.45	16.14	20.00	55.07	−1.88
Lohardaga	15.28	25.61	16.80	48.88	−1.40
Pakaur	14.49	2.50	14.63	42.87	−4.04
Palamau	19.27	22.41	21.28	49.58	1.40
Pashchimi Singhbhum	10.02	17.01	9.30	59.74	−8.09
Purbi Singhbum	9.26	−1.05	8.31	59.35	−9.81
Ranchi	12.52	20.96	13.42	51.29	−3.21
Sahibganj	12.37	3.51	13.96	40.30	1.05

Table 5. Percent change of Annual and Seasonal rainfall (1952–2002).

1952–2002	Annual	Pre-Monsoon	Monsoon	Post-Monsoon	Winter
Stations	% Change	% Change	% Change	% Change	% Change
Bokaro	−3.22	16.86	−16.27	5.44	−8.82
Chatra	−16.18	55.12	−16.43	21.82	−8.58
Deogarh	−13.75	35.58	−22.10	14.79	21.84
Dhanbad	−4.77	42.65	−23.28	4.24	17.29
Dumka	−11.25	19.49	−16.39	10.78	−11.40
Garhwa	−18.00	20.36	−23.17	28.61	−3.16
Giridih	−13.26	9.93	−22.73	21.51	16.41
Godda	−17.98	45.31	−24.88	3.89	−14.90
Gumla	−11.26	27.65	−15.25	7.42	16.52
Hazaribagh	−9.35	41.60	−15.33	19.48	−8.00
Kodarma	−16.07	42.67	−22.78	17.31	−3.49
Lohardaga	−10.56	33.40	−16.05	15.82	8.23
Pakaur	−10.63	47.21	−18.46	1.28	−6.46
Palamau	−18.15	29.16	−22.16	21.29	14.92
Pashchimi Singhbhum	0.41	41.20	−2.91	−0.58	−6.13
Purbi Singhbum	4.26	45.22	0.66	0.95	−2.18
Ranchi	−4.98	38.34	−9.81	10.43	−1.96
Sahibganj	−13.04	46.35	−21.20	3.15	5.29

Out of 18 stations, only two stations (Deogarh and Dumka) showed significant downward trends during winter rainfall (1952–2002). Before shift point (1901–1951), none of the station showed significant downward trends in winter rainfall.

3.6. Comparative Analysis of Mean Annual Precipitation between Two Time Period (1901–1951 and 1952–2002)

Percentage change in annual rainfall was computed by calculating percentage of second period average (1952–2002) from first period average (1901–1951) to demonstrate the decrease in rainfall over all the stations. Table 6 showed the decrease in rainfall which was found lowest in Ranchi (−2.59) and highest in Godda (−10.45). This shows a higher variation in change percentage in the state.

Table 6. Percentage change in the mean of 1901–1951 over the mean of 1952–2002.

Station Name	Change Percentage	Station Name	Change Percentage
Bokaro	−6.15	Hazaribagh	−7.44
Chatra	−9.03	Kodarma	−9.43
Deogarh	−9.65	Lohardaga	−7.93
Dhanbad	−6.62	Pakaur	−8.79
Dumka	−8.93	Palamau	−9.58
Garhwa	−9.46	Paschimi Singhbhum	−4.16
Giridih	−9.11	Purbi Singhbhum	−2.59
Godda	−10.45	Ranchi	−6.06
Gumla	−7.33	Sahibganj	−8.73

3.7. Spatial Analysis of Precipitation Series

To analyze the spatial behavioral changes in rainfall, the linear regression slope of each station was interpolated using Kriging in ArcGIS environment for the whole study period (1901–2002) and after shift (1952–2002) which is shown in Figures 7–11. The interpolated linear slopes for the annual precipitation for the period 1901 to 2002 (Figure 7a) indicated a negative slope all over the entire study area while for period 1952–2002 (Figure 7b), it varied from positive to negative. Positive linear slopes were found in Northern and South Eastern region that showed a decrease in rainfall from north to south east direction. For the whole study period, the negative slope value (which decreased up to −2.10 mm/year) was less in comparison to entire duration (up to −4.33 mm/year) in annual rainfall. While the rise in magnitude of trend is found only after the shift point (up to 0.84 mm/year). During monsoon season (Figure 8a,b) decrease in rainfall magnitude varied from −0.23 to −2.29 mm/year for the whole study period. The maximum negative slopes were shown by Deogarh, Godda, Dumka, Sahibganj, and Pakaur. The rainfall magnitude varied from −0.11 to −4.75 mm/year after shift point. Maximum negative slopes were shown by Chatra, Palamau, Deogarh, Kodarma, Godda, Dumka, and Sahibganj. Decline in rainfall increased during the period 1952-2002 as compared to the period 1901–2002.

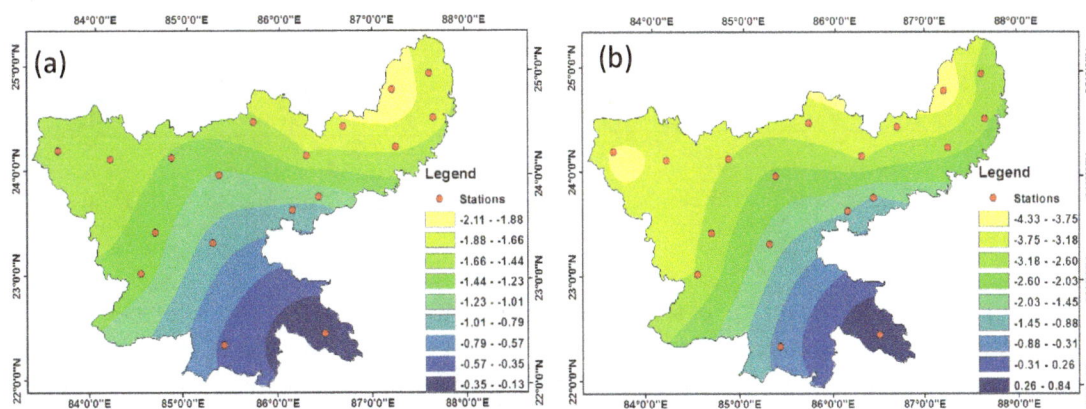

Figure 7. Spatial distribution of temporal change in Annual rainfall during 1901 to 2002 (**a**) and from 1952 to 2002 (**b**).

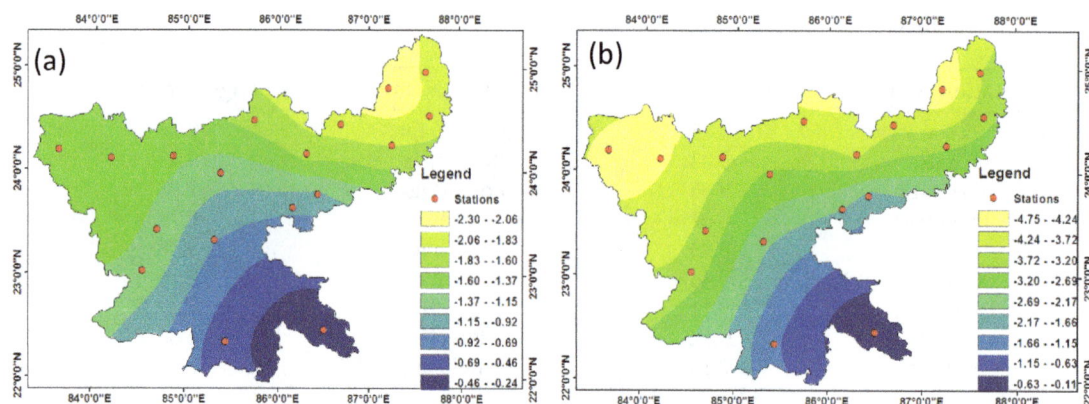

Figure 8. Spatial distribution of temporal change in Monsoon rainfall during 1901 to 2002 (**a**) and from 1952 to 2002 (**b**).

In the pre-monsoon season the rainfall magnitude varied between -0.0407 to 0.0823 mm/year during 1901–2002 (Figure 9a,b) and the negative slopes were prominent from western to the southern region. The stations that showed negative slopes were Garhwa, Palamau, Chatra Hazaribagh, Chatra, Kodarma, Giridih, Dhanbad, Bokaro, Ranchi, Gumla, Lohardaga and Purbi Singhbhum. During 1952–2002 the linear interpolated slopes for the pre-monsoon season varied between 0.13 and 1.16 mm/year during 1952–2002 that showed the absence of negative slopes in the region.

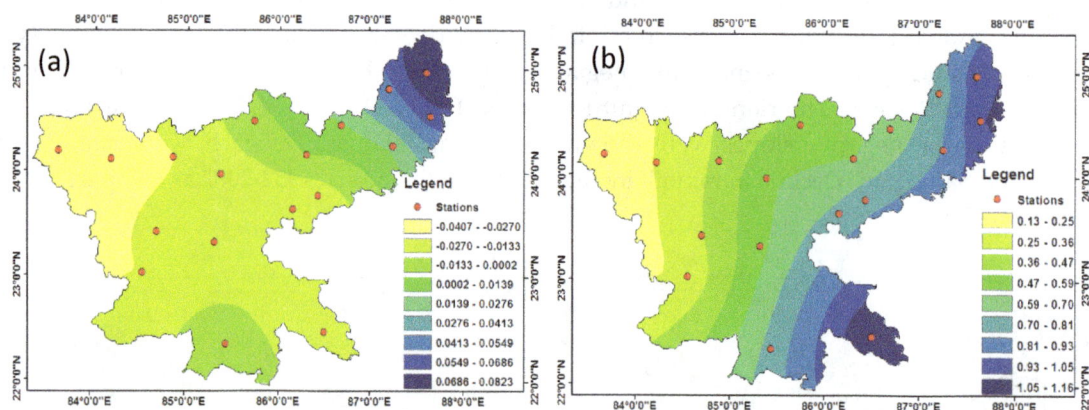

Figure 9. Spatial distribution of temporal change in Pre monsoon rainfall during 1901 to 2002 (**a**) and from 1952 to 2002 (**b**).

The interpolated slopes for the post monsoon (Figure 10a,b) precipitation for the period 1901 to 2002 indicated a prevalent positive slope (0.08 to 0.30 mm/year) all over the study area, it showed an increase in rainfall during post-monsoon. After shift point both negative and positive slopes were observed in the study area and it varied between -0.23 and 0.33 mm/year. The negative slopes were found only at northeastern and southern region of the study area which varied from -0.04 to -0.24 mm/year.

For the winter season, negative interpolated slopes were prevalent all over the study area during 1901–2002 (Figure 11a,b). The rainfall magnitude varies from -0.02 to -0.18 mm/year during this time period. The maximum negative slopes were observed at Deogarh, Dumka and Dhanbad. After shift point both negative and positive slopes, ranged between -0.24 and 0.12 mm/year were observed. However, widespread negative slopes (-0.03 to -0.024 mm/year) were depicted covering most of the state, except for western part of Jharkhand during 1952–2002.

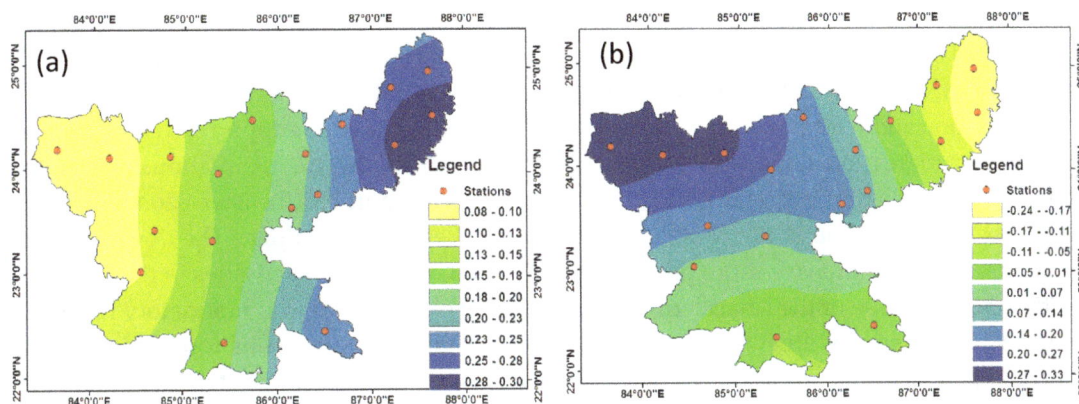

Figure 10. Spatial distribution of temporal change in Post monsoon rainfall during 1901 to 2002 (**a**) and from 1952 to 2002 (**b**).

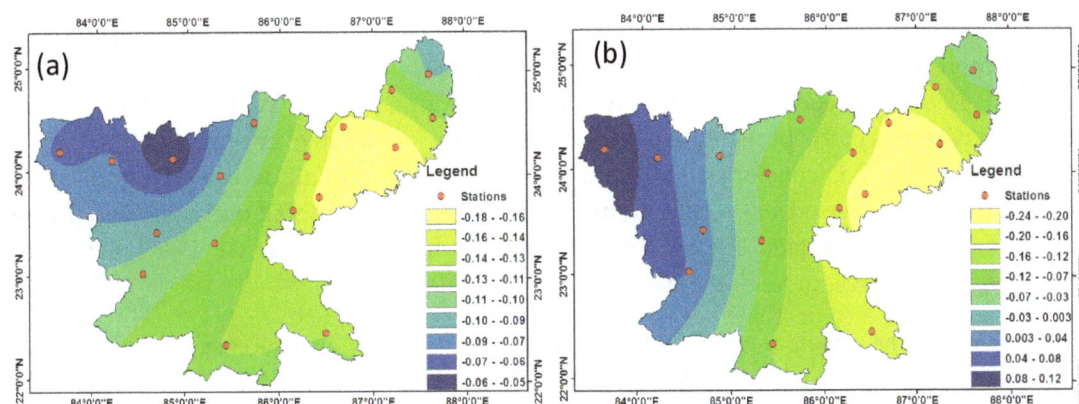

Figure 11. Spatial distribution of temporal change in winter rainfall during 1901 to 2002 (**a**) and from 1952 to 2002 (**b**).

3.8. Precipitation Trends over Entire Jharkhand

The results of trend analysis of mean annual and seasonal rainfall for whole Jharkhand is demonstrated in Table 7. The results showed statistically significant downward trend in annual and monsoon rainfall. However, insignificant upward trend was observed for pre-monsoon and post-monsoon season rainfall for the entire state. A decline of 14.11% and 15.65% was noticed for annual and monsoon season rainfall respectively. The winter season showed statistically insignificant downward trend, with a decline of 19.38%.

Table 7. Mann Kendall and Percent change analysis over entire Jharkhand.

Time Series	Test Z	β	% Change	Time Series	Test Z	β	% Change
Annual	−2.6 **	−1.8	−14.11	Post Monsoon	1.5	0.2	28.55
Premonsoon	0.2	0.03	3.38	Winter	−0.9	−0.1	−19.38
Monsoon	−2.7 **	−1.7	−15.65				

*** if trend at α = 0.001 level of significance ** if trend at α = 0.01 level of significance * if trend at α = 0.05 level of significance + if trend at α = 0.1 level of significance.

4. Discussion and Conclusions

In the present study, trends for annual and seasonal rainfall series were analyzed for Jharkhand during the period 1901–2002 using India water portal rainfall data. The western part of Jharkhand experiences lower rainfall as compared to eastern Jharkhand. The retreating monsoon enters from the eastern zone, so it experiences higher rainfall. Coefficient of variation was found higher in the western region of the Jharkhand state. Autocorrelation was absent in the dataset and the results of the trend analysis (Mann-Kendall) showed a downward trend of rainfall in almost over all stations for annual, monsoon and winter season. The results of Mann-Kendall were supported with interpolated maps of linear regression slope values. The slope of the downward trends in annual precipitation ranged between 0.25 mm per year (at Purbi Singhbhum station in Northeastern zone) to 2.49 mm per year (at Godda station in Southeastern zone) as per Sen's slope values. The downward trend in seasonal rainfall will have a more pronounced effect on agricultural activities in the area. It may affect the growth phase of the Kharif crops (May–October) and irrigation is mandatory to tackle the moisture stress. Mann Whitney Pettit and cumulative deviations test results found most probable year of change in the state to be year 1951. There was an upward trend in the state during the period 1901–1951(before change point), which got reversed during the period 1952–2002 (after change point). For whole Jharkhand a downward trend in annual rainfall was noticed in the study.

Sathiyamoorthy and Rao et al found a reduction in strength of Tropical Easterly Jet Stream during monsoon in the recent five decades, which are responsible for the formation of monsoon depressions during the southwest monsoon season, and that are the important rain bearing systems during the southwest monsoon season [31,32]. The decrease in frequency of cyclonic storms over Indian seas during 1981–1997 have been reported by Ray and Srivastava [33]. These weather systems declining frequency may be one of the probable reasons for the decline in rainfall over the area.

From the study, it is concluded that annual and monsoon rainfall decreased significantly in Jharkhand during the period 1901–2002. If this downward trend in rainfall persists, it would badly impact the economy of the state. There is a need to integrate the changing climate in the planning and management of water resources of the state.

Acknowledgments: The authors are thankful to India water portal for providing the essential data required for the present study. Sincere thanks to Arun Kumar Taxak (Research scholar, Indian Institute of Technology, Roorkee) for his help in data handling. Authors are sincerely grateful to Amit Ghosh (PhD. student, Jadavpur University) for his earnest help in using R software. Finally, we would like to thank the anonymous reviewers for their valuable comments and suggestions which helped us on improving this paper.

Author Contributions: The design of study, data collection and data analysis has been done by Shonam Sharma and Prasoon Kumar Singh contributed to the interpretation part of the study. While the manuscript writing has been done by both the authors.

Conflicts of Interest: The authors declare no conflict of interest.

References

1.   Cruz, R.; Lasco, R.; Pulhin, J.; Pulhin, F.; Garcia, K. *Climate Change Impact on Water Resources in Pantabangan Watershed, Philippines*; AIACC Final Technical Report; The International Start Secreteriat: Washington, DC, USA, 2006; pp. 9–107.

2.   Hu, Z.Z.; Yang, S.; Wu, R. Long-term climate variations in China and global warming signals. *J. Geophys. Res. Atmos.* **2003**, *108*. [CrossRef]

3.   Peterson, B.J.; Holmes, R.M.; McClelland, J.W.; Vörösmarty, C.J.; Lammers, R.B.; Shiklomanov, A.I.; Shiklomanov, I.A.; Rahmstorf, S. Increasing river discharge to the Arctic Ocean. *Science* **2002**, *298*, 2171–2173. [CrossRef] [PubMed]

4.   Dore, M.H. Climate change and changes in global precipitation patterns: What do we know? *Environ. Int.* **2005**, *31*, 1167–1181. [CrossRef] [PubMed]

5.   Stocker, T.F.; Qin, D.; Plattner, G.-K.; Tignor, M.; Allen, S.K.; Boschung, J.; Nauels, A.; Xia, Y.; Bex, V.; Midgley, P.M. *Climate Change 2013: The Physical Science Basis*; Cambridge University Press: Cambridge, UK; New York, NY, USA, 2014.

6.   Jones, P.; Hulme, M. Calculating regional climatic time series for temperature and precipitation: Methods and illustrations. *Int. J. Climatol.* **1996**, *16*, 361–377. [CrossRef]

7.   Hulme, M.; Osborn, T.J.; Johns, T.C. Precipitation sensitivity to global warming: Comparison of observations with HadCM2 simulations. *Geophys. Res. Lett.* **1998**, *25*, 3379–3382. [CrossRef]

8.   Nicholls, N. El nino-southern oscillation and rainfall variability. *J. Clim.* **1988**, *1*, 418–421. [CrossRef]

9.   Clark, C.O.; Cole, J.E.; Webster, P.J. Indian Ocean SST and Indian summer rainfall: Predictive relationships and their decadal variability. *J. Clim.* **2000**, *13*, 2503–2519. [CrossRef]

10. Times of India. Available online: http://timesofindia.indiatimes.com/india/40-of-India-still-banks-on-monsoon-for-agriculture/articleshow/47115057.cms (accessed on 7 February 2017).

11. Guhathakurta, P.; Rajeevan, M. Trends in the rainfall pattern over India. *Int. J. Climatol.* **2008**, *28*, 1453–1470. [CrossRef]

12. Rotstayn, L.D.; Lohmann, U. Tropical rainfall trends and the indirect aerosol effect. *J. Clim.* **2002**, *15*, 2103–2116. [CrossRef]

13. Krishnakumar, K.N.; Prasada Rao, G.S.L.H.V.; Gopakumar, C.S. Rainfall trends in twentieth century over Kerala, India. *Atmos. Environ.* **2009**, *43*, 1940–1944. [CrossRef]

14. Subash, N.; Singh, S.S.; Priya, N. Extreme rainfall indices and its impact on rice productivity—A case study over sub-humid climatic environment. *Agric. Water Manag.* **2011**, *98*, 1373–1387. [CrossRef]

15. Patra, J.P.; Mishra, A.; Singh, R.; Raghuwanshi, N. Detecting rainfall trends in twentieth century (1871–2006) over Orissa state, India. *Clim. Chang.* **2012**, *111*, 801–817. [CrossRef]

16. Basistha, A.; Arya, D.; Goel, N. Analysis of historical changes in rainfall in the Indian Himalayas. *Int. J. Climatol.* **2009**, *29*, 555–572. [CrossRef]

17. Kumar, V.; Jain, S.K. Trends in rainfall amount and number of rainy days in river basins of India (1951–2004). *Hydrol. Res.* **2011**, *42*, 290–306. [CrossRef]

18. Meshram, S.G.; Singh, V.P.; Meshram, C. Long-term trend and variability of precipitation in Chhattisgarh state, India. *Theor. Appl. Climatol.* **2016**. [CrossRef]

19. Duhan, D.; Pandey, A. Statistical analysis of long term spatial and temporal trends of precipitation during 1901–2002 at Madhya Pradesh, India. *Atmos. Res.* **2013**, *122*, 136–149. [CrossRef]

20. Barsugli, J.; Anderson, C.; Smith, J.; Vogel, J. *Options for Improving Climate Modeling to Assist Water Utility Planning for Climate Change*; Water Utility Climate Alliance: Clearwater, FL, USA, 2009.

21. Brekke, L.D. *Climate Change and Water Resources Management: A Federal Perspective*; DIANE Publishing: Darby, PA, USA, 2009.

22. Agricultural Drought Assessment Report. Available online: http://bhuvan.nrsc.gov.in/bhuvan/PDF/NADAMS_July12_Report.pdf (accessed on 20 August 2016).

23. India Water Portal. Available online: http://www.indiawaterportal.org/metdata (accessed on 10 January 2016).

24. Pettitt, A. A non-parametric approach to the change-point problem. *Appl. Stat.* **1979**, *28*, 126–135. [CrossRef]

25. Buishand, T.A. Some methods for testing the homogeneity of rainfall records. *J. Hydrol.* **1982**, *58*, 11–27. [CrossRef]

26. Anderson, R.L. Distribution of the serial correlation coefficient. *Ann. Math. Stat.* **1942**, *13*, 1–13. [CrossRef]

27. Sen, P.K. Estimates of the regression coefficient based on Kendall's Tau. *J. Am. Stat. Assoc.* **1968**, *63*, 1379–1389. [CrossRef]

28. Cleveland, W.S. Robust locally weighted regression and smoothing scatterplots. *J. Am. Stat. Assoc.* **1979**, *74*, 829–836. [CrossRef]

29. Cleveland, W.S.; Grosse, E. Computational methods for local regression. *Stat. Comput.* **1991**, *1*, 47–62. [CrossRef]

30. Helsel, D.R.; Hirsch, R.M. *Statistical Methods in Water Resources*; US Geological Survey: Reston, VA, USA, 2002.

31. Sathiyamoorthy, V. Large scale reduction in the size of the tropical easterly jet. *Geophys. Res. Lett.* **2005**, *32*. [CrossRef]

32. Rao, B.; Rao, D.; Rao, V.B. Decreasing trend in the strength of tropical easterly jet during the Asian summer monsoon season and the number of tropical cyclonic systems over Bay of Bengal. *Geophys. Res. Lett.* **2004**, *31*, L14103. [CrossRef]

33. Ray, K.S.; Srivastava, A. Is there any change in extreme events like heavy rainfall? *Curr. Sci.* **2000**, *79*, 155–158.

3

Assessing River Low-Flow Uncertainties Related to Hydrological Model Calibration and Structure under Climate Change Conditions

Mélanie Trudel *, Pierre-Louis Doucet-Généreux and Robert Leconte

Université de Sherbrooke, Department of Civil Engineering, 2500, boul de l'Université, Sherbrooke, QC J1K 2R1, Canada; pierre-louis.doucet-genereux@usherbrooke.ca (P.-L.D.-G.); robert.leconte@usherbrooke.ca (R.L.)
* Correspondence: melanie.trudel@usherbrooke.ca

Academic Editor: Yang Zhang

Abstract: Low-flow is the flow of water in a river during prolonged dry weather. This paper investigated the uncertainty originating from hydrological model calibration and structure in low-flow simulations under climate change conditions. Two hydrological models of contrasting complexity, GR4J and SWAT, were applied to four sub-watersheds of the Yamaska River, Canada. The two models were calibrated using seven different objective functions including the Nash-Sutcliffe coefficient (NSE_Q) and six other objective functions more related to low flows. The uncertainty in the model parameters was evaluated using a PARAmeter SOLutions procedure (PARASOL). Twelve climate projections from different combinations of General Circulation Models (GCMs) and Regional Circulation Models (RCMs) were used to simulate low-flow indices in a reference (1970–2000) and future (2040–2070) horizon. Results indicate that the NSE_Q objective function does not properly represent low-flow indices for either model. The NSE objective function applied to the log of the flows shows the lowest total variance for all sub-watersheds. In addition, these hydrological models should be used with care for low-flow studies, since they both show some inconsistent results. The uncertainty is higher for SWAT than for GR4J. With GR4J, the uncertainties in the simulations for the 7Q2 index (the 7-day low-flow value with a 2-year return period) are lower for the future period than for the reference period. This can be explained by the analysis of hydrological processes. In the future horizon, a significant worsening of low-flow conditions was projected.

Keywords: low flows; hydrological uncertainty; climate change; calibration

1. Introduction

Climate change is expected to have a direct impact on water resources. The increase in extreme weather events will particularly intensify both the severity and frequency of low flows (e.g., [1–4]). A wide range of applications such as water supply, irrigation, navigation, and hydropower production can be strongly affected by water shortages in rivers. Therefore, it is crucial for low-flow management that prediction about low flows and the impacts of climate change become available. However, significant uncertainties remain in climate change studies on water resources. Typically, studies evaluating the impact of climate change on low flows apply outputs (temperature and precipitation) from global or regional climate circulation models to hydrological models. Thus, uncertainties have many sources related to emission scenarios, General Circulation Models (GCMs), downscaling methods such as Regional Circulation Models (RCMs), and hydrological models (structure and parameters). Assessing different sources of uncertainty in river flow projections has become a topic of interest (e.g., [5–9]). Nevertheless, only a few aspects of the uncertainty related to hydrological model structure and parameters in low-flow projections have been studied.

Najifi et al. [10] used eight GCMs and two emission scenarios with four hydrological models of different complexity calibrated with three objective functions to evaluate the uncertainties associated with hydrological models in a climate change impact study of a watershed in Oregon, USA. The results showed that total uncertainty and uncertainties related to hydrological models are higher for the dry season than for the wet season. Velàzquez et al. [11] analyzed the hydrological response of four hydrological models driven by five members of a RCM for a watershed in Canada and three members of a RCM for a watershed in Germany. Results showed that low-flow projections are strongly affected by the structure of the hydrological model. Vansteenkiste et al. [12] applied more than 50 runs of global/regional climate models to five hydrological models (lumped conceptual to physically based). They also compared a manual and an automatic calibration for a watershed in Belgium. They found that the structure and the calibration of the model highly influenced low-flow projections resulting in higher uncertainties related to the hydrological model for low flows than for peak flows. Dams et al. [13] studied three climate change scenarios with four physically-based hydrological models calibrated with four objective functions for a watershed in Belgium. They concluded that, although uncertainty related to climate change scenarios carry the largest source of uncertainties, the hydrological model structure contributes considerable uncertainty to low-flow projections. Parajka et al. [14] used 4 climate change scenarios with a conceptual semi-distributed model calibrated with 11 objective functions in 3 different decades to assess the uncertainty in low-flow projections for 262 watersheds in Austria. The results show that climate scenarios have the most impact on the total uncertainty; although the uncertainty related to hydrological model calibration is not negligible. They also found that relative contributions to the uncertainty vary with the hydrological regime (summer low-flow or winter low-flow). In these studies, many authors pointed out that calibrated models have difficulty in simulating low flows compared to observations. This may be related to uncertainties in the parameters of the hydrological model or in the hydrological processes and state variables simulated by the model. The objective function used to calibrate the model can also introduce uncertainty on low-flow simulation.

To our knowledge, no studies assessed parameter uncertainty related to choosing the objective function of the calibration for low-flow simulations. Parameter uncertainty stems from the equifinality of the model, where multiple hydrological model parameter ensembles produce similar or acceptable model outputs (example river flows) [15], but also from the calibration method used, which includes the objective function used [14], as well as model input and output errors [16]. Different methods, each with their strengths and weaknesses [17,18], allow the estimation of the parameter uncertainties for the hydrological model during calibration, including the Bayesian approach [19], Generalized Likelihood Uncertainty Estimation (GLUE) [20,21], Markov chain Monte Carlo (MCMC) [19], sequential uncertainty fitting (SUFI-2) [22], and parameter solution (PARASOL) [23].

This study aims to answer three questions. Are hydrological model parameter uncertainties influenced by using an objective function targeted to low flows? Are the impacts of the objective function on the overall hydrological model uncertainty (model parameters, structure and calibration) the same for a global model as for a distributed model? What is the order of magnitude of this uncertainty compared with the uncertainty of GCMs in current and future climates?

To address these questions, a global hydrological model, GR4J, and a distributed model, SWAT, were used to simulate low flows for the Yamaska River watershed, in Québec, Canada. Four sub-watersheds were studied since they display differences in area and land cover. Some of these sub-watersheds currently have substantial low-flow issues. Six objective functions related to low flows were used, along with the Nash-Sutcliffe efficiency (NSE_Q) [24], which is most often used to calibrate hydrological models. A method similar to the Parameter Solution (PARASOL) [23] was used to evaluate parameter uncertainty. This method selects "behavioral" results from simulations generated through a calibration process based on an objective threshold. A group of "behavioral" parameters is then selected for each calibration process. The uncertainty in streamflows, hydrological processes and state variables resulting from this group of parameters was analyzed over a reference (1970–2000) and a future (2040–2070) period. Twelve climatic projections from different combinations of RCMs/GCMs were used to evaluate the uncertainty in climate projections. An analysis of variance (ANOVA) [8] was used to evaluate the relative contributions to overall uncertainty of climate model, hydrological model, objective function, hydrological model parameters, as well as their interactions.

2. Study Area

The Yamaska River watershed is located in the southeast of the province of Quebec, Canada [25]. It covers an area of 4784 km^2, divided between the Appalachian Mountain region upstream and the lowlands of the St. Lawrence downstream. The main rivers of the watershed are the Yamaska River, Southeast Yamaska River and Noire River. Three reservoirs are located in the Yamaska River watershed: Choiniere (4.70 km^2), Waterloo (1.50 km^2) and Brome Lake (14.53 km^2). These reservoirs are operated to mitigate spring flooding and to ensure a minimum streamflow for the municipal water supply. Agriculture plays a predominant role in the economy of the Yamaska River watershed. Indeed, about 52% of the territory is dedicated to agriculture and pasture land. The forested area covers about 43% of its territory, urban areas 3%, and water areas and wetlands cover 2% (Figure 1). The soil in the Yamaska River watershed consists largely of sandy silt in its upstream part and silty sand to clay in the downstream part.

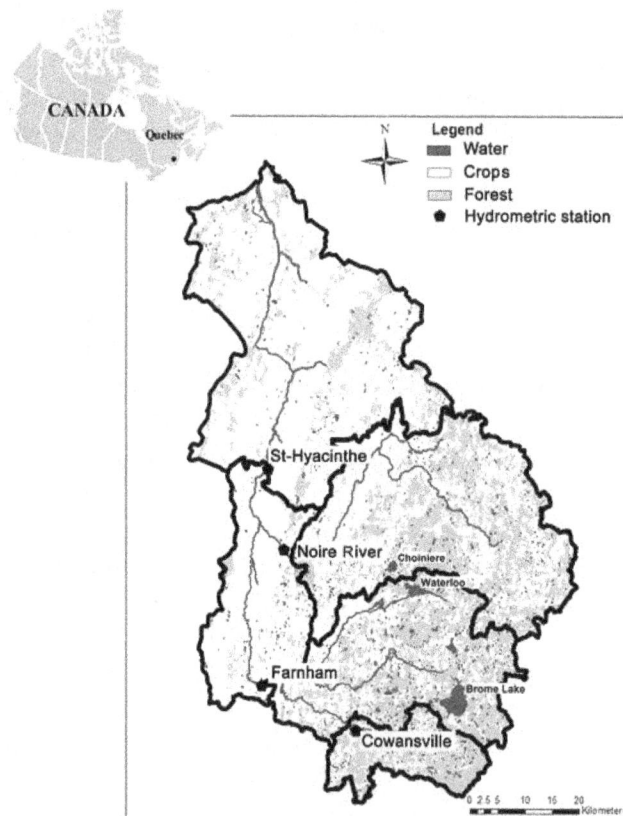

Figure 1. Yamaska River watershed.

Water quality is a serious issue for municipalities. Indeed, given the high percentage of farmland, various pollutants are released and reach the hydrographic network of the watershed. This problem becomes even more acute during periods of low flow as the dilution of contaminants is reduced.

In general, the climate of southern Quebec is humid continental with large seasonal variations. Annual precipitation averages (computed from 1980–2010) around 1100 mm. For summer months (May to October), rainfall averages around 650 mm. This region receives close to 200 cm of snowfall over an average winter, and snow usually remains on the ground from December to March. Summers are fairly warm and humid in this region with average daily temperatures near 20 °C and average monthly precipitation around 100 mm. Spring and fall are changeable seasons, prone to extremes in temperature and precipitation.

Four hydrometric stations are located in the watershed at the outlet of each sub-watershed (Figure 1). Noire River (1414 km²) and Cowansville (214 km²) are upstream sub-watersheds. The Cowansville sub-watershed drains into the Farnham sub-watershed (1202 km²). They all drain into the St-Hyacinthe sub-watershed (3289 km²). Characteristics of each sub-watershed are presented in Table 1.

Table 1. Physiographic characteristics of the study sub-watersheds.

Sub-Watershed	Drainage Area (km²)	Slope (%)	Land Cover—Forest (%)	Land Cover—Crops (%)
Cowansville	214	8.42	73.0	19.6
Farnham	1202	5.42	51.6	33.6
Noire River	1414	2.77	40.7	53.2
St-Hyacinthe	3289	3.49	43.2	52.1

3. Methods

The study was carried out using two continuous hydrological models, GR4J and SWAT, which are briefly described below.

3.1. GR4J and SWAT Model Descriptions

The GR4J model (Figure 2a) is a lumped rainfall-runoff model [26] using two conceptual reservoirs (S and R) and four calibration parameters: X1, the exchange term coefficient; X2, the maximum capacity of the routing reservoir; X3, the characteristic time of the unit hydrographs; and X4, the maximum capacity of the soil reservoir. Daily time series of precipitation (P) and potential evapotranspiration (PET) are required as inputs. In this study, PET (Equation (1)) was computed using the Hamon model [27], and P includes the snowmelt using a simple degree-day model. In addition to the four calibration parameters of the GR4J model, three parameters were added: X5, a multiplicative adjustable factor to calculate the actual evapotranspiration (ET) from PET, and two parameters (X6, a melt temperature threshold (°C); and X7, a melt rate factor (mm/day)) for the snowmelt model based on a degree-day approach.

$$\text{PET} = \frac{2.1\, H_t^2 e_s}{(T_{mean} + 273.2)}, \tag{1}$$

where H_t (day) is the average number of daylight hours per day, e_s is the saturation vapor pressure (kPa), and T_{mean} is the daily mean temperature (°C).

The Soil and Water Assessment Tool (SWAT) [28] (Figure 2b) is a semi-distributed and semi-conceptual model that simulates water, nutrient and pesticide transport at the watershed scale on a daily time step. The watershed is subdivided into sub-watersheds, river reaches and hydrological response units (HRUs). HRUs are generated by combining soil type, land use and slope classes within sub-watersheds to obtain areas which are considered hydrologically homogeneous. The SWAT simulates the hydrologic cycle based on the water balance equation (Equation (2)) [28]:

$$SW_t = SW_0 + \sum_{i=1}^{t} \left(R_{day} - Q_{surf} - ET - w_{seep} - Q_{gw} \right), \tag{2}$$

where SW_t is the soil water content (mm) at time t, SW_0 is the initial soil water content (mm) at time i, t is the time (days), R_{day} is the amount of precipitation (mm), Q_{surf} is the amount of surface runoff (mm), ET is the amount of actual evapotranspiration, w_{seep} is the amount of water percolating from the unsaturated to the saturated zone (mm), and Q_{gw} is the amount of return flow (mm). Snowmelt is calculated as a function of air temperature, snow cover area, a melt factor, a base temperature threshold, and the temperature of the snow pack. Snowmelt is added to rainfall to be infiltrated and/or to contribute to surface runoff. The Soil Conservation Service (SCS) runoff equation is used to compute runoff Q_{surf}. The Penman-Monteith method is used to compute PET. Actual evapotranspiration varies depending on PET, interception by the canopy, and the amount of water in the unsaturated soil. Routing is simulated with the variable storage coefficient method. SWAT can also simulate nutrient and pesticide transport, but these model components were not used in this study. Details on the hydrological processes in SWAT can be found in the SWAT manual [29].

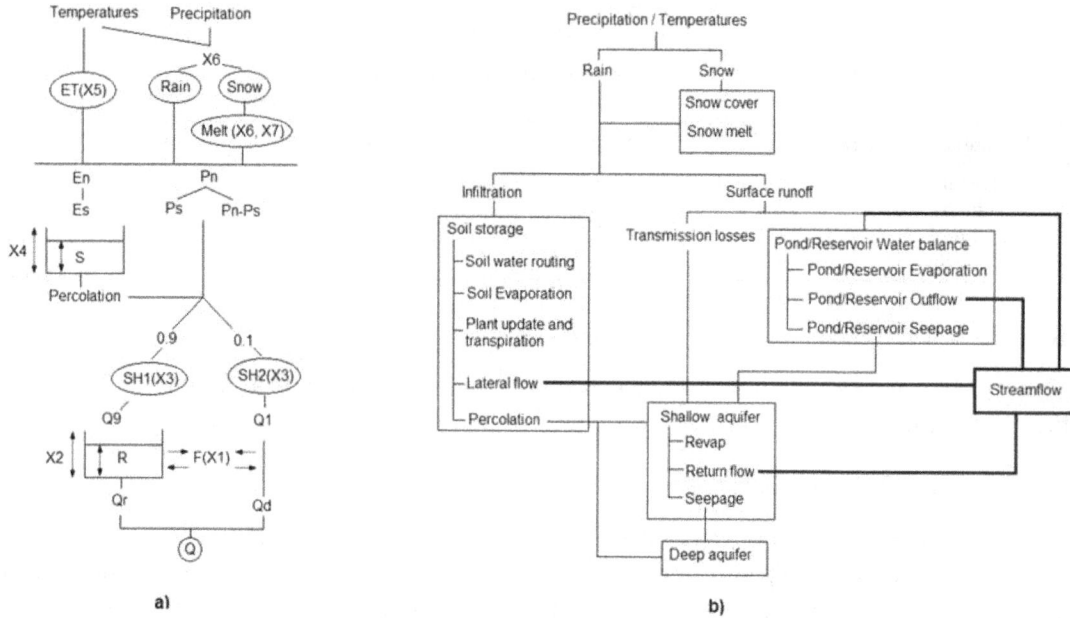

Figure 2. Schematic diagrams of the structure of the (**a**) GR4J model; and (**b**) SWAT model.

3.2. Low-Flow Indices

Low-flow indices calculated from streamflow measurements were used to evaluate the impact of climate change on low flows. In the USA and Canada, the most widely used indices are the lowest average flows that occur over a consecutive n-day period for m-year return periods (nQm). 7Q2, 7Q10 and 30Q5 indices are used by the *Ministère du Développement Durable, de l'Environnement et de la Lutte contre les changements climatiques* (MDDELCC) of Quebec [30] for the evaluation of low flows in environmental studies. These low-flow indices are related to river conditions associated with environmental issues, such as pollutant dilution or fish habitat quality. In addition, the MDDELCC limits river water withdrawals by municipalities to 15% of 7Q2 [31]. The 7Q2 and 7Q10 indices used in this study were determined by fitting the 7-day lowest streamflow for each time period to a Log Pearson Type III distribution for 2-year and 10-year return periods respectively. The Log Pearson Type III distribution was found to better represent the observed distribution compared to other distributions (e.g., Gumbel, lognormal).

3.3. Calibration, Validation and Statistical Tests

In this study, the GR4J and SWAT models were calibrated using historical river flow observations for the period from 1 January 1998, to 31 December 2009. Snow water equivalent (SWE) data were available at only one station near Brome Lake (see Figure 1) at a bi-weekly frequency during the months of January-April and consequently were not used in the calibration process. The validation period spans from 1 January 1985 to 31 December 1997. A method based on the PARAmeter SOLutions procedure (PARASOL) [23] was used to perform optimization and parameter uncertainty analyses. The objective function was minimized using the shuffled complex evolution method developed at The University of Arizona (SCE-UA) [32]. The simulations performed by the SCE-UA optimization were divided into "behavioral" simulations and "non-behavioral" simulations similar to the GLUE methodology [21]. According to [21], behavioral models are those models that provide good fits to any observables available, while non-behavioral models do not. The SCE-UA samples over the entire parameter space with a focus on solutions near the optimum. A threshold was used to select "behavioral" simulations. This study used a bootstrap method [33] applied to the best solution to define a distribution of the objective function. The threshold for the behavioral simulations was set to a confidence interval of 95%.

The calibration of the GR4J model was performed for each of the four sub-watersheds individually since the GR4J model does not simulate river routing from one sub-watershed to another as it is a lumped (i.e., global) model. For the SWAT model, the calibration was done from upstream to downstream. The Cowansville and Noire River sub-watersheds were calibrated first. The "best" parameters were selected and set to these two sub-watersheds before resuming the calibration process of the Farnham sub-watershed. The St-Hyacinthe sub-watershed was then calibrated with the "best" parameters of the Farnham, Noire River and Cowansville sub-watersheds. As the SWAT model is spatialized, streamflows coming from the Choiniere and Lac Brome reservoirs influence downstream flows. To account for reservoir management, the daily median was calculated based on streamflow observations from 1980–2010. These values were used as the reservoir output in the SWAT model. The same values were kept for reference and future climate conditions, underlying that reservoir operating rules will remain the same in a future climate. Regulation of these reservoirs depends on both the water supply and conservation of aquatic habitat and is relatively constant over the years.

A large variety of goodness-of-fit criteria have been proposed to evaluate the performance of hydrological models. These goodness-of-fit criteria can also be used as an objective function for the calibration of hydrological models. The NSE_Q goodness-of-fit criterion is the widely used [34,35]. Like many other goodness-of-fit criteria based on the mean square error, it puts greater weight on high flows, and therefore can adequately simulate snow accumulation and melt. A few studies have been carried out on the meaning and interpretation of goodness-of-fit criteria used to evaluate models in low-flow conditions [36] and concluded that the NSE_Q is not suitable to evaluate model performance for low flows. Pushpalatha et al. [36] proposed other goodness-of-fit criteria suitable for evaluating low-flow simulations. However, they used a root mean square error calculated on root square streamflows as objective function, and the impact of the choice of the objective function on the performance of low-flow simulations was not examined. Based on these findings, six goodness-of-fit criteria were selected in this study and used as objective functions for the calibration of hydrological models. A seventh goodness-of-fit criterion based on the normalized 7Q2 (p7Q2) was also developed and tested, since this criterion is important for Yamaska River watershed managers. However, using only the p7Q2, calibration of both models tended towards a solution where there was not much accumulation of snow, resulting in low melting of the snow cover. In addition, models tended to simulate low flows during the late winter instead of summer. This can be explained by lower precipitation observed between February and March combined with low melting during this period. Thus, a combination of NSE_Q and p7Q2 was tested to bring the model to better simulate the spring melt and still maintaining a good fit for low-flow. Using a threshold of 0.3 on the NSE_Q and then optimizing the p7Q2, visually it was found that both calibrated models were able simulate reasonable accumulation and melting of snow. Table 2 summarizes these goodness-of-fit criteria. All goodness-of-fit criteria tend to 1 when the calibration is good. They were grouped into four categories. Group 1 is the NSE_Q that focuses on high flows; group 2 focuses both on high flows and low flows (NSE_{sqrtQ} and NSE_{lnQ}); group 3 applies to the summer period only ($NSE_{Q\text{-summer}}$ and $NSE_{lnQ\text{-summer}}$); and group 4 focuses on very low flows (NSE_{iQ} and p7Q2).

Table 2. Description of the seven goodness-of-fit criteria used as objective functions for the model calibration and to evaluate model performance.

Goodness-of-Fit (Group)	Description	Equation
NSE_Q (1)	NSE calculated on streamflows	$1 - \dfrac{\sum_{t=1}^{T}\left(Q_{obs}^{t}-Q_{sim}^{t}\right)^2}{\sum_{t=1}^{T}\left(Q_{obs}^{t}-\overline{Q}_{obs}\right)^2}$
NSE_{sqrtQ} (2)	NSE calculated on root squared transformed streamflows	$1 - \dfrac{\sum_{t=1}^{T}\left(sqrt(Q_{sim}^{t}+\eta)-sqrt(Q_{sim}^{t}+\eta)\right)^2}{\sum_{t=1}^{T}\left(sqrt(Q_{obs}^{t}+\eta)\eta-sqrt\overline{Q}_{obs+\eta}\right)^2}$
NSE_{lnQ} (2)	NSE calculated on log-transformed streamflows	$1 - \dfrac{\sum_{t=1}^{T}\left(\ln(Q_{sim}^{t}+\eta)-\ln(Q_{sim}^{t}+\eta)\right)^2}{\sum_{t=1}^{T}\left(\ln(Q_{obs}^{t}+\eta)\eta-\ln\overline{Q}_{obs+\eta}\right)^2}$
$NSE_{Q\text{-}summer}$ (3)	NSE calculated on streamflows between May and October	$1 - \dfrac{\sum_{t=1}^{T}\left(Q_{obs-summer}^{t}-Q_{sim-summer}^{t}\right)^2}{\sum_{t=1}^{T}\left(Q_{obs-summer}^{t}-\overline{Q}_{obs-summer}\right)^2}$
$NSE_{lnQ\text{-}summer}$ (3)	NSE calculated on log transformed streamflows between May and October	$1 -$ $\dfrac{\sum_{t=1}^{T}\left(\ln(Q_{sim-summer}^{t}+\eta)-\ln(Q_{sim-summer}^{t}+\eta)\right)^2}{\sum_{t=1}^{T}\left(\ln(Q_{obs-summer}^{t}+\eta)\eta-\ln\overline{Q}_{obs-summer+\eta}\right)^2}$
NSE_{iQ} (4)	NSE calculated on inverse transformed streamflows	$1 - \dfrac{\sum_{t=1}^{T}\left(1/Q_{obs}^{t}-1/Q_{sim}^{t}\right)^2}{\sum_{t=1}^{T}\left(1/Q_{obs}^{t}-\overline{1/Q}_{obs}\right)^2}$
NSE_Q and $p7Q2$ (4)	The 7-day low-flow value with a 2-year return period combined with a threshold on NSE calculated on streamflows	$1 - \dfrac{abs(7Q2_{obs}-7Q2_{sim})}{7Q2_{obs}}$ and $1 - \dfrac{\sum_{t=1}^{T}\left(Q_{obs}^{t}-Q_{sim}^{t}\right)^2}{\sum_{t=1}^{T}\left(Q_{obs}^{t}-\overline{Q}_{obs}\right)^2} > 0.3$

Analysis of variance (ANOVA) was performed to quantify the contribution of each source of uncertainty [8,9,14] on the computed 7Q2 index as follows:

$$\sigma^2_{7Q2} = \sigma^2_{CM} + \sigma^2_{HM} + \sigma^2_{PAR} + \sigma^2_{OF} + \sigma^2_{INT} + \sigma^2_{\varepsilon} \qquad (3)$$

where σ^2_{7Q2} is the total variance on 7Q2 index, σ^2_{CM}, σ^2_{HM}, σ^2_{PAR}, σ^2_{OF}, σ^2_{INT}, and σ^2_{ε} are the variances linked to climate models, hydrological models, hydrological model parameters, objective functions, their interactions and the residual variance, respectively. Only first-order interactions (between two-factors) were considered. The ANOVA was performed for each sub-watershed in both reference and future periods.

3.4. Climate Change Projections

For the evaluation of the uncertainty related to climate models, 12 regional climate projections were used: 7 from the North America Regional Climate Change Assessment Program (NARCCAP) [37] and 5 from the Canadian Regional Climate Model version 4 (CRCM) [38] driven by the Canadian Global Climate Model, version 3 (CGCM3) [39]. A bias correction on temperature and precipitation was performed to adjust the reference and future periods with the observations. This method assumes that the relative bias of climate models is the same for the future period and the reference period. For temperatures, a bias correction based on monthly averages was performed. The difference between the average monthly temperatures observed and the average monthly temperatures simulated for the reference period was applied to daily data for the reference and future periods. For precipitation, a bias correction was applied for both the monthly average precipitation occurrence (ratio of the number of days above a threshold of precipitation on the number of days in the month) and the average monthly rainfall intensity (average daily precipitation) using the Local Intensity method (LOCI). Details of this methodology can be found in of Schmidli et al. [40].

Figure 3 shows a statistically significant increase in atmospheric temperatures between the reference and future horizons. There is a mean difference of 2.5 °C for the minimal temperature and of 2.7 °C for the maximum temperature between the reference and future climate for the summer period (May to October) (Figure 3a,b). Precipitation is also significantly influenced by climate change.

Firstly, we can observe an increase in the variability of future climatic projections. Secondly, there is a decrease of the mean precipitation for the months of June and July of 5 mm (4.7%) and 9 mm (7.5%) respectively; and an increase for May, August, September and October, of 13 mm (13.7%), 8 mm (6.5%), 2 mm (1.8%, not significant) and 10 mm (10.3%) respectively (Figure 3c). Other studies support the increase in precipitation in late spring and early fall, and a decrease of precipitation during summer for southern Quebec (e.g., [41]).

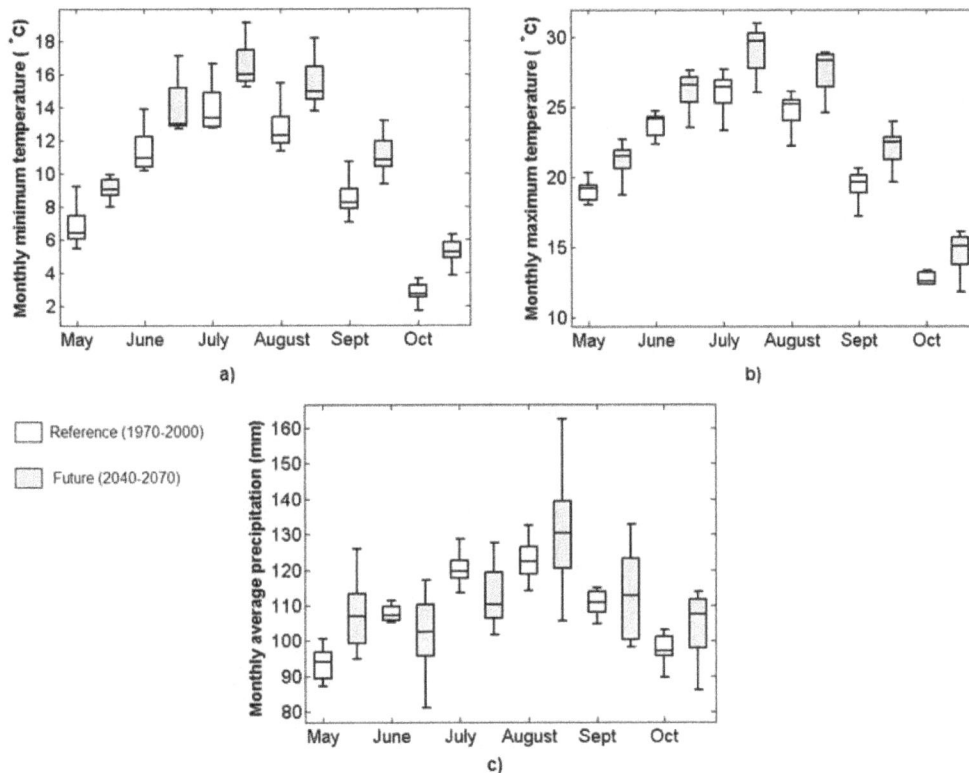

Figure 3. Uncertainties related to the climate model for the (**a**) monthly minimal temperature; (**b**) monthly maximal temperature; and (**c**) precipitation for the Yamaska River watershed. Boxes represent the 25–75 percentile values. Whiskers correspond to the 5–95 percentile values.

4. Results and Analysis

4.1. Model Calibration and Validation

Both GR4J and SWAT models were calibrated using alternately each of the 7 goodness-of-fit criteria listed in Table 2 as objective functions, while the remaining six other criteria were computed to assess model performance. Figure 4 presents the relationship between the objective function values and other goodness-of-fit criteria for the best simulation (parameter set corresponding to the highest value of the objective function) for each of the four sub-watersheds for the model calibration and validation (i.e., 8 points per model for each graph in Figure 4). The closer the points are to the upper right corner of the graphs, the better the model. Values of NSE < 0 (which means that a model is a worse predictor than the average of the observations) were set to 0 for better visualization of the results. Data points located along the diagonal of a graph mean that the hydrological model performs equally well according to a given goodness-of-fit criterion as compared to the objective function used for the calibration/validation phases. Data points located above the diagonal mean that the model performs better with the objective functions in the calibration/validation phases as compared to the selected goodness-of fit criterion.

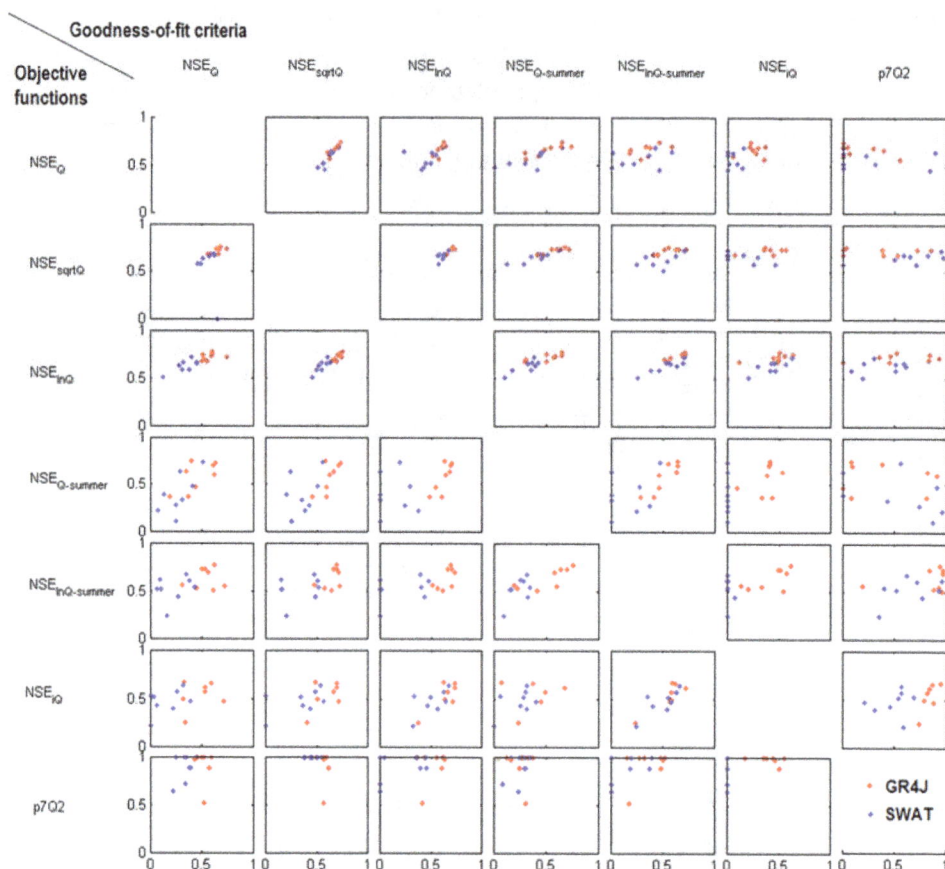

Figure 4. Relationship between objective function values and goodness-of-fit criteria for the best simulation.

In general, the GR4J model (red) performed better than the SWAT model (blue), especially for the smallest sub-watershed, Cowansville, where low flows can be very low. The calibration using the NSE_Q objective function shows values higher than 0.45 for both the calibration and validation period, but poorly represents the p7Q2 index and NSE_{iQ}. Nevertheless, the seven objective functions for both models were less successful for both calibration and validation for this sub-watershed. The calibration using the p7Q2 converged to a value near one (perfect match between simulated and observed 7Q2). However, the other goodness-of fit criteria were low. It is possible that the equifinality was greater for this objective function, and that the calibration did not converge to a global optimization.

Calibrating the GR4J model using the NSE_{lnQ} objective function gave better results for low flows than using the other objective functions for each of the four sub-watersheds simulated. This can be seen in Figure 4 (line 3) where the red points are closest to the upper right corner of the graphs. The objective functions $NSE_{lnQ\text{-summer}}$, NSE_{iQ} and the one based on a combination of NSE_Q and p7Q2 also gave good results for very low flows.

For the SWAT model, the calibration using NSE_Q and NSE_{sqrtQ} objective functions well represented these three goodness-of-fit criteria, but did not simulate well the very low flows, as shown by the poor goodness-of-fit criteria in the validation (group 4—NSE_{iQ} and p7Q2). Only the NSE_{lnQ} and NSE_{iQ} objective functions were able to successfully represent the NSE_{iQ} goodness-of-fit. The p7Q2 value simulated using the $NSE_{Q\text{-summer}}$ objective function (line 4, column 7) was equal to zero for both the calibration and validation mode for the Noire River sub-watershed. This indicates that the simulated 7Q2 value was equal to 0 m^3/s for both calibration and validation. The $NSE_{Q\text{-summer}}$ objective function also gave poor results for the NSE goodness-of-fit including a logarithm or inverse flow. It is likely that the model often simulated flows equal to 0 m^3/s with the NSE$_{Q\text{-summer}}$ objective function for the Noire River sub-watershed.

4.2. Future Low-Flow and Uncertainty Analysis

The 7Q2 index was used to evaluate the impacts of climate change and their uncertainties for low flows. An ANOVA was performed according to Bosshard et al. [8] and Andor et al. [9] (see Equation (3)) to identify major sources of uncertainty on the computed 7Q2 index (Figure 5).

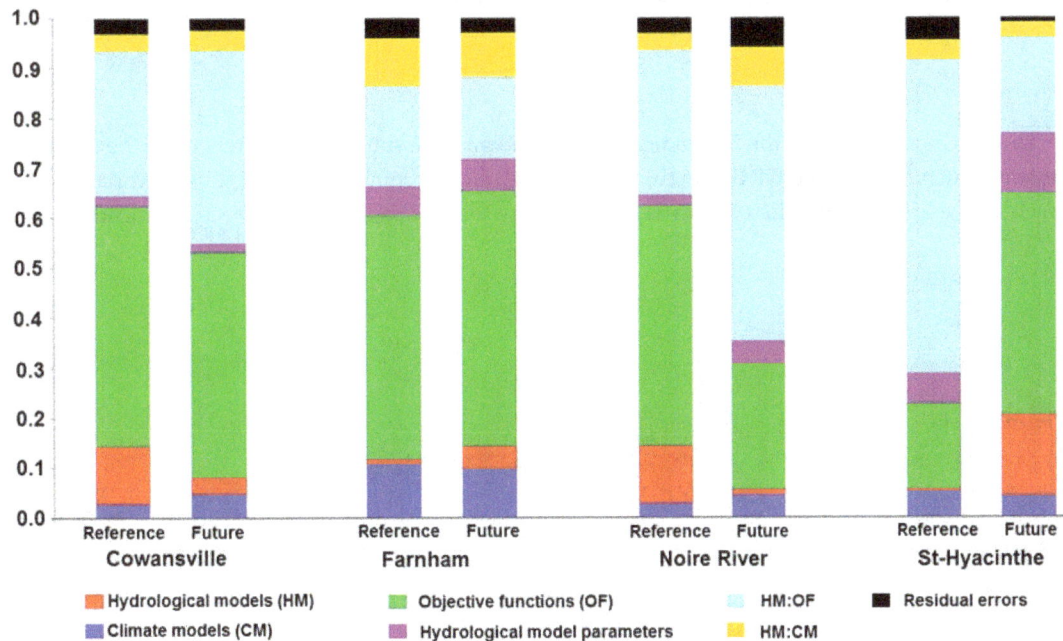

Figure 5. ANOVA partitioning among the sources of uncertainty computed on the 7Q2 index for each sub-watershed in reference and future periods.

The major source of uncertainty in computing the 7Q2 index was the objective functions used to calibrate the hydrological model. The interaction-term between objective functions and hydrological models was also high. It shows the importance of the choice of the objective function for the calibration in low-flow analysis. The contribution of the hydrological model to the uncertainty increased in the future period for the Farnham and St-Hyacinthe sub-watersheds, while it decreased for Cowansville and Noire River sub-watersheds. It is interesting to note that both Farnham and St-Hyacinthe sub-watersheds are regulated, while Cowansville and Noire River sub-watersheds are natural. Note that SWAT can explicitly incorporate flow regulating structures such as dams, while GR4J cannot. This difference in model structure explains why hydrological model uncertainty increases in the future, as natural inflow variability is augmenting in a future climate, while hydraulic structures dampen this variability. An increase of the contribution of the hydrological models in the 2040–2070 period was also observed, along with a decrease of the contribution of the climate models in the future, and vice versa. As demonstrated by other authors [9,14], our results show that the relative importance of sources of uncertainties varies in time and with watersheds characteristics.

For each objective function and sub-watershed the ANOVA was applied to identify the contribution of the sources of uncertainty in the reference and future periods. The median, the 25%–75% interquartile range and the minimum and maximums values were also computed. Figures 6–9 present the results for the Cowansville (Figure 6), Farnham (Figure 7), Noire River (Figure 8) and St-Hyacinthe (Figure 9) sub-watersheds. The dotted line shows the observed value of the low-flow indices for the reference period.

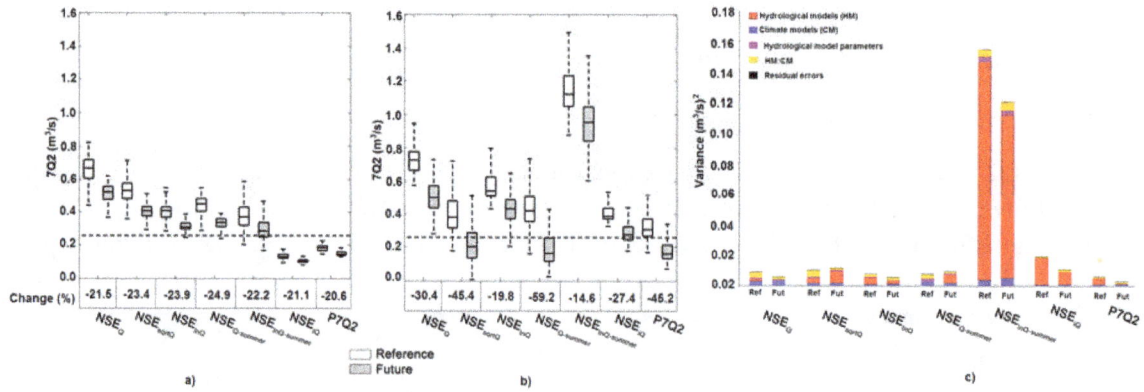

Figure 6. Uncertainties for the 7Q2 indices for the Cowansville sub-watershed related to the climate model structure for the (**a**) GR4J; and (**b**) SWAT model; and (**c**) comparison of 7Q2 index variance for each source of uncertainty in reference (Ref) and future (Fut) periods for each objective function.

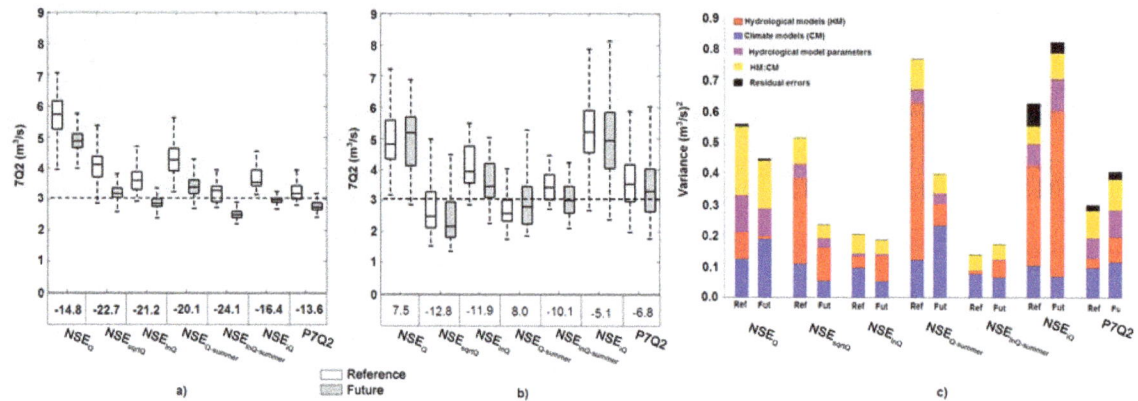

Figure 7. Uncertainties for the 7Q2 indices for the Farnham sub-watershed related to the climate model structure for the (**a**) GR4J; and (**b**) SWAT model; and (**c**) comparison of 7Q2 index variance for each source of uncertainty in reference (Ref) and future (Fut) periods for each objective functions.

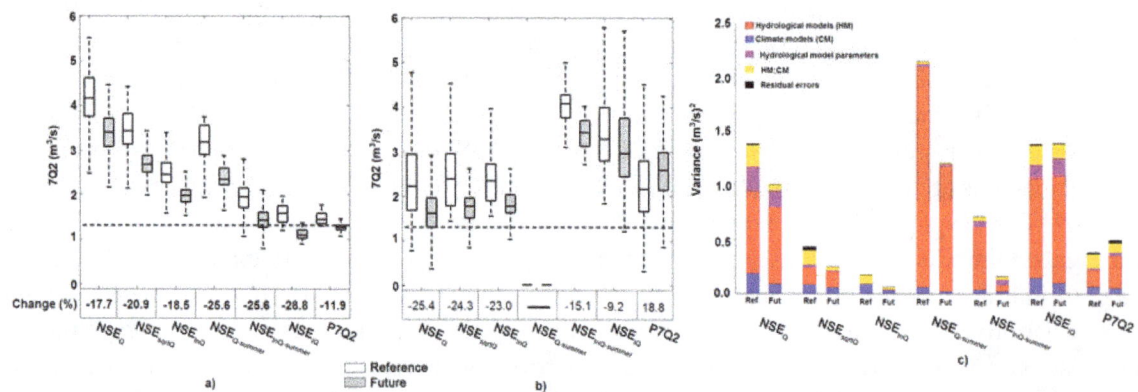

Figure 8. Uncertainties for the 7Q2 indices for the Noire River sub-watershed related to the climate model structure for the (**a**) GR4J; and (**b**) SWAT model; and (**c**) comparison of 7Q2 index variance for each source of uncertainty in reference (Ref) and future (Fut) periods for each objective functions. For the St-Hyacinthe sub-watershed, the GR4J model overestimated the 7Q2 values for the reference period for all objective functions (Figure 9a).

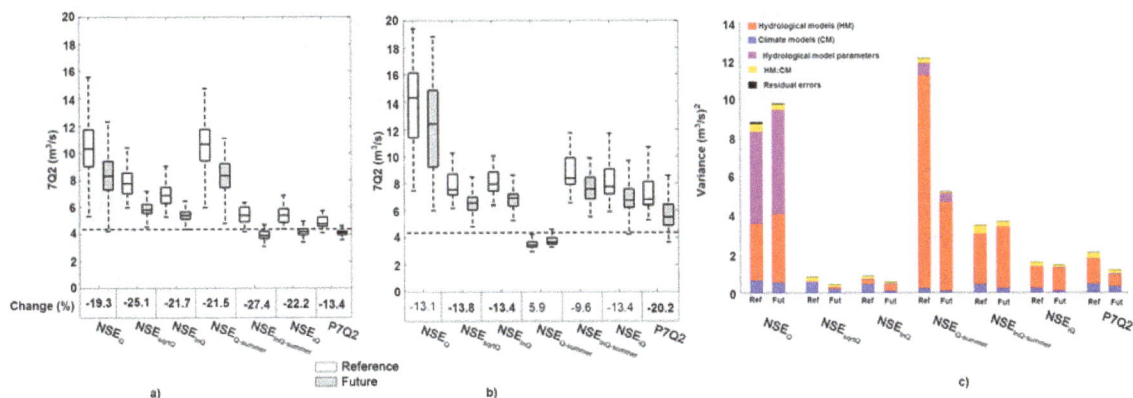

Figure 9. Uncertainties for the 7Q2 indices for the St-Hyacinthe sub-watershed related to the climate model structure for the (**a**) GR4J; and (**b**) SWAT model; and (**c**) comparison of 7Q2 index variance for each source of uncertainty in reference (Ref) and future (Fut) periods for each objective functions.

The 7Q2 indices simulated in the reference period with climate models were compared against the 7Q2 indices observed over the same period. For the Cowansville sub-watershed with the GR4J model (Figure 6a), the 7Q2 indices were overestimated compared to the observed value of 7Q2 (dotted line) with the objective functions of groups 1, 2 and 3 for the reference period; and they were underestimated with the objective functions of group 4.

For the simulations with the SWAT model, each objective function overestimated the 7Q2 values in the reference period when compared to the observed values (dotted line) for this period (Figure 6b). The variance (Figure 6c) was higher for the $NSE_{lnQ-summer}$ objective function, where the contribution of the hydrological model to overall uncertainty is large. This is because SWAT produced much larger 7Q2 values compared to GR4J when calibrated with this objective function. In general, a lower uncertainty for the 7Q2 index was observed for the future period than for the reference period. For the NSE_Q objective function, climate model is the main source of uncertainty, while for objective functions more related to low-flow, the main contribution to the total uncertainty is the hydrological model.

At the Farnham sub-watershed, using the GR4J model, all objective functions overestimated the 7Q2 using climatic projections of the reference period compared to the observed value (dotted line) for the same period (Figure 7a). When the SWAT model was used (Figure 7b), the NSE_Q, NSE_{lnQ}, NSE_{iQ}, $NSE_{lnQ-summer}$ objective functions overestimated the 7Q2 value observed in the same period (while the NSE_{sqrtQ}, $NSE_{Q-summer}$, p7Q2 functions underestimated this 7Q2 value). The main source of uncertainty for the NSE_Q, NSE_{lnQ}, $NSE_{lnQ-summer}$, and P7Q2 objective functions was the climate model, while the hydrological model was the main source of uncertainty for other objective functions. For the NSE_Q, NSE_{sqrtQ}, NSE_{lnQ}, and $NSE_{Q-summer}$ objective functions, the total variance in the future period was lower than the total variance in the reference period, while the opposite was observed when the $NSE_{lnQ-summer}$, NSE_{iQ}, and P7Q2 objective functions were used to calibrate the models. This difference in the total variance came from the variance related to the hydrological model. The interaction term between climate models and hydrological models was not negligible, but was similar between the reference and future periods.

For the Noire River sub-watershed, the GR4J model using the climatic projections for the reference period overestimated the observed values of 7Q2 for all objective functions (Figure 8a). Similarly, the SWAT model overestimated the 7Q2 observed values using the climatic projections in the reference period for all objective functions, with the NSE_Q objective function being worse than the others (Figure 7b). Model calibration with the $NSE_{Q-summer}$ objective function was problematic since all the simulated 7Q2 values were equal to zero. Calibrating the model using this objective function simulated very low contribution of groundwater to streamflow (see Section 4.3.2), resulting of values of streamflow equal to zero. The main source of uncertainty (Figure 8c) was the hydrological model for all objective functions except for the NSE_{lnQ} objective function, where the main contribution was the climate model. This objective function also shown the lower total variance compared to the other objective functions. Furthermore, total variance was lower for the future period than for the reference period.

The SWAT model behaved similar to the GR4J model. Each objective function, except $NSE_{Qsummer}$, overestimated the 7Q2 values in the reference period. Note that the simulations for the St-Hyacinthe sub-watershed were influenced by the simulations of the upstream sub-watersheds. As the $NSE_{Qsummer}$ objective function often simulated streamflows equal to zero for the Noire River sub-watershed, the streamflows simulated for the St-Hyacinthe sub-watershed were lower with this objective function. The NSE_Q is the objective function that overestimated the 7Q2 values the most (Figure 9b). The total variance was lower for the future period than for the reference period for all objective functions except the NSE_Q. The variance of NSE_{sqrtQ} and NSE_{lnQ} objective functions were the lowest. For all objective functions, except for these two objective functions, the main source of uncertainty was the hydrological model.

For all sub-watersheds, the NSE_{lnQ} shown the lowest total variance and the lowest variance related to hydrological models. In general, the SWAT model shown higher 25%–75% interquartile range than the GR4J model.

Comparison of the 7Q2 index for the reference and future periods was used to evaluate whether low flows will worsen with climate change. Relative changes in the median between the reference and future periods are indicated under the x-axis for each sub-watershed (Figures 6–8). A positive change indicates that future 7Q2 is higher than reference 7Q2. Significant changes (p-value < 0.05) are shown in bold.

For the GR4J model, the changes shown in Figures 6a, 7a, 8a and 9a are all negative and statistically significant, meaning that there is a worsening, or decrease, of low flows for all objective functions and sub-watersheds assessed in the study. Current observed values of 7Q2 indices are also indicated in these figures by a dotted line. The changes vary from −11.9% to −28.8% depending on the objective functions and the sub-watersheds considered in the analysis. The Cowansville sub-watershed showed the smallest range of change depending of objective functions (from −20.6% to −24.9%), while the Noire River sub-watershed shown the largest range (from −11.9% to −28.1%). The Cowansville sub-watershed is the most forested, while the Noire River sub-watershed is the most agricultural sub-watershed in this study. The simulation using the p7Q2 objective function to calibrate the GR4J model shown the smallest change for 7Q2 between the reference and future periods, depending on the sub-watershed analyzed.

The changes in 7Q2 between current and future climates simulated with the SWAT model were not always significant (Figures 6b, 7b, 8b and 9b). Changes that were significant were almost always negative, meaning a worsening in low-flow conditions. The changes simulated for the Cowansville sub-watershed were mostly significant and were all negative. The changes in 7Q2 simulated with the SWAT model for this sub-watershed were much larger than the ones simulated with the GR4J model. The changes greatly varied according to the objective function used, from −19.8% to −59.2%.

For the Farnham sub-watershed, there were only a few significant changes in the simulated 7Q2, which were all negative. Some objective functions gave positive differences, although not statistically significant. Reservoir flow regulation for this sub-watershed (Choinière, Waterloo, and Brome Lake, see Figure 1) simulated with the SWAT model explains these results as these reservoirs control downstream flows by reducing high flows and increasing low flows. Therefore, the presence of reservoirs should help to alleviate the negative impacts of climate change on low flows. Note that the GR4J model resulted in significant negative changes for the Farnham sub-watershed. This is because GR4J does not explicitly simulate reservoir storage and routing. This illustrates the importance of a hydrological model capable of explicitly handling storage structures for low-flow impact studies in regulated watersheds. For the Noire River and St-Hyacinthe sub-watersheds, the $NSE_{Q-summer}$ objective function used to calibrate SWAT did not give conclusive results. The NSE_Q, NSE_{sqrtQ}, NSE_{lnQ}

and NSE$_{\text{lnQ-summer}}$ objective functions all resulted in a decrease of the 7Q2 value. However, the change in the 7Q2 values between the reference and future periods was not significant. This may be caused by the difficulty in simulating very low flows with SWAT for those sub-watersheds (zero flow values were obtained), which possibly related to the high number of parameters in the model, as compared to GR4J.

4.3. Hydrological Processes and State Variable Analysis

To better understand the results of the uncertainty analysis, different hydrological processes and state variables for the GR4J and SWAT models were analyzed according to the objective functions used to calibrate the models.

4.3.1. The GR4J Model

For the GR4J model, the hydrological processes analyzed were actual evapotranspiration (ET) and percolation (PERC). The model state variables analyzed included the soil moisture reservoir (S) and the routing reservoir (R). For the GR4J model, the dynamics of hydrological processes and temporal characteristics of state variables were found to be similar for all four sub-watersheds.

Overall, GR4J calibrated using NSE$_Q$ simulated less ET than using other objective functions (Figure 10). Potential evapotranspiration was simulated with the Hamon model and was identical for all objective functions used to calibrate the model. Actual evapotranspiration depends of the X5 parameter and the on the amount of water in the soil, represented by the soil moisture reservoir S, that depends of the X4 parameter (Section 3.1). These parameters changed depending of the objective function used to calibrate the model. The simulations obtained with GR4J calibrated with all objective functions, except p7Q2, showed a significant increase of ET during the summer-fall period (May–October). For the p7Q2, ET remained unaffected during that period between reference and future horizons. On a seasonal basis, the simulations showed an increase in ET between May and June in the future period compared to the reference period, while simulated ET remained similar or decreased slightly from July to October depending on the objective function selected for model calibration. A decrease in ET between June and October is explained by a lack of water in the soil moisture reservoir to satisfy potential evapotranspiration. The increase in ET between May and June thus can lead to a decrease in the amount of water present in the soil during the summer. The sum of actual ET from May to October significantly decreased between 8% and 11% depending on the sub-watershed and objective functions, except for NSE$_{iQ}$ and p7Q2, while the difference was not significant. For all objective functions, uncertainties in ET were larger in the future than in the reference period. Larger uncertainty in temperatures in the future period explains this phenomenon (see Equation (1)). Since the evapotranspiration uncertainty increased in the future period, it cannot explain the decreasing uncertainty for low-flow indices obtained for the future period (Figures 6a, 7a, 8a and 9a).

Total percolation simulated over the entire year varied between 40 mm to 85 mm depending on the sub-watershed and objective function. The overall evolution of percolation during the year was similar for all calibrations and sub-watersheds, showing lower percolation from May to October. The sums of percolation for the summer (May to October) are presented in Figure 11. The percolation is dependent on the X4 parameter, the maximum capacity of the soil reservoir, which relates to soil properties and land use/land cover characteristics. For all objective functions, uncertainty in percolation decreased between May and October in the future period compared to the reference period (Figure 11). The low percolation simulated is explained by a shortage of water in the soil moisture reservoir. Thus, the overall decrease in the uncertainty for the 7Q2 index in the future period can be explained by a decrease in the uncertainty for percolation during the summer period.

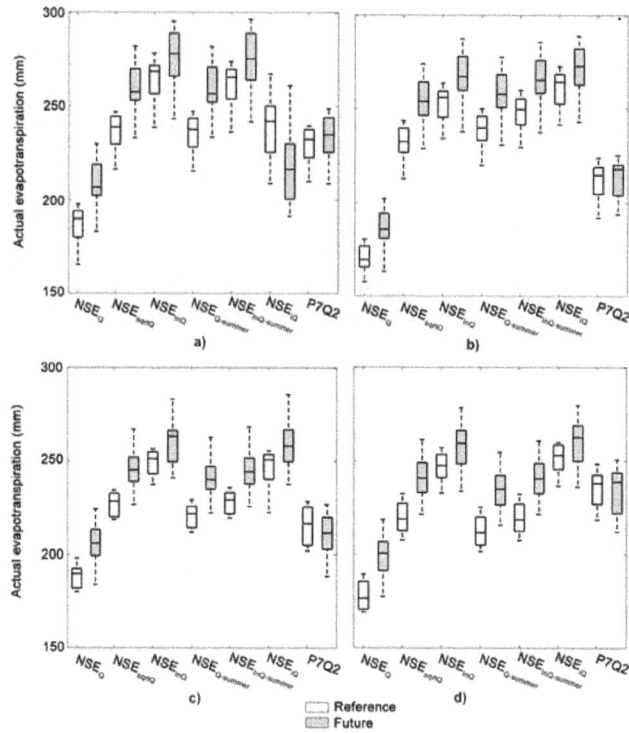

Figure 10. Sum of actual evapotranspiration simulated with the GR4J model for the summer (May to October) for the (**a**) Cowansville; (**b**) Farnham; (**c**) Noire River; and (**d**) St-Hyacinthe sub-watersheds.

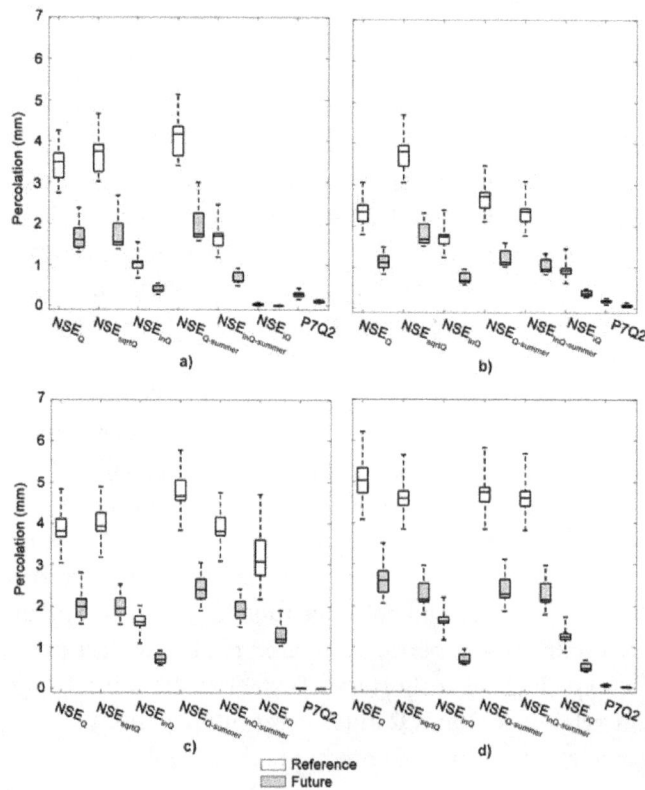

Figure 11. Sum of percolation simulated with the GR4J model for the summer (May to October) for the (**a**) Cowansville; (**b**) Farnham; (**c**) Noire River; and (**d**) St-Hyacinthe sub-watersheds.

The soil reservoir (S) was analyzed based on the percentage to which the reservoir is filled. Generally, it was found that the level of the reservoir S was high during the period from November to March, reaching 86% to 90% capacity, depending on the objective function used to calibrate the model. It gradually decreased until July when it reached a minimum and then gradually increased. The simulation with the NSE_Q objective function had the smallest variation between the maximum and the minimum level of reservoir S, while simulations with NSE_{iQ} and p7Q2 objective functions had the largest variation (Table 3). Low levels of reservoir S observed with the NSE_{iQ} and p7Q2 objective functions mean that there was a small amount of water in the soil during the summer period. This small amount of water in the soil limited the ET and percolation between mid-June and late August. In addition, the difference between the minimum and maximum level of reservoir S in the future period was significantly lower than in the reference period. The water content in the soil was significantly lower in the 2040–2070 period.

Table 3. Variation of the levels of the soil reservoir S for the GR4J model.

Level of S (%)—Period	NSE_Q	NSE_{sqrtQ}	$NSEln_Q$	$NSE_{Q-summer}$	$NSE_{lnQ-summer}$	NSE_{iQ}	P7Q2
Cowansville							
Maximum—reference	90.8	87.0	86.2	86.0	87.3	89.3	89.0
Minimum—reference	52.2	46.8	32.8	47.9	37.6	12.4	26.4
Maximum—future	91.6	88.4	87.8	87.2	86.1	90.8	90.6
Minimum—future	37.3	32.2	19.9	33.2	23.7	6.5	16.2
Farnham							
Maximum—reference	92.2	86.5	85.9	86.4	85.9	85.7	90.3
Minimum—reference	50.5	46.6	38.1	43.5	41.2	31.9	24.3
Maximum—future	92.7	87.8	87.5	87.7	87.4	87.4	91.6
Minimum—future	36.1	32.9	24.8	29.3	27.1	19.6	15.0
Noire River							
Maximum—reference	89.1	85.6	84.3	85.7	85.7	85.3	85.1
Minimum—reference	51.7	46.8	36.7	49.1	46.7	41.7	10.8
Maximum—future	90.2	86.6	85.6	86.9	86.3	86.5	87.3
Minimum—future	38.0	33.3	23.8	35.7	33.3	27.7	6.0
St-Hyacinthe							
Maximum—reference	89.8	86.8	85.3	87.0	87.0	85.4	86.6
Minimum—reference	56.0	49.4	37.3	50.6	49.4	34.5	17.8
Maximum—future	90.7	87.9	86.7	88.3	88.3	86.9	88.4
Minimum—future	41.4	35.4	23.6	36.4	35.4	21.6	10.0

The routing reservoir (R) was also analyzed based on the percentage to which the reservoir was filled. The level of the reservoir R was typically high during the period from November to March, reaching 60% to 80% capacity depending on the sub-watershed and objective functions. It gradually decreased until July when it reached a minimum followed by a gradual rise (Table 4). The difference between the maximum and minimum level was significantly lower for the NSE_Q objective function than for other objective functions. In the future period, the maximum level did not show a significant difference from the current period, whereas the minimum levels during the summer decreased significantly. In addition, in the reference period, the minimum level was reached during the winter, whereas for the future period, the minimum level was reached in the summer.

Table 4. Variation of the levels of the soil reservoir R for the GR4J model.

Level of R (%)—Period	NSE_Q	NSE_{sqrtQ}	$NSElnQ$	$NSE_{Q\text{-summer}}$	$NSE_{lnQ\text{-summer}}$	NSE_{iQ}	P7Q2
Cowansville							
Maximum—reference	68.4	71.1	70.4	74.3	77.9	78.6	68.4
Minimum—reference	52.1	50.8	46.0	55.6	48.4	49.0	44.5
Maximum—future	69.4	71.7	71.0	76.0	73.3	81.4	69.9
Minimum—future	47.6	46.4	43.0	50.3	44.4	46.1	41.6
Farnham							
Maximum—reference	57.9	63.3	63.8	68.8	67.7	59.5	58.3
Minimum—reference	45.1	45.4	43.8	49.8	47.0	40.0	40.5
Maximum—future	58.5	63.7	64.1	69.3	68.1	60.0	58.6
Minimum—future	42.8	42.0	40.7	46.2	43.4	37.7	38.7
Noire River							
Maximum—reference	64.6	68.8	71.6	75.2	79.1	89.1	63.1
Minimum—reference	49.0	49.7	49.2	54.6	54.2	57.9	38.2
Maximum—future	65.2	70.0	73.1	76.7	77.9	87.2	63.2
Minimum—future	45.1	45.4	45.2	49.3	48.8	52.2	36.5
St-Hyacinthe							
Maximum—reference	63.7	67.0	67.0	64.8	73.2	71.3	63.9
Minimum—reference	48.3	47.7	45.2	48.7	51.2	47.0	39.3
Maximum—future	64.0	66.9	67.3	65.2	73.5	72.0	63.3
Minimum—future	44.4	43.4	41.6	44.6	46.1	43.2	37.2

4.3.2. The SWAT Model

For the SWAT model, the hydrological processes analyzed were actual evapotranspiration in mm (ET), and groundwater contribution to streamflow in mm (GW). The soil water content in mm (SW), a model state variable, was also analyzed. For the SWAT model, the response of the hydrological processes differed for the four sub-watersheds and the objective function used to calibrate the model. This can be explained by the higher number of parameters in the SWAT model leading a different combination of parameters to a similar final result (streamflow). Although, intermediate results, such as hydrological processes, changed depending on the combination of parameters. As another possibility, differences in land cover/land use, drainage area, and slope between sub-watersheds may partly explain the differences in simulation results (Table 1).

The hydrological response of the Farnham sub-watershed is influenced by the regulation of the upstream reservoirs simulated by the SWAT model. In addition, the hydrological response of the St-Hyacinthe sub-watershed is highly influenced by upstream sub-watersheds. Thus, the hydrological processes and state variables of these two sub-watersheds were not analyzed here and only the Cowansville and Noire River sub-watersheds were analyzed.

Total actual evapotranspiration was computed for the summer/fall season (May to October). Potential evapotranspiration was simulated using the Penman-Monteith method and was the same for all objective functions used to calibrate SWAT. Actual evapotranspiration depends on the amount of water in the soil and a specific parameter (ESCO). The amount of water in the soil simulated by SWAT depends of parameters related to the characteristics of the soil. For the Cowansville sub-watershed, ET simulated with the $NSE_{lnQ\text{-summer}}$ and NSE_{iQ} objective functions was statistically significantly higher than ET simulated with the NSE_Q objective function (Figure 12a) for both current and future periods. For the Noire River sub-watershed (Figure 12b), ET simulated with the NSE_{lnQ} objective function was significantly lower than with the other objective functions, again for current and future climates. In the future period, a statistically significant increase of median values of ET between 13% and 14% was simulated for the Cowansville sub-watershed and between 11% and 12% for the Noire River sub-watershed. The objective function used to calibrate the model therefore had little impact on the magnitude of the change of ET attributable to climate change.

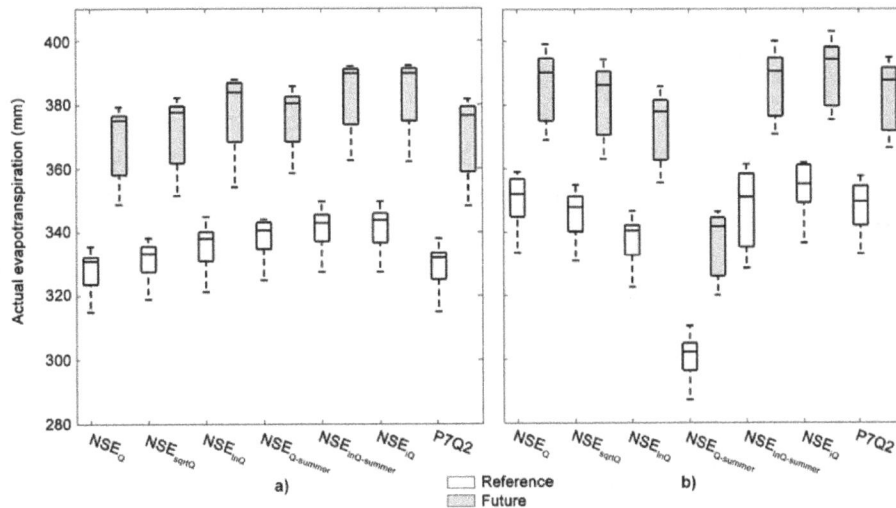

Figure 12. Sum of actual evapotranspiration simulated with the SWAT model for the summer (May to October) for the (**a**) Cowansville sub-watershed; and (**b**) Noire River sub-watershed.

The total GW for the summer/fall period (May to October) was also analyzed (Figure 13). For the Cowansville sub-watershed (Figure 13a), a very low contribution from groundwater was observed for the simulation using the NSE_{iQ} and p7Q2 objective functions. A significant difference was observed between values of GW for these two objective functions and all other objective functions. The simulation using the NSE_Q objective function showed the highest contribution from groundwater to streamflow. A significant difference was observed between the value of GW for the simulation using NSE_Q and all other objective functions except NSE_{Q_summer}. For the Noire River sub-watershed a very low contribution of groundwater was simulated with the $NSE_{Q-summer}$ objective function. The low-flow indices simulated with the $NSE_{Q-summer}$ objective function were therefore equal to 0 (Figure 8b). For the future period, a significant decrease of the groundwater contribution to streamflow was observed with the NSE_Q, NSE_{sqrtQ} and NSE_{lnQ} objective function for the Cowansville sub-watershed. No significant difference was observed between the reference and future periods for the Noire River sub-watershed, as the corresponding boxplots overlap, see Figure 13b.

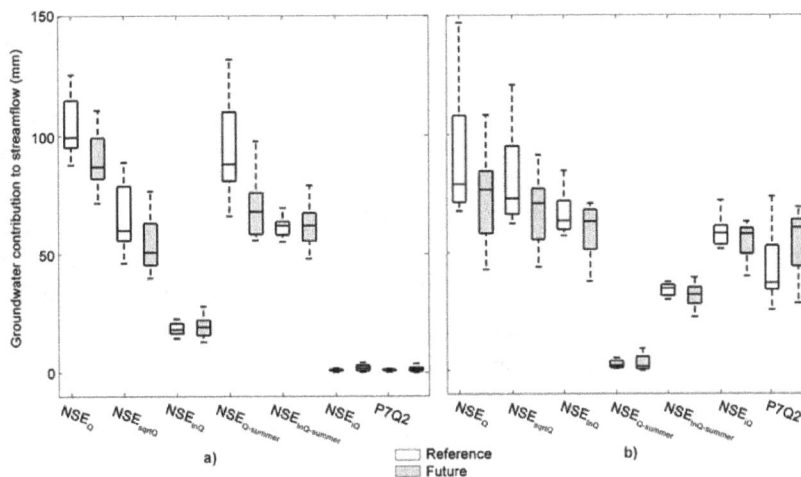

Figure 13. Sum of groundwater contribution to streamflow during the summer simulated with the SWAT model for the (**a**) Cowansville sub-watershed; and (**b**) Noire River sub-watershed.

The median value of SW for the summer/fall period was analyzed (Figure 14). SW for the Cowansville sub-watershed was significantly higher than in the Noire River sub-watershed by approximately 90–100 mm for both current and future climates. These differences are attributed to differences in land cover/land use of these watersheds, the Cowansville being predominantly forested, while the Noire is mostly agricultural. For the future period, a significant decrease of SW between 1% and 2% was simulated for the Cowansville sub-watershed and between 2% and 3% for the Noire River sub-watershed for all objective functions considered. Differences attributed the choice of the objective function for model calibration were also noticed. For the Cowansville sub-watershed (Figure 14a), significantly lower values of SW were simulated with the NSE_{iQ}, $NSE_{Q-summer}$ and p7Q2 objective functions. For the Noire River sub-watershed (Figure 14b), the $NSE_{Q-summer}$ this time showed a significantly higher value of SW. The lower ET simulated with this objective function resulted in a higher soil water content.

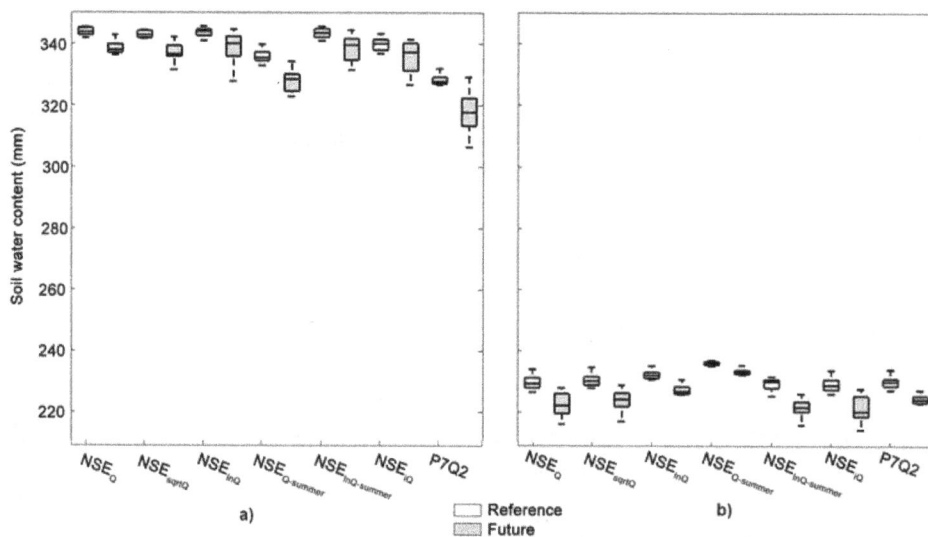

Figure 14. Median of soil water content during the summer simulated with the SWAT model for the (**a**) Cowansville sub-watershed; and (**b**) Noire River sub-watershed.

To summarize, analysis of the SWAT results did not clearly demonstrate a causal relationship between the objective function related to low flows used to calibrate the model and hydrological processes influencing low-flow values. Indeed, the different processes analyzed did not follow a particular trend. For example, groundwater flow in the Cowansville sub-watershed was very low for the NSE_{iQ} and p7Q2 objective functions (see Figure 14a) and corresponding 7Q2 (see Figure 6b) were among the lowest for these objective functions. However, a similar consistency was not observed in the Noire River sub-watershed, where objective functions producing the lowest groundwater flow (NSE_{lnQ} and $NSE_{ln-Q-summer}$) did not produce values of 7Q2which were among the lowest (see Figure 8b). Thus, the low flows produced by the SWAT model did not result from the same hydrological processes (e.g., groundwater flow), but rather from different hydrological processes (e.g., groundwater, interflow, evapotranspiration, etc.), acting individually or in combination and which differed from one watershed to another. Only further analysis of the model results, together with observations, could help to better understand the behavior of the model as a function of different objective functions and also of land use/land cover types. Finally, the question of whether modeling of these processes was faithful to what is actually observed in the field remains unresolved due to the lack of data to make such an assessment. For example, although the overall seasonal variation of ET, SW and GW is acceptable, measurements of these processes and state variables will allow quantitatively evaluating simulation results. It is also possible that equifinality, causing several sets of parameters to produce a similar response from the model, may be the cause of the lack of consistency observed between the modeled hydrological processes and low-flow indices.

5. Discussion and Conclusions

This study investigated the impact of the objective function on the uncertainty of hydrological model results for simulating low flows in a climate change context. Two models were evaluated: the GR4J lumped conceptual model and the SWAT spatially distributed model. A PARASOL-based method was used to evaluate uncertainties related to hydrological model parameters. An ANOVA method was used to quantify the contribution of climate models, hydrological models, hydrological model parameters, objective functions and their interaction to the total uncertainty of an index related to low-flow (7Q2).

Our results shown that the objective function is an important source of uncertainty for low-flow simulations. This finding is partially supported by Parajka et al. [14] who concluded that uncertainty from the objective function is not negligible, but not the main contribution to total uncertainty. However, they used 11 different combinations of a weighted average between two criteria, namely NSE_Q and NSE_{lnQ}, along with a single hydrological model. Our study used a larger spectrum of objective functions related to low-flow with two hydrological models of different complexity. Our results also showed that the interaction-term between the hydrological models and the objective functions is important. The contribution of hydrological model parameters is the lowest contribution to the total uncertainty of low-flow projections. When the ANOVA was applied for each objective function, the relative contribution of climate models and hydrological models was shown to vary with watershed characteristics and time horizon (reference and future periods). This is consistent with previous studies [9,14].

The model calibrated with the NSE_Q objective function, one of the seven objective functions tested in this study, and which is commonly used for model calibration in climate changes studies, was found to poorly simulate low flows. Besides, many studies assessing the impact of climate change on the hydrological response highlight that calibrated models have difficulty in simulating low flows compared to observations (e.g., [11,13]). Furthermore, total variance on the 7Q2 index was large using this objective function. Studies on low flows using NSE_Q may be producing high uncertainties and therefore should be avoided. Using the logarithmic transformed NSE (NSE_{lnQ}) objective function, both models succeeded in adequately representing the other's goodness-of-fit related to low flows. This finding confirms results reported by other studies [14,36]. In addition, using the NSE_{lnQ} objective function, the total variance on 7Q2 index was the lowest for all sub-watersheds.

As observed by Najifi et al. [10] and Chen et al. [7], the structure of the hydrological model can influence estimates of climate change impacts on low flows. Using 12 regional climate projections from different climate models, the GR4J model calibrated with objective functions related to low flows better reproduced the observed 7Q2 index for the reference (i.e., 1970–2000) period. The SWAT model hardly reproduced the observed indices for the reference period. This problem was also noticed by Dams et al. [13] using SWAT. Moreover, the 25%–75% interquartile range on 7Q2 index was found to be higher for the SWAT model than for the GR4J model. SWAT includes a large number of parameters compared to GR4J, which may amplify the effect of the model equifinality, and therefore be responsible for the larger parameter uncertainty obtained with SWAT. For the GR4J model, the 25%–75% interquartile range on 7Q2 index was lowest with the NSE_{iQ} and p7Q2 objective functions, which are more related to low flows.

In general, the total variance was lower for the future period than the reference period. . For high flows, precipitation is an important factor in calculating streamflow. Increasing the uncertainty of precipitation caused by the climate model structure led to an increase in the uncertainty for high flows for the future period [8]. In the dry period, the hydrological model structure can plays a major role and an increase in the uncertainty of precipitation did not lead necessarily to increasing uncertainty for low flows. The analysis of hydrological processes can help understanding why the uncertainty of low-flow indices for the future period was lower than for the reference period. There is no upper limit to streamflow during high flows, while streamflows are physically bounded to zero. In the future period, more streamflow values are simulated close to zero, which reduces the uncertainty on low-flow indices, such as the 7Q2.

Figures 9 and 11 show that uncertainties for actual evapotranspiration increased in the future horizon compared to the reference period for both models, even when simulated ET is different between the GR4J (from 150 to 300 mm) and the SWAT model (from 300 to 400 mm). The difference may be due to the model used to calculate potential evapotranspiration (ETP) (Hamon for GR4J, Penman-Monteith for SWAT), since comparing ETP produced by these two models shown differences. However, ETP simulated by the Hamon model for each sub-watershed was higher than ETP simulated with Penman-Monteith model, which cannot explain higher ET simulated by the SWAT model. Therefore, the difference must come from the model structure used to convert ETP to ET, which is very different between GR4J and SWAT. In addition, soil moisture is conceptualized differently in each model, which may contribute to the difference in simulated ET values. The increase of uncertainty on ET is difficult to tie to the decrease of uncertainty on the 7Q2 index, which, as mentioned above, is related to longer periods of very low flow values in a future climate.

For the GR4J model, uncertainties in percolation decrease since the available water in the soil also decreases. This can explain the lower uncertainties observed for the 7Q2 index in the future period compared to the reference period, despite the increase in the uncertainty for precipitation.

Hydrological processes in the SWAT model are more complex and the objective function used for the calibration may change the dominant hydrological processes leading to low-flow situations. Different land cover/land use between sub-watersheds, which are explicitly represented in SWAT, but not in GR4J, may also partly explain the differences in hydrological processes driving the production of low-flow. The nested method used for the calibration can also introduce uncertainty, since model calibration of upstream sub-watersheds can influence the calibration of downstream sub-watersheds. These findings highlight the need to further investigate hydrological processes involved in low-flow simulations, and hydrological models should be used with care for low-flow studies, since they show some inconsistent results.

For the Cowansville sub-watershed, there is a clear signal from the two hydrological models that low flows will be worsening in the future. It is difficult to make conclusions about the impact of climate change on lows flows in the Farnham sub-watershed since streamflows are regulated. In particular, while a significant decrease in low flows in the future is expected with the GR4J model, which does not handle reservoir storage, conflicting results are obtained depending on the objective function used to calibrate the SWAT model. Additional studies should be conducted to determine whether water released from the reservoir can be maintained in the future without affecting other usages and violating reservoir constraints. For the Noire River and St-Hyacinthe sub-watersheds, results differ depending on the hydrological model used to conduct the impact study. For example, a decrease in low flows was obtained with GR4J as shown by the decrease in both 7Q2 and 7Q10 indices for all objective functions used to calibrate the model. However, these results neglect the effect of reservoir storage in the Farnham sub-watershed located upstream, and are therefore considered suspect since GR4J cannot simulate reservoir storage. On the other hand, the SWAT model, which explicitly considers storage regulation, generated no significant changes in the 7Q2 index, while at the same time a significant worsening of low-flow conditions was obtained through a lowering of the 7Q10 index in the St-Hyacinthe watershed. Again, possible equifinality problems with SWAT may explain these contradictory results.

Great care therefore must be taken in selecting an objective function to calibrate a hydrological model which is well adapted to simulate low flows. This is particularly striking in climate change studies for watersheds which expect a worsening of low-flow conditions in the future. Moreover, this study highlights the importance of model structure in simulating low flows in climate change impact assessments. While models must explicitly consider storage retention for regulated watersheds, increasing model complexity must be addressed with care as low flows are the result of hydrological processes such as infiltration, evapotranspiration and percolation, which interact to produce subtle changes in low-flow regimes in a way that is very dependent on watershed physiographic characteristics. More than ever, climate change impact studies on hydrological regimes require conducting hydrological model uncertainty assessments.

Acknowledgments: Funds from the 2006-2012 Québec's Climate Change Action Plan. Natural Sciences and Engineering Research Council of Canada. We wish to thank the North American Regional Climate Change Assessment Program (NARCCAP) for providing the data used in this paper. NARCCAP is funded by the National Science Foundation (NSF), the U.S. Department of Energy (DoE), the National Oceanic and Atmospheric Administration (NOAA), and the U.S. Environmental Protection Agency Office of Research and Development (EPA).

Author Contributions: Robert Leconte coordinated the project, was involved in most simulations and finalized the paper. Melanie Trudel and Pierre-Louis Doucet-Généreux performed the simulations and analyzed the data. Melanie Trudel wrote the paper.

Conflicts of Interest: The authors declare no conflict of interest. The founding sponsors had no role in the design of the study; in the collection, analyses, or interpretation of data; in the writing of the manuscript, and in the decision to publish the results.

References

1. Mauser, W.; Marke, T.; Stoeber, W. Climate Change and water resources: Scenarios of low-flow conditions in the Upper Danube River Basin. *IOP Conf. Ser. Earth Environ. Sci.* **2008**. [CrossRef]
2. Rahman, M.; Bolisetti, T.; Balachandar, R. Effect of climate change on low-flow conditions in the Ruscom River Watershed, Ontario. *Trans. ASABE* **2010**, *53*, 1521–1532. [CrossRef]
3. Ryu, J.H.; Lee, J.H.; Jeong, S.; Park, S.K.; Han, K. The impacts of climate change on local hydrology and low flow frequency in the Geum River Basin, Korea. *Hydrol. Process.* **2011**, *25*, 3437–3447. [CrossRef]
4. Centre d'expertise hydrique du Québec (CEHQ). *Atlas Hydroclimatique du Québec Méridional—Impact des Changements Climatiques sur les Régimes de Crue, D'étiage et D'hydraulicité à L'horizon 2050*; Centre D'expertise Hydrique du Québec (CEHQ): Ville de Québec, QC, Canada, 2015; p. 81.
5. Wilby, R.L.; Harris, I. A framework for assessing uncertainties in climate change impacts: Low-flow scenarios for the River Thames, UK. *Water Resour. Res.* **2006**, *42*, W02419. [CrossRef]
6. Kay, A.L.; Davies, H.N.; Bell, V.A.; Jones, R.G. Comparison of uncertainty sources for climate change impacts: Flood frequency in England. *Clim. Chang.* **2009**, *92*, 41–63. [CrossRef]
7. Chen, J.; Brissette, F.P.; Leconte, R. Uncertainty of downscaling method in quantifying the impact of climate change on hydrology. *J. Hydrol.* **2011**, *401*, 190–202. [CrossRef]
8. Bosshard, T.; Carambia, M.; Goergen, K.; Kotlarski, S.; Krahe, P.; Zappa, M.; Schär, C. Quantifying uncertainty sources in an ensemble of hydrological climate-impact projections. *Water Resour. Res.* **2013**, *49*, 1523–1536. [CrossRef]
9. Andor, N.; Rössler, O.; Köplin, N.; Huss, M.; Weingartner, R.; Seibert, J. Robust changes and sources of uncertainty in the projected hydrological regimes of Swiss Catchments. *Water Resour. Res.* **2014**, *50*, 7541–7562.
10. Najafi, M.R.; Moradkhani, H.; Jung, I.W. Assessing the uncertainties of hydrologic model selection in climate change impact studies. *Hydrol. Process.* **2011**, *25*, 2814–2826. [CrossRef]
11. Velàzquez, J.A.; Schmid, J.; Ricard, S.; Muerth, M.J.; Gauvin St-Denis, B.; Minville, M.; Chaumont, D.; Caya, D.; Ludwig, R.; Turcotte, R. An ensemble approach to assess hydrological models' contribution to uncertainties in the analysis of climate change impact on water resources. *Hydrol. Earth Syst. Sci.* **2013**, *17*, 565–578. [CrossRef]
12. Vansteenkiste, T.; Tavakoli, M.; Ntegeka, V.; de Smedt, F.; Batelaan, O.; Pereira, F.; Willems, P. Intercomparison of hydrological model structures and calibration approaches in climate scenario impact projections. *J. Hydrol.* **2014**, *519*, 743–755. [CrossRef]
13. Dams, J.; Nossent, J.; Senbeta, T.B.; Willems, P.; Batelaan, O. Multi-model approach to assess the impact of climate change on runoff. *J. Hydrol.* **2015**, *529*, 1601–1616. [CrossRef]

14. Parajka, J.; Blaschke, A.P.; Blöschl, G.; Haslinger, K.; Hepp, G.; Laaha, G.; Schöner, W.; Trautvetter, H.; Viglione, A.; Zessner, M. Uncertainty contributions to low-flow projections in Austria. *Hydrol. Earth Syst. Sci.* **2016**, *20*, 2085–2101. [CrossRef]

15. Beven, K. A manifesto for the equifinality thesis. *J. Hydrol.* **2006**, *320*, 18–36. [CrossRef]

16. Her, Y.; Chaubey, I. Impact of the numbers of observations and calibration parameters on equifinality, model performance, and output and parameter uncertainty. *Hydrol. Process.* **2015**, *29*, 4220–4237. [CrossRef]

17. Yang, J.; Reichert, P.; Abbaspour, K.C.; Xia, J.; Yang, H. Comparing uncertainty analysis techniques for a SWAT application to the Chaohe Basin in China. *J. Hydrol.* **2008**, *358*, 1–23. [CrossRef]

18. Zhang, J.; Li, Q.; Guo, B.; Gong, H. The comparative study of multi-site uncertainty evaluation method based on SWAT model. *Hydrol. Process.* **2015**, *29*, 2994–3009. [CrossRef]

19. Vrugt, J.A.; Gupta, H.V.; Bouten, W.; Sorooshian, S. A shuffled complex evolution metropolis algorithm for optimization and uncertainty assessment of hydrologic model parameters. *Water Resour. Res.* **2003**. [CrossRef]

20. Beven, K.; Binley, A. The future of distributed models: Model calibration and uncertainty prediction. *Hydrol. Process.* **1992**, *6*, 279–298. [CrossRef]

21. Beven, K.; Binley, A. GLUE: 20 years on. *Hydrol. Process.* **2014**, *28*, 5897–5918. [CrossRef]

22. Abbaspour, K.C.; Yang, J.; Maximov, I.; Siber, R.; Bogner, K.; Mieleitner, J.; Zobrist, J.; Srinivasan, R. Spatially-distributed modelling of hydrology and water quality in the pre-alpine/alpine Thur watershed using SWAT. *J. Hydrol.* **2007**, *333*, 413–430. [CrossRef]

23. Van Griensven, A.; Meixner, T. A global and efficient multi-objective auto-calibration and uncertainty estimation method for water quality catchment models. *J. Hydroinform.* **2007**, *9*, 277–291. [CrossRef]

24. Nash, J.E.; Sutcliffe, J.V. River flow forecasting through conceptual models part I—A discussion of principles. *J. Hydrol.* **1970**, *10*, 282–290. [CrossRef]

25. OBV Yamaska. *Plan directeur de L'eau, 2e Version*; Organisme de Basin Versant de la Yamaska: Granby, QC, Canada, 2014.

26. Perrin, C.; Michel, C.; Andréassian, V. Improvement of a parsimonious model for streamflow simulation. *J. Hydrol.* **2003**, *279*, 275–289. [CrossRef]

27. Hamon, W.R. Estimating potential evapotranspiration. *J. Hydraul. Eng. Div.* **1961**, *87*, 107–120.

28. Arnold, J.G.; Srinivasan, R.; Muttiah, R.S.; Williams, J.R. Large area hydrologic modeling and assessment Part I: Model development. *J. Am. Water Resour. Assoc.* **1998**, *34*, 73–89. [CrossRef]

29. Neitsch, S.L.; Arnold, J.G.; Kiniry, J.R.; Williams, J.R. *Soil and Water Assessment Tool Theoretical Documentation Version 2009*; Grassland, Soil and Water Research Laboratory, Agricultural ResearchService and Blackland Research Center, Texas Agricultural Experiment Station: College Station, TX, USA, 2011.

30. Ministère de développement durable, de l'environnement et des parcs (MDDEP). *Calcul et Interprétation des Objectifs Environnementaux de Rejet Pour les Contaminants du Milieu Aquatique*, 2nd ed.; Direction du Suivi de L'état de L'environnement; Québec, Ministère du Développement Durable, de L'environnement et des Parcs: Rimouski, QC, Canada, 2007; p. 56.

31. Ministère du Développement Durable, de L'environnement et de la Lutte Contre les Changements Climatiques (MDDELCC). *Guide de Production des Installations de Production d'eau Potable*; Ministère du Développement Durable, de L'environnement et de la Lutte Contre les Changements Climatiques: Rimouski, QC, Canada, 2015.

32. Duan, Q.; Sorooshian, S.; Gupta, V. Effective and efficient global optimization for conceptual rainfall-runoff models. *Water Resour. Res.* **1992**, *28*, 1015–1031. [CrossRef]

33. Ritter, A.; Munoz-Carpena, R. Performance evaluation of hydrological models: Statistical significance for reducing subjectivity in goodness-of-fit assessments. *J. Hydrol.* **2013**, *480*, 33–45. [CrossRef]

34. McCuen, R.H.; Knight, Z.; Cutter, A.G. Evaluation of the Nash-Sutcliffe Efficiency Index. *J. Hydrol. Eng.* **2006**, *11*, 597–602. [CrossRef]

35. Gupta, H.V.; Kling, H. On typical range, sensitivity, and normalization of Mean Squared Error and Nash-Sutcliffe Efficiency type metrics. *Water Resour. Res.* **2011**, *47*, 1–3. [CrossRef]

36. Pushpalatha, R.; Perrin, C.; Le Moine, N.; Andréassian, V. A review of efficiency criteria suitable for evaluating low-flow simulations. *J. Hydrol.* **2012**, *420*, 171–182. [CrossRef]

37. Mearns, L.; McGinnin, S.; Arritt, R.; Biner, S.; Duffy, P.; Gutowski, W.; Isaac, H.; Richard, J.; Leung, R.; Nunes, A.; et al. *The North American Regional Climate Change Assessment Program Dataset*; National Center for Atmospheric Research Earth System Grid Data Portal: Boulder, CO, USA, 2007.

38. Caya, D.; Laprise, R. A semi-implicit semi-lagrangian regional climate model: The Canadian RCM. *Mon. Weather Rev.* **1999**, *127*, 341–362. [CrossRef]

39. Flato, G. The Third Generation Coupled Global Climate Model (CGCM3). Available online: http://www.ec.gc.ca/ccmac-cccma/default.asp?n=1299529F-1 (accessed on 28 February 2017).

40. Schmidli, L.; Frei, C.; Vidal, P. Downscaling from GCM precipitation: A benchmark for dynamical and statistical downscaling methods. *Int. J. Climatol.* **2006**, *26*, 679–686. [CrossRef]

41. Ouranos. *Vers L'adaptation. Synthèse des Connaissances sur les Changements Climatiques au Québec. Partie 1: Évolution Climatique au Québec*; Ouranos: Montréal, QC, Canada, 2015.

4

Rising Precipitation Extremes across Nepal

Ramchandra Karki [1,2,*], Shabeh ul Hasson [1,3], Udo Schickhoff [1], Thomas Scholten [4] and Jürgen Böhner [1]

[1] Center for Earth System Research and Sustainability, Institute of Geography, University of Hamburg, Bundesstraße 55, 20146 Hamburg, Germany; shabeh.hasson@uni-hamburg.de (S.H.); udo.schickhoff@uni-hamburg.de (U.S.); juergen.boehner@uni-hamburg.de (J.B.)

[2] Department of Hydrology and Meteorology, Government of Nepal, 406 Naxal, Kathmandu, Nepal

[3] Department of Space Sciences, Institute of Space Technology, Islamabad 44000, Pakistan

[4] Soil Science and Geomorphology, University of Tübingen, Department of Geosciences, Rümelinstrasse 19-23, 72070 Tübingen, Germany; thomas.scholten@uni-tuebingen.de

[*] Correspondence: Ramchandra.Karki@studium.uni-hamburg.de

Academic Editor: Christina Anagnostopoulou

Abstract: As a mountainous country, Nepal is most susceptible to precipitation extremes and related hazards, including severe floods, landslides and droughts that cause huge losses of life and property, impact the Himalayan environment, and hinder the socioeconomic development of the country. Given that the countrywide assessment of such extremes is still lacking, we present a comprehensive picture of prevailing precipitation extremes observed across Nepal. First, we present the spatial distribution of daily extreme precipitation indices as defined by the Expert Team on Climate Change Detection, Monitoring and Indices (ETCCDMI) from 210 stations over the period of 1981–2010. Then, we analyze the temporal changes in the computed extremes from 76 stations, featuring long-term continuous records for the period of 1970–2012, by applying a non-parametric Mann−Kendall test to identify the existence of a trend and Sen's slope method to calculate the true magnitude of this trend. Further, the local trends in precipitation extremes have been tested for their field significance over the distinct physio-geographical regions of Nepal, such as the lowlands, middle mountains and hills and high mountains in the west (WL, WM and WH, respectively), and likewise, in central (CL, CM and CH) and eastern (EL, EM and EH) Nepal. Our results suggest that the spatial patterns of high-intensity precipitation extremes are quite different to that of annual or monsoonal precipitation. Lowlands (Terai and Siwaliks) that feature relatively low precipitation and less wet days (rainy days) are exposed to high-intensity precipitation extremes. Our trend analysis suggests that the pre-monsoonal precipitation is significantly increasing over the lowlands and CH, while monsoonal precipitation is increasing in WM and CH and decreasing in CM, CL and EL. On the other hand, post-monsoonal precipitation is significantly decreasing across all of Nepal while winter precipitation is decreasing only over the WM region. Both high-intensity precipitation extremes and annual precipitation trends feature east−west contrast, suggesting significant increase over the WM and CH region but decrease over the EM and CM regions. Further, a significant positive trend in the number of consecutive dry days but significant negative trend in the number of wet (rainy) days are observed over the whole of Nepal, implying the prolongation of the dry spell across the country. Overall, the intensification of different precipitation indices over distinct parts of the country indicates region-specific risks of floods, landslides and droughts. The presented findings, in combination with population and environmental pressures, can support in devising the adequate region-specific adaptation strategies for different sectors and in improving the livelihood of the rural communities in Nepal.

Keywords: Nepal; spatial precipitation pattern; precipitation extremes; consecutive dry days; high-intensity precipitation

1. Introduction

Precipitation extremes are one of the major factors that trigger natural disasters, such as droughts, floods and landslides, which subsequently cause the loss of property and life, and deteriorate socioeconomic development. Under the prevailing anthropogenic warming, the precipitation extremes are observed to be intensified globally, exacerbating the existing problems of food and water security as well as disaster management [1–11].

Consistent with the global pattern [4,10], the world's most disaster-prone region of South Asia [12] has also experienced an overall increase in precipitation extremes [11], though such a pattern is heterogeneous across the region [5,11,13,14]. For instance, studies have found a rise in the summer monsoonal precipitation extremes over central India and northeastern Pakistan [15–20]. In contrast, precipitation extremes feature a falling trend over southwestern Pakistan [20], the eastern Gangetic plains and some parts of Uttaranchal, India [21]. Further, contrary to the extremes observed at low altitudes or over the plains, extremes observed in the high-altitude mountainous regions exhibit quite an opposite sign of change due to the influence of local factors, and are thus less predictable [22].

Situated in the steep terrain of the central Himalayan range, Nepal is likewise more susceptible to the developments of heavy rainfall events and subsequent flooding and droughts severely impacting the marginalized mountain communities, as was impressively illustrated by recent events. For instance, the cloudburst of 14–17 June 2013 in the northwestern mountainous region near the Nepalese border killed around 5700 people and affected more than 100,000, extensively damaging the property in both Nepal and India [23]. A heavy rain event of 14–16 August 2014 likewise caused massive flooding and triggered a number of landslides, resulting in huge losses of life and property, affecting around 35,000 households [24]. Similarly, one of the worst winter droughts of the country in 2008/2009 reduced yield of wheat and barley by 14% and 17%, respectively, leading to severe food shortage in 66% of rural households in the worst hit far- and mid-western hill and mountain regions [25]. Such intense precipitation and extreme dryness (droughts) negatively impact the yield of both cash and cereal crops [26], and in turn, the livelihood of around 60% of the total Nepalese population directly dependent on agriculture [27]. Therefore, analyzing the precipitation extremes and their time evolution under prevailing climatic changes is of paramount importance for ensuring food and water security in Nepal and developing a region-specific disaster management strategy.

As compared to the rest of South Asia, studies on the observed precipitation extremes over Nepal are rare and sporadic, lacking a comprehensive picture across the country. For instance, computing various extreme precipitation indices from only 26 stations for the period of 1961–2006, Baidya et al. [28] have found an increasing trend in total events and heavy precipitation events from most of the stations. In contrast, analyzing only a subset of extreme precipitation indices used in Baidya et al. [28], from the daily interpolated gridded precipitation of APHRODITE (1951–2007) Duncan et al. [29] concludes that the monsoonal and annual precipitation extremes are unlikely to worsen over Nepal. Since precipitation extremes are more localized and can be smoothed over when limited gauge data is interpolated [30], employing APHRODITE may have significantly affected the computation of extreme statistics as it hardly incorporates any real observations from the early 1950s west of Kathmandu [31]. Recently, analyzing the extreme precipitation indices only within the Koshi River basin in eastern Nepal for the 1975–2010 period, Shrestha et al. [32] have reported an overall increase in the precipitation total and intensity, though such trends were statistically insignificant. These discordant findings of changes in the precipitation extremes over Nepal may be attributed to employing varying datasets, analyzing different time periods, and focusing on distinct study regions, hence lacking a countrywide picture.

In view of these limited countrywide studies with contrasting findings—and because an understanding of the extreme precipitation events is crucial to socioeconomic development—, this study presents an exploratory analysis of the widely adopted daily precipitation extremes across the whole of Nepal based on the maximum number of high-quality long-term station observations. For this, the extreme indices from the Expert Team on Climate Change Detection, Monitoring and

Indices (ETCCDMI) along with a few additional indices are computed from the daily precipitation observations. First, we have analyzed the spatial distribution of the most relevant precipitation extremes as well as the seasonal precipitation patterns from 210 surface weather stations across Nepal over the most recent period, 1981–2010. Moreover, time evolution of the computed precipitation extremes has been analyzed by ascertaining the monotonic trends from the long-term continuous record available at 76 stations over the 1970–2012 period. In this regard, a robust non-parametric Mann–Kendall [33,34] trend test along with the trend free pre-whitening (TFPW) procedure has been applied. The trends at the local stations are further assessed for their field significance over the distinct physio-geographic regions of Nepal, in order to establish the dominant patterns of changes in precipitation extremes.

2. Study Area

Nepal is a mountainous country that stretches between 26.36° N–30.45° N and 80.06° E–88.2° E, encompassing an area of 147,181 km^2 and an elevation range of 60–8848 m above sea level (asl). Along the cardinal directions and altitude, the country is divided into five standard physiographic regions, such as Terai, Siwaliks, Middle Mountains, High Mountains, and High Himalaya [35,36], which according to Duncan and Biggs [37] are further categorized into three broader zones, namely, Lowlands (Terai and Siwaliks), Mid-Mountains and Hills (Middle and High Mountains) and High Mountains (High Himalayas). Owing to such unique physiographical and topographical distribution, the country features a variety of climates that range from the tropical savannah over the southern plains to the polar frost in the northern mountains within a short horizontal distance of less than 200 km [38]. The population density is highest in the eastern lowlands region, followed by the regions of central lowlands, middle and eastern middle mountains, and western lowlands, respectively, while such density is lowest for the rest of Nepal.

There are four seasons in Nepal, namely, pre-monsoon (March–May), monsoon (June–September), post-monsoon (October–November) and winter (December–February). Pre-monsoon is characterized by hot, dry and westerly windy weather with mostly localized precipitation in a narrow band, whereas the monsoon is characterized by moist southeasterly monsoonal winds coming from the Bay of Bengal and occasionally from the Arabian Sea with widespread precipitation. Post-monsoon refers to a dry season with sunny days featuring a driest month, November. Winter is a cold season with precipitation mostly in the form of snow in high-altitude mountainous regions. Precipitation over Nepal is received by two major weather systems; the southwest monsoon greatly impacts the southeastern parts of the country during the monsoon season while the western disturbances predominantly affect the northwestern high mountainous parts during the winter season [39–43]. Similar to the monsoon season, precipitation during pre- and post-monsoon seasons is also generally higher towards the east [44]. In the Marsyangdi River basin (MRB) of Nepal, observed precipitation at the stations below 2000 m is mostly received in the form of rain while at the stations above this height, snowfall accounts for 17% ± 11% of the annual totals, where such a fraction rises with altitude [45].

Classifying the precipitation regimes of Nepal based on the shape and magnitude of monthly precipitation from 222 stations, Kansakar et al. [35] have illustrated that the precipitation patterns are mainly controlled by the orographic effect of the complex central Himalayan terrain and the east—west progression of the summer monsoon. Thus, owing to the intricate interaction amid the weather systems and their alteration by the extreme topographical variations (high mountains, valleys and river catchments), spatial distribution of precipitation in Nepal is highly heterogeneous (Figure 1a). For instance, the annual precipitation varies from less than 200 mm for the driest regions (Mustang, Manang, and Dolpa, located at the leeward-side north of the Annapurna) to above 5000 mm in and around the Lumle region. Two additional wetter regions with annual precipitation greater than 3500 mm are Num and Gumthang. On the other hand, regions of low precipitation typically also reside in the leeward-side of the Khumbu, Everest and other high mountainous regions [38,41,46]. Along the altitudinal extent of the central Himalayan region (~74°–88° E), the Tropical Rainfall Measuring

Mission (TRMM) precipitation dataset (1998–2005) indicates two parallel significant peak precipitation zones [47]; the first zone is at the mean elevation of ~0.95 km (mean relief of ~1.2 km) while the second one is at ~2.1 km (mean relief of ~2 km). Barros et al. [45] have suggested a weak altitudinal gradient of precipitation between 1000 and 4500 m altitude, whereas in deep river valleys with steep slopes, such a gradient of rainstorm is very strong. In general, they have found the maximum precipitation along the ridges and a strong east−west ridge-to-ridge precipitation gradient. Similar results were obtained from station-based observations in different regions of Nepal where precipitation peaks around 2500–3600 mm and decreases further with increase in altitude in high mountain regions [48–51].

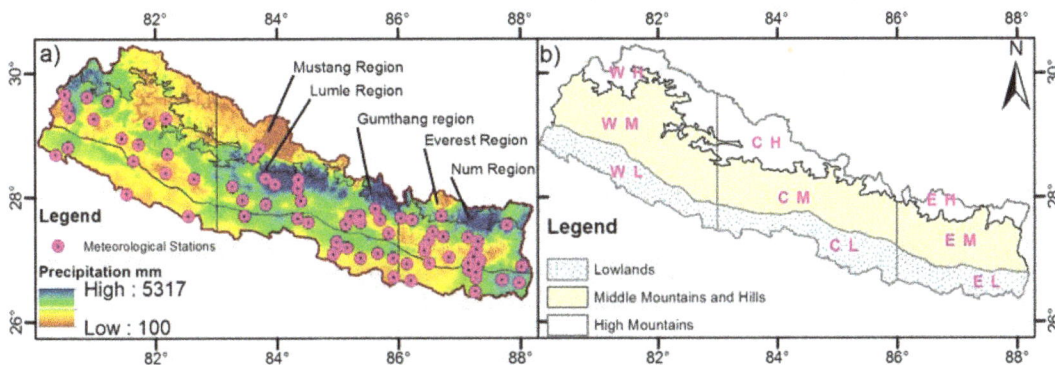

Figure 1. (a) Annual precipitation distribution (mm) over Nepal with precipitation pocket areas and dry areas delineated, and the meteorological stations used for the daily extreme analysis overlaid; **(b)** The three broader physiographic zones of Nepal and nine sub-regions (WL—Western Lowlands, WM—Western Middle Mountains and Hills, WH—Western High Mountains, CL—Central Lowlands, CM—Central Middle Mountains and Hills, CH—Central High Mountains, EL—Eastern Lowlands, EM—Eastern Middle Mountains and Hills, EH—Eastern High Mountains).

In addition to the horizontal and altitudinal precipitation gradients, a large seasonal precipitation gradient (~factor of 4) has also been observed over a short horizontal distance of ~10 km in MRB [52].

Precipitation does not feature long-term trends at seasonal and annual scales, except a localized trend in some parts of the Koshi River basin [44,53,54]. However, it features a significant relationship with the Southern Oscillation Index (SOI) [55]. Changes in the monsoonal precipitation regimes indicate the extension of the active monsoon duration mainly due to significantly delayed withdrawal, though the onset timing has been observed unchanged [56,57].

As the precipitation distribution is highly heterogeneous across the country, characterizing strong north−south and east−west gradients, the whole country is divided into nine sub-regions (Figure 1b) for regional field significance study. The latitudinal extent has been divided based on the demarcation of three broader physiographic regions, while for the longitudinal division, longitudes of 83° E and 86° E have been taken as the demarcation points (as used in [44]), yielding western, central and eastern regions, each containing three physio-geographic regions. For instance, the sub-regions within the western longitudinal belt are the western lowlands (WL), western middle mountains and hills (WM) and western high mountains (WH). The case for the central (CL, CM and CH) and eastern (EL, EM and EH) longitudinal belts is similar.

3. Data

We have obtained daily precipitation data from all available surface weather stations in Nepal that are being maintained by the Department of Hydrology and Meteorology (DHM), Nepal. These observations are consistent in terms of the measurement method as the obtained stations use the same type of US-standard 8-inch diameter manual precipitation gauges [58]. Though underestimated as in other types of gauges, these gauges can also measure snow water equivalent. For that, the snow deposited in the gauge is at first melted by pouring hot water and then measured as normal rainfall measurement. In the DHM database, the longest precipitation record available since 1946 is from only three stations, although there are records from 23 stations since 1947, and from around 40 stations since 1950. Until 1956, the precipitation observations were available only from the stations located east of Kathmandu and particularly within the Koshi River basin. Afterwards, the number of precipitation stations considerably increased, reaching up to 100, 190, 240, 250, 370 and 410 by 1961, 1971, 1981, 1991, 2001 and 2010, respectively. However, all the available precipitation gauges do not feature regular data since the time of their inception [31]. For instance, among 450 stations that have been operational until recently, regular data of varying lengths from only around 400 stations either due to short-term discontinuity of the stations or their relocation are available (Figure 2). In order to ensure a balanced spatial distribution of stations across Nepal and employing the maximum common length of the high-quality continuous data available from the maximum number of stations, we have restricted the period of our daily extreme analysis to 1970–2012. Within such a period, the stations that feature data gaps for more than (1) a fortnight within a year; (2) four consecutive years or (3) for six years in total were excluded from the analysis.

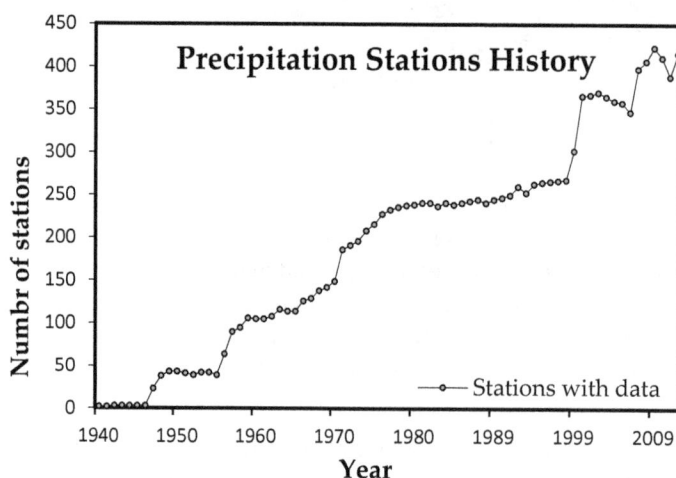

Figure 2. The number of precipitation gauges and their age in the Department of Hydrology and Meteorology (DHM) database.

The quality control of the data from the considered stations was performed using the RClimDex toolkit [59], which can identify potential outliers and negative precipitation values [4]. After the quality control, testing the homogeneity is the most important step [60] as it identifies the variations that occurred due to purely non-climatic factors, such as faults in the instruments or changes in the measurement method, aggregation method, station location, station exposure and observational practice [61]. The homogeneity test was performed for each station by monthly time series using the RHtest toolkit, which can statistically identify the multiple step changes by using a two-phase regression model with a linear trend of the entire time series [62]. Since the stations observe large Euclidean distances in complex mountainous terrains, we have performed a relative homogeneity test, without using a reference time series [63,64]. The inhomogeneity of a station has been decided based on the RHtest results, graphical examination and coincidence of known ENSO or localized precipitation events. The stations featuring any inhomogeneity were excluded from the analysis. Such strict station selection criteria have yielded the continuous, homogeneous, high-quality daily observations from only 76 stations for the 1970–2012 period. These 76 stations have been used for the computation of daily precipitation extremes (Figure 1a and Table 1) and seasonal precipitation total.

For sub-regions considered for field significance study, these stations ensure an adequate spatial distribution of stations at least across the middle mountain and lowland sub-regions. The number of stations that fall in each sub-region of WM, WL, CH, CM, CL, EM and EL are 14, 4, 3, 16, 11, 19 and 9, respectively. Since the stations within the western and central regions are relatively younger, there are more stations within the eastern region that fulfill the selection criteria. As for high mountain sub-regions, long-term data was available only from CH, due to the low density of the high-altitude station network. Hence, we limit our sub-regional analysis to seven sub-regions only.

Table 1. List of meteorological stations.

Region	ID	Name	Lat (°)	Lon (°)	Height (m)
WL	106	Belauri santipur	28.683	80.35	159
	209	Dhangadhi (atariya)	28.8	80.55	187
	416	Nepalgunj Reg. off.	28.052	81.523	144
	510	Koilabas	27.7	82.533	320
WM	101	Kakerpakha	29.65	80.5	842
	103	Patan (west)	29.467	80.533	1266
	104	Dadeldhura	29.3	80.583	1848
	201	Pipalkot	29.617	80.867	1456
	202	Chainpur (west)	29.55	81.217	1304
	203	Silgadhi doti	29.267	80.983	1360
	206	Asara ghat	28.953	81.442	650
	303	Jumla	29.275	82.18	2366
	308	Nagma	29.2	81.9	1905
	402	Dailekh	28.85	81.717	1402
	404	Jajarkot	28.7	82.2	1231
	406	Surkhet	28.587	81.635	720
	504	Libang gaun	28.3	82.633	1270
	511	Salyan bazar	28.383	82.167	1457
CL	703	Butwal	27.694	83.466	205
	902	Rampur	27.654	84.351	169
	903	Jhawani	27.591	84.522	177
	907	Amlekhganj	27.281	84.992	310
	909	Simara airport	27.164	84.98	137
	910	Nijgadh	27.183	85.167	244
	911	Parwanipur	27.079	84.933	115
	912	Ramoli bairiya	27.017	85.383	152
	1109	Pattharkot (east)	27.1	85.66	162
	1110	Tulsi	27.013	85.921	251
	1111	Janakpur airport	26.711	85.924	78
CM	701	Ridi bazar	27.95	83.433	442
	722	Musikot	28.167	83.267	1280
	802	Khudi bazar	28.283	84.367	823
	804	Pokhara airport	28.2	83.979	827
	807	Kunchha	28.133	84.35	855
	808	Bandipur	27.942	84.406	995
	810	Chapkot	27.883	83.817	460
	814	Lumle	28.297	83.818	1740
	904	Chisapani gadhi	27.56	85.139	1729
	1008	Nawalpur	27.813	85.625	1457
	1015	Thankot	27.688	85.221	1457
	1022	Godavari	27.593	85.379	1527
	1023	Dolal ghat	27.639	85.705	659
	1029	Khumaltar	27.652	85.326	1334
	1030	Kathmanduairport	27.704	85.373	1337
	1115	Nepalthok	27.42	85.849	698
CH	601	Jomsom	28.78	83.72	2744
	604	Thakmarpha	28.739	83.681	2655
	607	Lete	28.633	83.609	2490

Table 1. *Cont.*

Region	ID	Name	Lat (°)	Lon (°)	Height (m)
	1112	Chisapani bazar	26.93	86.145	107
	1213	Udayapur gadhi	26.933	86.517	1175
	1216	Siraha	26.656	86.212	102
	1311	Dharan bazar	26.792	87.285	310
EL	1316	Chatara	26.82	87.159	105
	1319	Biratnagar airport	26.481	87.264	72
	1320	Tarahara	26.699	87.279	121
	1408	Damak	26.671	87.703	119
	1409	Anarmani birta	26.625	87.989	122
	1102	Charikot	27.667	86.05	1940
	1103	Jiri	27.633	86.233	2003
	1202	Chaurikhark	27.7	86.717	2619
	1203	Pakarnas	27.443	86.569	1944
	1204	Aisealukhark	27.36	86.749	2063
	1206	Okhaldhunga	27.308	86.504	1731
	1207	Mane bhanjyang	27.215	86.444	1528
	1210	Kurule ghat	27.136	86.43	341
	1211	Khotang bazar	27.029	86.843	1305
EM	1303	Chainpur (east)	27.292	87.317	1262
	1305	Leguwa ghat	27.154	87.289	444
	1306	Munga	27.05	87.244	1457
	1307	Dhankuta	26.983	87.346	1192
	1308	Mul ghat	26.932	87.32	286
	1309	Tribeni	26.914	87.16	146
	1322	Machuwaghat	26.938	87.155	168
	1325	Dingla	27.353	87.146	1169
	1403	Lungthung	27.55	87.783	1780
	1410	Himali gaun	26.887	88.027	1654

Note: WL (western lowlands), WM (western middle mountains and hills), WH (western high mountains), CL (central lowlands), CM (central middle mountains and hills), CH (central high mountains, EL (eastern lowlands), EM (eastern middle mountains and hills) and EH (eastern high mountains)).

In addition to a daily extreme precipitation trend analysis, we present spatial variability maps of mean seasonal and of physically relevant daily precipitation extreme indices from the maximum number of stations that have the data for at least 20 years in the recent normal period (1981–2010) but two stations from high altitude (>3500 m) with only five years of data are also included to represent the high-altitude spatial pattern. Since these indices do not require the high-quality data, around 210 precipitation stations were considered for this analysis.

4. Methodology

4.1. Precipitation Indices

We have considered the extreme precipitation indices that are developed and recommended by the Expert Team on Climate Change Detection, Monitoring and Indices (ETCCDMI), jointly established by the World Meteorological Organization (WMO) Commission for Climatology and the Research Programme on Climate Variability and Predictability (CLIVAR) (Table 2). Based on the calculation method, these indices fall into four groups [65,66], namely, absolute indices, threshold indices, duration indices and percentile-based threshold indices.

Table 2. The description of ETCCDMI (Expert Team on Climate Change Detection, Monitoring and Indices).

Category	ID	Name of Index	Definition	Unit
HIP	R95	Very wet days	Annual total precipitation of days in >95th percentile	mm
HIP	RX1day	Max 1-day precipitation amount	Annual maximum 1-day precipitation	mm
HIP	RX5day	Max 5-day precipitation amount	Annual maximum consecutive 5-day precipitation	mm
HIP	R99	Extremely wet days	Annual total precipitation of days in >99th percentile	mm
FP	R10	Number of heavy precipitation days	Annual count of days when precipitation is $\geq$10 mm	Days
FP	R20	Number of very heavy precipitation days	Annual count of days when precipitation is $\geq$20 mm	Days
DWS	CDD	Consecutive dry days	Maximum number of consecutive dry days (precipitation <1 mm)	Days
DWS	CWD	Consecutive wet days	Maximum number of consecutive wet days (precipitation $\geq$1 mm)	Days
EX	WD	Annual wet/rainy days	Annual count of days when precipitation is $\geq$1 mm	Days
EX	PRCPTOT	Annual total wet-day precipitation	Annual total from days $\geq$1 mm precipitation	mm
EX	SDII	Simple daily intensity index	Ratio of annual total to WD in a year	mm/day

Note: HIP: High-intensity-related precipitation extreme, FP: Frequency-related precipitation extreme, DWS: Dry and wet spell or duration, EX: Extra.

The absolute indices include the annual maximum of 1-day and 5-day precipitation referred to as RX1day and RX5day, respectively. Threshold indices comprise of numbers of (1) heavy precipitation days (R10) and (2) very heavy precipitation days (R20). R10 (R20) denotes the counts of days in a year when precipitation was $\geq$10 mm ($\geq$20 mm). The duration indices encompass the consecutive wet and dry days (CWD and CDD), which describe the maximum lengths of consecutive wet and dry days, respectively. Unlike the absolute threshold-based indices, percentile-based indices better represent the spatial aspects of precipitation extremes, accounting for spatial heterogeneity of precipitation. However, different researchers have used different percentiles as thresholds for defining such extremes. For example, Krishnamurthy et al. [18] and Bookhagen [67] have used the 90th percentile, Duncan et al. [29] and Casanueva et al. [68] have used the 95th percentile and Alexander et al. [4], Klein Tank et al. [5], Donat et al. [14], and Sheikh et al. [11] have used the 95th and 99th percentiles. Here, the percentile-based threshold indices include two; the yearly total precipitation from very wet days (R95) and extremely wet days (R99) observing above than 95th and 99th precipitation percentiles, respectively. Since R95 and R99 indices are based on the long-term percentile thresholds, they may differ from station to station but not on an inter-annual scale. Three additional indices that do not fall in any of the above categories are the annual wet days (WD) described as the annual count of days when precipitation was $\geq$1 mm, annual total precipitation from WD (PRCPTOT) and the simple daily intensity index (SDII) defined as the ratio of PRCPTOT to WD within a year. These indices were computed using the RClimDex toolkit [59] while the regionalization of the physically relevant indices has been performed using Global Kriging interpolation (See [38] for details).

The considered extreme precipitation indices are mainly the indicators of floods and droughts or indirect indicators which in combination with the main indicators provide valuable information. For instance, R95, R99, RX1day and RX5day indices characterize the magnitude of very intense precipitation events that trigger flash floods and landslides. R10 and R20 assess the frequency of heavy and very heavy precipitation events. On the other hand, CDD and CWD indirectly provide the indication of droughts, which is quite important for the agricultural activities [68]. The SDII, PRCPTOT and WD indirectly provide useful information when combined with other extreme indices. Therefore, we have presented and discussed the results in four broad categories, namely, high-intensity-related precipitation extreme (HIP) indices which consist of absolute and percentile-based threshold indices, frequency-related precipitation extremes or threshold-based (FP) indices, dry and wet spell or duration (DWS) indices and extra indices (EX) (Table 2).

In addition to the computation of extreme precipitation indices, the temporal changes in these indices and seasonal total precipitation have also been assessed. For this, trends in the considered indices (except R99) were analyzed using the Mann−Kendall (MK) trend test [33,34] while the magnitude of trend was estimated using the Theil−Sen's (TS) slope method [69,70]. Trend assessment for R99 was possible, but no trends were found for more than half of the stations due to large inter-annual variability that resulted in zero values for more than half of the R99 time series. They were therefore, excluded from the analysis.

4.2. Trend Analysis

4.2.1. Mann−Kendall Trend

The MK trend test [33,34] has been widely used to assess the significance of a trend in the time series as the test does not require normally distributed data sets [71,72] and can cope with missing data records and extremes.

4.2.2. Theil−Sen's Slope

If a linear trend is present in a time series, the true slope of the existing trend can be computed using the non-parametric TS approach. This test is widely used and robust as it is less sensitive to outliers and missing values in data [69,70].

4.2.3. Trend-Free Pre-Whitening

The MK test is based on the assumption that the time series is serially independent. However, often the hydro-meteorological time series contain a serial correlation [73–75], affecting the MK test results. For instance, existence of a positive (negative) serial correlation in a time series overestimates (underestimates) the significance of the MK test (e.g., [74,76]). To limit such effect of serial correlation on the MK test results, several pre-whitening (PW) procedures have been proposed [71,75–77] including trend-free pre-whitening (TFPW). TFPW more effectively reduces the effect of a serial correlation present within the hydro-meteorological time series on the MK test results [78–80]. Here, we have used TFPW as proposed by Yue et al. [74].

In TFPW, the initial step is to estimate the true slope of a trend using Sen's slope method, unitize the time series by dividing each sample with the sample mean and de-trend the time series. The lag-1 auto-correlation is then estimated and removed if existing and the time series is subsequently blended back to the pre-identified trend component. Finally, the MK test is applied to pre-whitened time series (see [74,75] for further information).

4.2.4. Field Significance

A certain region can feature a number of stations with positive or negative trends. Thus, the field significance test is used to identify regions with trends of a consistent sign, independent of statistical significance of the individual station trends [81,82].

Various methods are available for assessing the field significance of local trends [71,73,74,83–86]. We have adopted the method of Yue et al. [74] to assess the field significance in the seven sub-regions (Figure 1b) with sufficient data. In this method, the original station network has been resampled 1000 times with the bootstrapping approach [87], distorting (preserving) the auto-(cross-)correlation to avoid its influence on the field significance analysis. The MK test is then applied to synthetic time series of each site. At the given significance level, the numbers of sites with significant upward trends and downward trends, respectively, have been counted using Equation (1) for each resampled network.

$$C_f = \sum_{i=1}^{n} C_i \tag{1}$$

where, n, indicates the number of stations within a region of analysis and C_i refers to a count of statistically significant trends (at 0.1 level) at the station, i. This procedure has been repeated 1000 times for each resampled network. Ranking the corresponding 1000 values of counts of significant positive (negative) trends in an ascending order using the Weibull [88] plotting position formula yields the empirical cumulative distributions, C_f, as:

$$P\left(C_f \leq C'^r{}_f\right) = r/(N+1) \tag{2}$$

where, r, is the rank of C and N is the number of resampled networks. The probability of a number of significant positive (negative) trends in the original network has been estimated by comparing with Cf of significant positive (negative) trends obtained from the resampled networks (Equation (3)).

$$P_{\text{obs}} = P\left(C_{f,\,\text{obs}} \leq C'^r{}_f\right) \text{ where } P_f = \begin{cases} P_{obs}, & \text{for } P_{obs} \leq 0.5 \\ 1 - P_{obs}, & P_{obs} > 0.5 \end{cases} \tag{3}$$

At the significance level of 0.1, if $P_f \leq 0.1$, then the trend over a region was considered significant. A similar approach has been employed by Petrow and Merz [89] for assessing the field significance of flood time series in Germany and by Hasson et al. [63] for hydro-meteorological time series in the Hindukush-Karakoram-Himalayan region of the upper Indus basin.

5. Results and Discussion

5.1. Spatial Distribution of Mean Seasonal and Daily Precipitation Indices

In addition to the seasonal mean precipitation distribution, spatial patterns of mainly the high-intensity- and frequency-related extremes (R95/R99 percentile precipitation, RX1day and WD, R10, R20, respectively), which are relevant for water resources, as well as flood and agriculture management, are computed from 210 stations over the period of 1981–2010 (Figures 3 and 4).

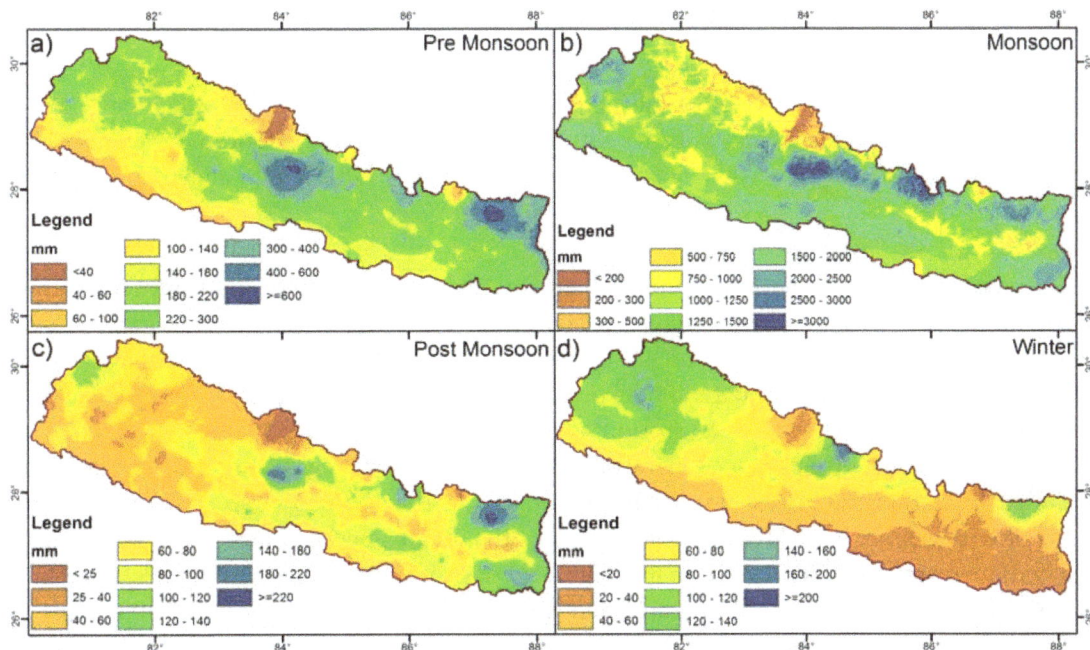

Figure 3. Spatial distribution of mean seasonal precipitation (mm) for (**a**) Pre monsoon; (**b**) Monsoon; (**c**) Post monsoon; and (**d**) winter season over the period of 1981–2010. Note: Legend scale of all four seasonal maps are different.

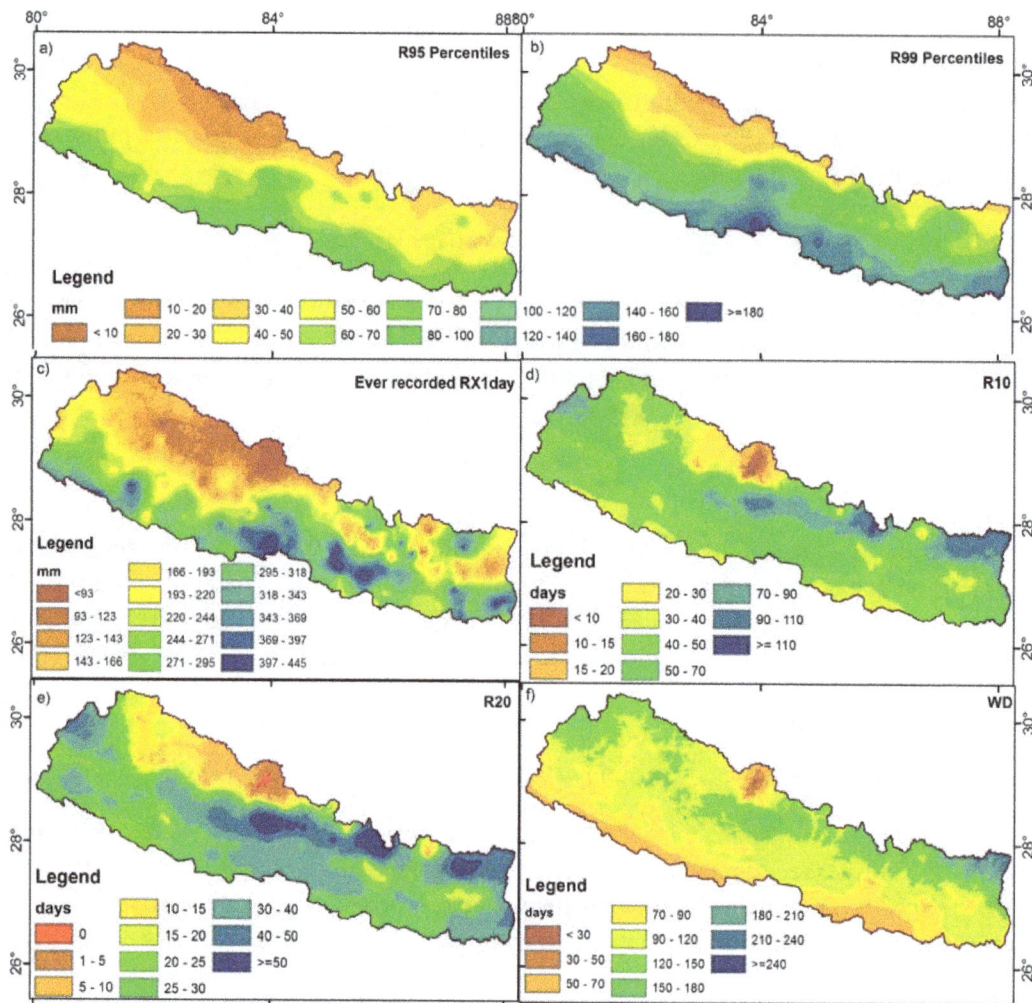

Figure 4. Spatial distribution of (**a**) 95th percentile; (**b**) 99th percentile mean values of daily precipitation; (**c**) ever-recorded one-day extreme precipitation; (**d**) mean annual number of days with precipitation ≥10 mm; (**e**) mean annual number of days with precipitation ≥20 mm (**f**) mean annual number of days with precipitation ≥1 mm (wet/rainy days) over the period of 1981–2010. Note: Legend scales are different.

The monsoonal precipitation dominates annual precipitation with contribution of around 80% of the annual precipitation, whereas precipitation during winter, pre- and post-monsoon seasons contribute only 3.5%, 12.5% and 4.0%, respectively [90]. Therefore, monsoonal precipitation distribution (Figure 3b) is similar to the annual precipitation distribution (Figure 1a and [38]), indicating three high-precipitation pocket areas around Lumle, Gumthang and Num that receive more than 3000 mm of monsoonal precipitation. On the other hand, dry leeward regions are in Mustang, Manang and Dolpa, featuring the lowest precipitation amounts of less than 150 mm. These findings are consistent with Böhner et al. [42], who reported that the regions north of high mountains are drier (<500 mm) in general and, depending on the alignment of surrounding mountain ranges, river valleys and the valleys lying in middle mountains and hills are also relatively drier (500–1000 mm) than surrounding mountain slopes and ridges. Such a pattern is more pronounced in the lower reaches of the river valleys, which lie between the southern frontal mountains and northern elevated mountain range. Such comparatively dry valleys are distinct in the surroundings of Tamor at Mulghat, Arun at Leguwaghat, DudhKoshi at Kuruleghat, Tamakoshi at Manthali, Sunkoshi at Nepalthok, Pachuwarght and Dolalghat, Bheri at Rakam and Dunai, Karnali at many upper river valleys (Jumla, Thirpu, Nagma, Gamshreenagar, etc.), Seti at Dipayal, Mahakali at Binayak and all other river valleys that lie north of tall mountain ranges (Figure 3b). On the other hand, as demonstrated earlier based on satellite data [47], precipitation is high (1500–2500 mm) on the windward side of both mountain ranges that lie north and south of such river valleys, creating double-peaked rainfall bands from south to north. In contrast to the river valleys lying within the northern elevated mountain range and southern frontal mountains, the majority of river valleys in Pokhara (near Lumle) and the surrounding region receive abundant (>1250 mm) monsoonal precipitation.

The precipitation in pre-monsoon, monsoon and post-monsoon seasons, more or less, follows the same spatial pattern in terms of representing three-peak precipitation pocket areas, as well as an east to west gradient. Pre-monsoon precipitation, mostly associated with thunderstorms, is very low in CH and the western half of the lowlands. There is a variation from less than 100 mm in WL, CH and in some areas on the leeward side of high mountains in the eastern region, to more than 700 mm in precipitation pocket areas and eastern mountains. Post-monsoon is the transition season between monsoon and winter. Therefore, following the retreat of monsoon from west to east, the western half of the country remains very dry, receiving below 40 mm of precipitation, whereas it is more than 200 mm in the eastern mountainous region. In contrast to other seasons, winter precipitation is higher over WM (>200 mm) and in a few isolated wet pockets in the central and eastern high mountainous regions. Winter precipitation is lowest over eastern lowlands (<20 mm) and feature a clear west to east as well as north to south gradient (Figure 3d). Our seasonal spatial maps are broadly similar to the previous studies for all of Nepal [29,44,90]; however, the inclusion of a large number of stations with the improved spatial interpolation technique has resulted in a much finer and realistic representation.

Contrary to the distribution of annual precipitation (Figure 1a), suggesting three rainfall pockets in Nepal, high values of extreme percentile thresholds are found in the lowlands (Terai and Siwalik), while low values are found in the highland regions (Figure 4a,b). For instance, annual mean of R99 percentile threshold in the lowlands is around 150 mm, whereas it is only about 30 mm in the highlands. Interestingly, the spatial pattern of R99 is quite similar to R95, and to some extent, to ever-recorded RX1day (Figure 4c). The high values of extreme thresholds in lowlands are associated with fewer heavy rainfall events (Figure 4a,b), while their relatively lower values observed in the middle mountains—despite the high annual precipitation (Figure 1a)—are due to weak but persistent rainfall [45,91]. The ever-recorded one-day maximum precipitation (RX1day) is found lowest in the central highlands followed by western highlands, whereas it is highest in the central lowlands followed by eastern and western lowlands. These spatial patterns are broadly similar to RX1day mapping using ground observation [92,93] and 90th percentile threshold mapping using TRMM data set [67]. The overall high values of R95, R99 and RX1day over lowlands suggest that these regions are more prone to soil erosion, landslides, flash flooding and subsequent inundations.

Unlike R95, R99 percentiles, and ever-recorded RX1day, spatial patterns of WD, R10 and R20 are largely similar to the annual distribution of precipitation. WD is lowest (below 40 days) in the very dry region of Mustang while it is highest (above 180 days) in EH followed by CM, WH and WM regions (Figure 4f). On average, WD is around 100 days over the lowlands and around 140 days over high and middle mountains. Similar to WD, R10 is lowest over Mustang, featuring only less than 11 days whereas R10 is above 110 days for the EH region (Figure 4d). On average, R10 is around 70 days over most of Nepal. R20 is almost zero over the dry Mustang region and at the minimum over WH and CH as these regions rarely experience very heavy precipitation events. However, over EH and CM regions and particularly over the high precipitation pockets, R20 is more than 50 days (Figure 4e). On average, R20 is around 40 days across the whole country. Interestingly, from Figure 4, it can be clearly noted that the frequency-related extreme indices of WD, R10 and R20 feature quite an opposite north–south gradient compared to high-intensity-related extreme indices of R95/R99 percentile precipitation and ever-recorded RX1day, with few sporadic exceptions.

The large spatial heterogeneity of mean precipitation across the country in different seasons indicates the requirement of localized information for water resources management and planning. Terai, Siwalik and river valleys, for example, are more prone to flood disaster, while mountains and hills face a higher landslide risk. The R95 and R99 thresholds computed here can also be directly utilizable for fixing the thresholds for flood warnings in different regions of Nepal.

5.2. Trend Analysis

In addition to the spatial distribution of mean precipitation and extreme indices, their time evolutions have been analyzed in order to see how these indices are changing over time. For this, trend slopes from the individual stations, their field significance for the seven geophysical sub-regions of Nepal along with the summary of these trend features are shown in Figures 5–9. The stationwise trends and percentages of the negative/positive trends along with their significance are also additionally included in Supplementary Materials Table S1.

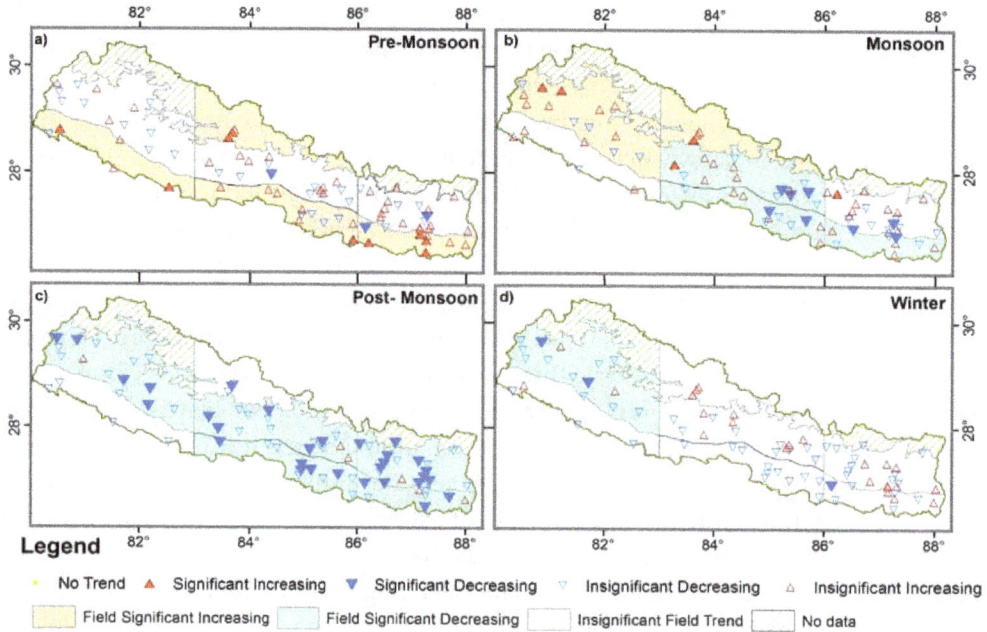

Figure 5. Stationwise and field significant trends of seasonal precipitation total for (**a**) Pre monsoon; (**b**) Monsoon; (**c**) Post monsoon; and (**d**) Winter season (significance at 0.1).

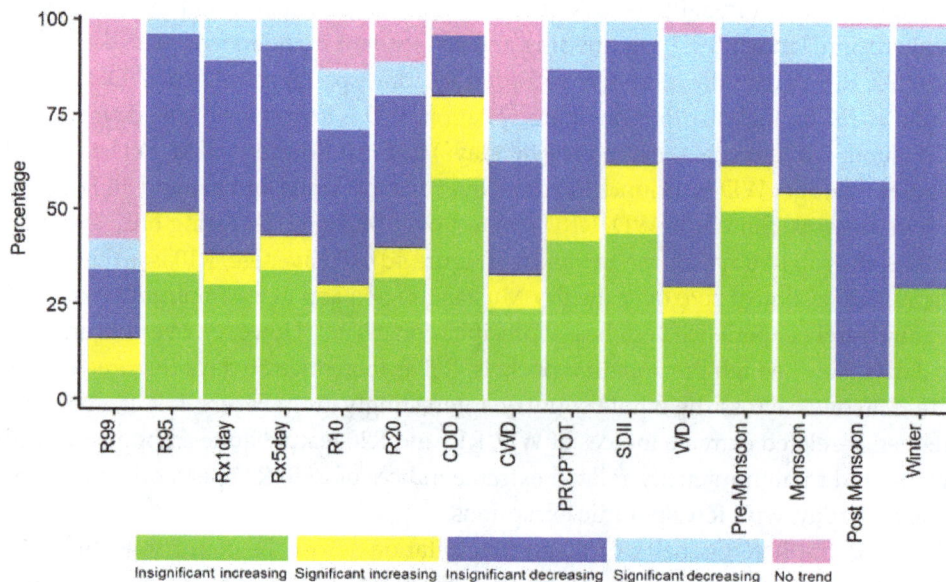

Figure 6. Percentage (from total stations) of stations with different trend features in Nepal.

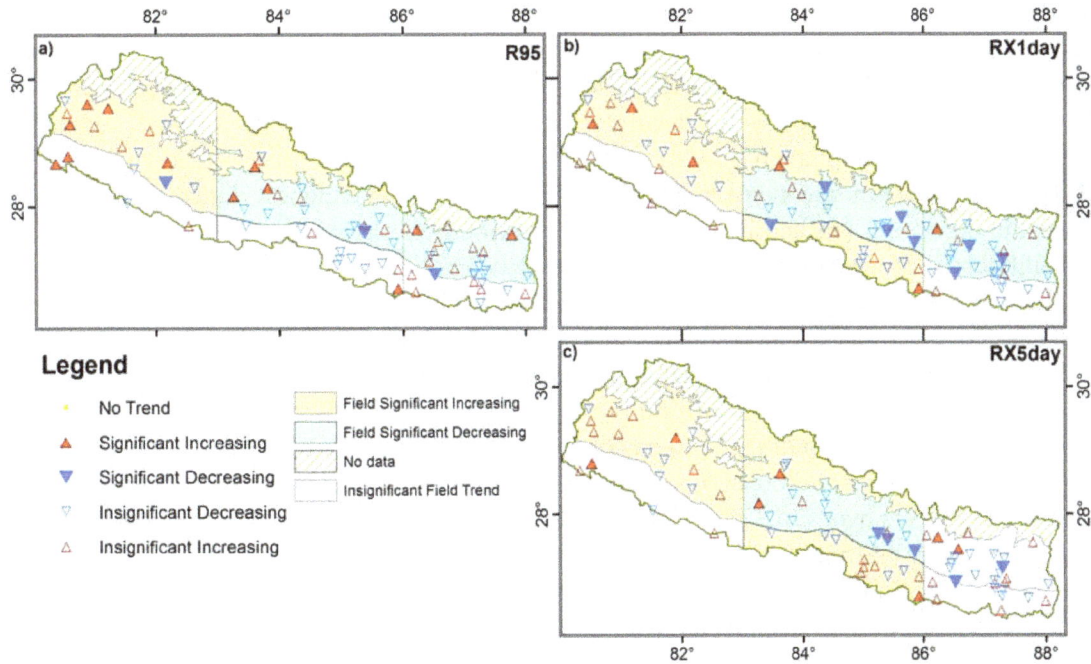

Figure 7. Stationwise and field significant trends for (**a**) R95; (**b**) RX1day; and (**c**) RX5day extreme precipitation indices (significance at 0.1).

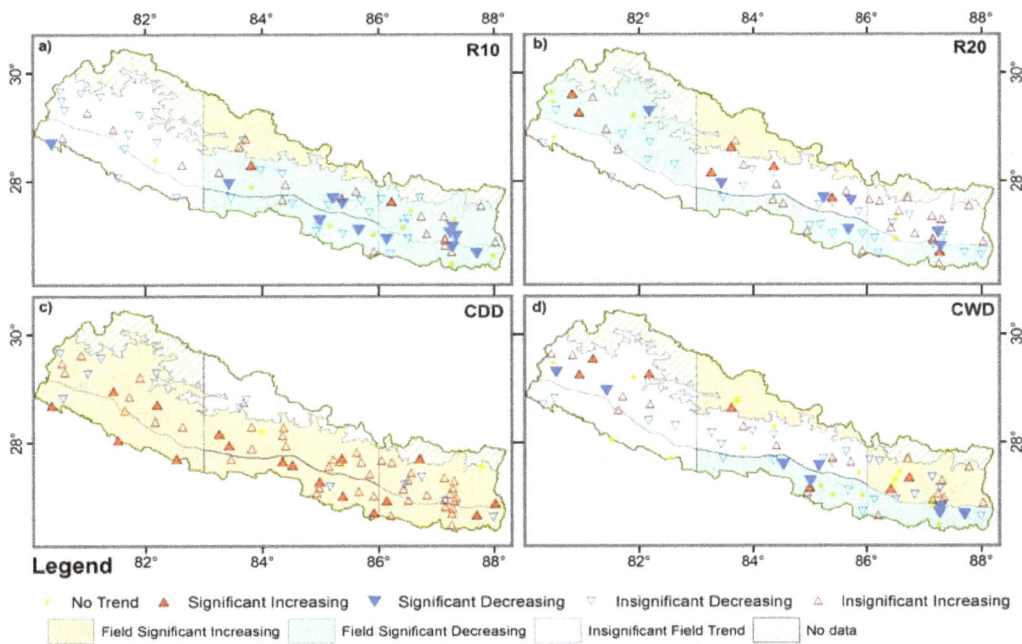

Figure 8. Stationwise and field significant trends for (**a**) R10; (**b**) R20; (**c**) CDD; and (**d**) CWD indices (significance at 0.1).

5.2.1. Seasonal Precipitation

All stations show a mixed pattern of increasing and decreasing trends for the pre-monsoon precipitation across Nepal (Figure 5a). However, around 62% of the total stations feature a rising trend, where such a trend is significant only at around 12%. Most of these stations are mainly concentrated within EL, CL, WL and CH regions. Only 4% of the total stations suggest a significant decreasing trend in pre-monsoon precipitation (Figure 6). The results of field significance analysis are also congruent with the stationwise trends, indicating a significant rise in pre-monsoon precipitation over WL, CL, EL and CH. These findings are mostly consistent with Duncan et al. [29] who have shown a countrywide precipitation rise. For the field insignificant middle mountain regions, differences with Duncan et al. [29] may arise due to employing distinct observational datasets and methodology. Since pre-monsoon precipitation is mostly accompanied with thunderstorms as evident heavily over EL [94], rise in the pre-monsoon precipitation over lowlands and CH regions will increase the extremely intense thunderstorms. Further, an increase in the pre-monsoon precipitation indicates changes in the seasonality of the precipitation regime over such regions [95,96].

The monsoon precipitation features a mixture of drying and wetting trends (Figure 5b). About one-fifth of the total stations exhibit significant trends with around 11% negative and 7% positive trends (Figure 6). The significant negative trends are concentrated in the central and eastern parts while significant positive trends are found in WM and CH regions. Field significance analysis more clearly suggests the rising and falling trends of monsoonal precipitation over the designated regions. For instance, increase in monsoonal precipitation is significant over the WM and CH regions whereas decrease is significant over CM, CL and EL regions, largely consistent with the signals observed at the local stations. Since the monsoonal precipitation is very important for summer crops (paddy, maize and millet), which constitute around 80% of the total national cereal production in Nepal [56], decreasing monsoonal precipitation may significantly affect the yield of cereal crops, as a significant decrease in the yield of rice has already been reported for the years of below-normal monsoonal precipitation [97].

Interestingly, most of the stations (92%) show a decrease in the post-monsoon precipitation, where such a signal is statistically significant at 41% of the total number of station. Decreasing post-monsoon precipitation is further suggested by the field significant decrease in all regions except for WL and CH (Figure 5c). This is in agreement with the findings of Khatiwada et al. [98], who have also indicated a decreasing post-monsoon precipitation over the Karnali basin in western Nepal for the 1981–2012 period. Over the same period, a significant decrease in precipitation during the post-monsoon dry months of November and winter season month December was also noticed in the Gandaki river basin of Nepal [57,99]. The observed decreasing post-monsoon precipitation may adversely affect the production of paddy crop, as it enters into sensitive stage of spikelet formation, fruiting and ripening, requiring more water during the post monsoon season [100].

Similar to post-monsoon precipitation, most of the stations feature a negative trend (68%) for winter precipitation over Nepal. However, such a negative trend is statistically significant at only 4% of the total stations, mainly lying within the WM region. Field significance analysis also suggests a significant decreasing trend for the winter precipitation over WM (Figure 5d). Khatiwada et al. [98] have also reported a decreasing winter precipitation over the Karnali basin in western Nepal for the 1981–2012 period. Based on GPCP and satellite-based datasets, Wang et al. [101] have likewise identified declining winter precipitation over the western region of Nepal in recent decades. Consistently, weakening influence of the western disturbances over the central Himalayas has also been found [102], and in line with this, decreasing winter precipitation has been reported in the adjoining western Himalayan region in India [103]. Further, Wang et al. [101] have attributed this decline to three main factors: (1) decadal trend towards negative phase of arctic oscillation in recent decades that has created a local mass flux circulation with descending branch over western Nepal; (2) the Indian ocean warming, and; (3) the anthropogenic aerosol loading. It is pertinent to mention that the winter precipitation, though low in volume, plays an important role in meeting the water demand of the winter crops and in feeding the rivers through accumulating their headwaters with snow that melts during the dry pre-monsoon season [42]. Particularly for the western hills and mountainous regions where food production is largely dependent upon rain-fed agriculture, decreasing winter precipitation may affect the winter crop production [101] of wheat, barley and potatoes, a major crop of the hills and mountains. Moreover, the decreasing winter precipitation in the region, where winter precipitation is substantially higher than in other regions, could also lead to a reduction in pre-monsoon season river flows, which are largely dependent on the snow and glacier melt during the dry season.

5.2.2. High-Intensity-Related Precipitation Extremes

The stationwise trends of R95 (total annual precipitation from the days of a year featuring >95 percentile precipitation) and RX1day and RX5day (annual maximum 1-day and 5-day precipitation) indices, along with their field significance, are shown in Figure 7.

The analysis reveals a mixture of equally increasing and decreasing trends in R95 across Nepal with only one-fifth of the total stations featuring significant trends. For instance, around 16% of the total stations show a significant positive trend in R95 while only 4% show its significant negative trend (Figure 6). The stations featuring a statistically significant rise in R95 are mainly concentrated within the western part of the country. Rising trends in R95 at local stations are found field significant over the WM and CH regions, while falling trends in R95 are field significant over the CM and EM regions.

For RX1day indices, about 60% of the analyzed stations feature falling trends and around one-tenth of the total stations show such a falling trend as significant. A large number of stations showing falling RX1day are located in CM and EM regions. In contrast, stations in the western region (WL and WM) have a higher number of increasing trends for RX1day. Similar to RX1day, a higher number of negative trends in RX5day is observed within CM and EM regions, while more positive trends are found in the western regions (WL and WM). Field significance results are largely in agreement with the stationwise trends. Positive trends in RX1day and RX5day are field significant over the WM, CL and CH regions, while negative trends are field significant over the CM region and RX1day decrease is additionally field significant over the EM region. Such a pattern of change in RX1day is consistent with the previous studies [32,104] that also report a decreasing trend of RX1day from most of the stations, where such a trend is particularly significant above 100 m (asl) within the Koshi River basin—a basin spanning mainly over the eastern and partially over the central region of our study area.

In summary, all three indices of R95, RX1day and RX5day feature a field significant rising trend over the WM and CH regions, whereas over the CL region, only the latter two are field significant. Decreasing trends in R95, RX1day and RX5day are found field significant over the CM region, while the former two are additionally field significant over the EM region. Coherence amid field significance rising trends of all three intensity-related extreme precipitation indices together with the dominance of stationwise significant rising trends over WM and CH regions indicate that precipitation extremes might be more intense in the near future.

On the other hand, decreasing field significant trends of all three indices over the CM region and of two indices, R95 and RX1dy, over the EM region indicate to some extent the weakening of intense precipitation extremes over such regions.

Since our analysis period is only until 2012, occurrence of extreme events during 14–17 June 2013 in Uttarkhanda, India and the bordering areas of western Nepal, and the 14–16 August 2014 event in western Nepal indicates the continuation of this pattern in the western region of Nepal. Further, Cho et al. [105] attributed the increase in extreme precipitation events like that of Uttarkhanda and the bordering region of western Nepal in recent decades to an amplification of an upper tropospheric mid-latitude shortwave trough pattern in the northern region of South Asia due to the increase in greenhouse gases and aerosols. In general, this kind of amplification in association with west–northwestward migration of monsoon low creates the highly favorable environment for vigorous interaction of tropical (monsoon) and extra-tropical (mid-latitude) circulation resulting in extreme precipitation events in the western Himalayan region [106]. Thus, the western mountainous region of Nepal lying at a higher latitude and being an adjacent region of western Himalayas could have greater influence of such mid-latitude wave train pattern, whereas opposite or no influence of that pattern can occur towards eastern region. However, the shortwave train pattern was analyzed only for June and no conclusions could be made for whole monsoon season during which extreme precipitation events occur.

In addition, physical mechanism responsible for enhancement of monsoon precipitation and extremes towards the central and eastern region of foothills of Himalayan region are normally associated with break/active monsoon condition in mainland India/Himalayan foothills (north ward migration of monsoon trough towards foothills of Himalaya from its normal position) during which interaction of southward migration of extratropical westerly troughs (dry air subsidence in Indian subcontinent at mid-to-upper troposphere) and weak monsoon circulation takes place [107]. The detail analysis of changing pattern of break monsoon situation and other physical mechanism responsible for changing extreme precipitation pattern over the whole monsoon season for Nepal is still lacking and it could be far from simple as the complex interaction of global, synoptic scale weather systems and topography takes place in the monsoon dominated region producing localized extreme precipitation.

The increasing intense precipitation over western mountainous region indicates higher risks of soil erosion and landslides in the fragile mountainous regions which are extremely vulnerable to these disasters due to manmade activities—deforestation for settlement, road network and agricultural activities—or natural causes like earthquakes. Lacking adaptive capacities of the remote mountains and hills further aggravate the situation. Additionally, increasing intense precipitation events in these regions consequently increases the risk of floods and inundation in densely populated river valleys and southern lowlands, destroying life and property, and causing damages to the agricultural land thereby impacting the socioeconomic development.

Nevertheless, absence of increasing trend of extreme intense precipitation in the central region where land is highly ruptured with recent earthquake still requires suitable adaptive measures to reduce the risk of landslides in the regions, as continuous but low intensity and even normal extreme threshold precipitation values of rainfall can easily trigger such disasters in the region.

5.2.3. Frequency-Related Precipitation Extremes

Since days with 10 mm and 20 mm precipitation events are quite common during the monsoon season over many parts of Nepal (Figure 4d,e), these events which are typically defined as heavy and very heavy precipitation in fact represent only moderate precipitation events over large areas of Nepal. Our results show a mixed pattern of stationwise increasing and decreasing trends for R10 (annual number of days with $\geq$10 m) across Nepal. However, around 57% of the total stations show a negative trend, which is significant at around 16% of the stations mostly concentrated in the central and eastern regions (Figure 6). Field significance test results also suggest the same, exhibiting significant negative trends for the CM, EM, CL and EL regions (Figure 8a). The pattern of change in R20 (annual counts of days when Precipitation is $\geq$20 mm) likewise indicates a mixed response with approximately equal numbers of rising and falling trends (Figure 8b). However, R20 is significantly decreasing for the regions of CL, EL and WM and significantly increasing for the region of CH. Further, coherent decrease in R10 and R20 over CL and EL clearly suggests a decrease in the number of moderate precipitation events over the respective regions.

5.2.4. Dry and Wet Spells

The analysis of CDD (consecutive dry days) indices suggests a widespread increase in the dry period over the whole country (Figure 8c). Around 80% of the analyzed stations exhibit an increase in CDD over the period of 1970–2012, though this trend is significant only at 22% of the total (Figure 6). In line with the stationwise trends, field significance trends are also statistically significant for all the sub-regions, except for CH. This finding is consistent with previous studies of Shrestha et al. [32] and Sigdel and Ma [104] over Koshi basin of Nepal. A similar increase in CDD has also been reported for the Songhu River basin in China [108] and across Bangladesh [109], as CDD is mostly related to the large-scale weather systems rather than localized systems [68].

The CWD (consecutive wet days) also reveals a mixed pattern (Figure 8d). However, the field significance analysis suggests a significant decreasing trend for CL and EL regions but a significant increasing trend for EM and CH regions. The CWD changes are mainly observed during the monsoon season; hence, such changes do not necessarily corresponds to changes in CDD, which are observed mostly in the dry seasons.

It is worth mentioning that duration and occurrence-related indices of CDD and CWD can only indirectly characterize the drought, which is a complex phenomenon and depends upon many other factors besides precipitation. Nevertheless, increasing CDD observed over the study area is consistent with the most widespread and worst drought observed in recent decades across the country [99,101,110,111]. Rise in CDD clearly indicates the prolongation of the dry period across the country, implying certain changes in the seasonality of prevailing precipitation regimes [95,96]. Further, this increase in the dry period can negatively impact crop yield and hydropower generation and can, moreover, elevate respiratory-related health problems in Nepal by increasing the concentration of particulate matter in the air. Since the frequency and scale of the forest fires in Nepal and other regions are also strongly related to the length of dry spells [112], the rise in CDD will aggravate such events, endangering wildlife and causing huge socioeconomic losses.

5.2.5. Extra Indices (PRCPTOT, SDII and WD)

One-fifth of the total number of stations exhibit significant trend changes in PRCPTOT. Around 13% of stations suggest a significant falling trend while 7% of stations suggest a significant increasing trend (Figure 6). The majority of statistically significant negative trends are concentrated in the central and eastern parts while higher numbers of significant positive trends are observed in the western parts of the country (Figure 9a). Interestingly, the stations above 29° N reveal an increasing trend while those below this latitude features a decreasing trend. The field significance test also indicates a significant decrease in PRCPTOT over CM, CL, EM and EL regions and a significant increase over WM and CH regions. The increasing trend in WM is consistent with Baidya et al.'s [28] findings for the western region. The spatial pattern of trend changes in PRCPTOT is quite similar to that of the monsoonal precipitation as it dominates (about 80% of) the total annual precipitation in Nepal. Decreasing PRCPTOT over the eastern region that covers most of the Koshi River basin within Nepal is consistent with the reports of significant precipitation decrease over the Koshi River basin during the 1994–2013 period [50] and over the middle mountains and hills during the 1975–2010 period [32]. Based on the coarse resolution Global Precipitation Climatology Project (GPCP) dataset, Yao et al. [113] have also provided evidence of decreasing precipitation over all the Himalayas and an increase in the eastern Pamir regions during 1979–2010.

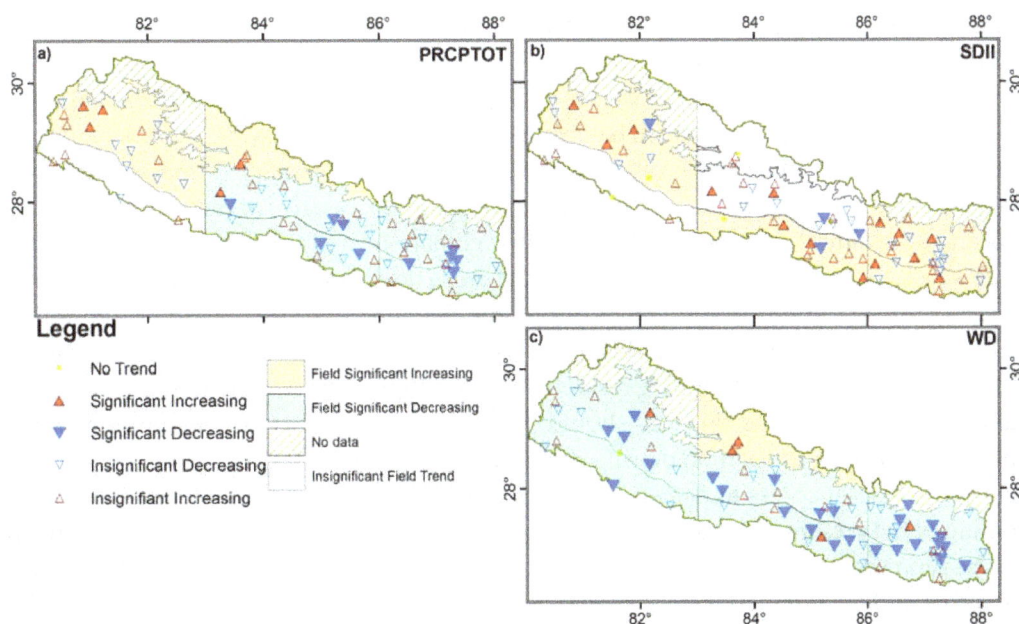

Figure 9. Stationwise and field significant trends for (**a**) PRCPTOT; (**b**) SDII; and (**c**) WD indices (significance at 0.1).

Notably, WD (wet days/rainy days) is decreasing (at 67% of the stations) across most of Nepal with a statistically significant decreasing pattern at one-third of the stations. Additionally, the individual stations' trends are also consistent with the field significant trends for all sub-regions, except for CH (Figure 9c). These findings are further consistent with the regional pattern of decreasing number of wet days over the whole of Southeast Asia [114].

The simple daily intensity index (SDII), defined as the ratio of PRCPTOT to WD, exhibits a positive trend (60% of the stations) across Nepal with statistically significant increasing trends at 14 stations (18%) and significant decreasing trends at four stations. Unlike significant decreasing trends in PRCPTOT over EM, EL and CL regions, in accord with Shrestha et al. [32], SDII features a significant increasing trend. An increase in SDII over such regions is mainly due to the higher decrease in WD than in PRCPTOT (Figure 9b). On the other hand, the reason for the increase in SDII over the western region is due to an increase in PRCPTOT but decrease in WD. Rising trends in R95, RX1day and RX5day over the WM region further indicate rising high-intensity precipitation extremes over the region. In contrast, decreasing trends in PRCPTOT, R95 and RX1day over the EM and CM region somewhat reinforce a decreasing trend of high-intensity precipitation extremes. Additionally, significant decreasing trends in WD, R10 and R20 indices over EL and CL regions indicate a decrease in the moderate daily precipitation events.

6. Conclusions

First, we have analyzed the spatial distribution of the seasonal precipitation and of a few physically relevant extreme precipitation indices from 210 stations across Nepal over the most recent period, 1981–2010. Then, the spatio-temporal variation of the computed daily precipitation extremes has been analyzed from long-term continuous records available from 76 stations over the period of 1970–2012 using a robust non-parametric Mann–Kendall [33,34] trend test and the trend-free pre-whitening (TFPW) procedure. Trends at local stations are further assessed for their field significance over the seven distinct physio-geographic regions of Nepal, in order to identify the most dominant patterns of changes in extreme precipitation indices over these regions.

By including more stations with updated records and employing robust test statistics as compared to previous studies, this study provides a detailed and coherent picture of spatio-temporal changes in the precipitation means and extremes across the whole of Nepal as well as over its seven physio-geographic sub-regions. Generally, we have found a prolongation of dry periods and a decrease in post-monsoon precipitation across the whole country, while annual and high-intensity precipitation extremes show contrasting evidence of increase in the western half and a moderately decreasing pattern in the east. Similarly, winter precipitation has significantly decreased across the western region, thereby indicating the weakened influence of western disturbances. Such information is anticipated to be useful in decision-making for the effective management of water resources, hydro-meteorologically induced disasters and agricultural practices, health and other climate-related livelihood activities in different regions as well as for the verification of climate model projections across Nepal. Similarly, the presented picture can support in strengthening the weather, climate and flood early warning and forecasting systems, in devising the landslide, flood risk and vulnerability maps, in adapting to region-specific drought-tolerant crops and in designing reservoir-type multipurpose hydropower projects. However, we caution that the presented findings should preferably be interpreted in connection with other relevant factors, such as population density, agricultural yield, food deficit regions, poverty, and the areas which have least access to post-disaster relief facilities and low coping capacity. Further, the shortcoming of the presented analysis is mainly the lack of high-altitude stations and thus the results may not be representative for the high mountainous regions, such as the western high mountains (WH) and eastern high mountains (EH). Hence, future research should focus on employing a broader database and larger set of extreme indices for the comprehensive analysis of changes in precipitation extremes at the seasonal scale in order to understand the physical mechanisms driving such changes. The main findings of the study are summarized below:

- Spatial distribution of monsoonal precipitation across Nepal indicates three high-precipitation pocket areas, such as surrounding regions of Lumle, Gumthang and Num, and four regions of low precipitation on the leeward side of high mountains, namely, Mustang, Manang, Dolpa, and Everest. Likewise, depending on the orientation of surrounding mountain ranges, lower precipitation is also found in the river valleys lying within the middle mountains and hill regions. Pre- and post-monsoonal precipitation more or less follow the spatial pattern of monsoonal precipitation in terms of representing three-peak precipitation pocket areas, as well as the east to west gradient. In contrast to other seasons, winter precipitation is higher over the western middle and high mountainous (WM) areas and lower over the eastern lowlands (EL), clearly depicting west to east and north to south gradients.

- Spatial distribution of precipitation extremes suggests that the high-intensity-related extremes (95th and 99th percentile thresholds and one-day extreme precipitation) are relatively more intense over the southern lowlands as compared to other regions suggesting higher chances of flooding and inundation in the region. However, the frequency-related extreme indices (WD—wet days, R10 and R20—heavy and very heavy precipitation days) generally feature low values.

- Long-term trends in the monsoonal and annual precipitation (PRCPTOT) indicate their significant increases over the middle mountains and hills within the western region (WM), and over the high mountains within the central region (CH). However, the monsoonal and annual precipitations feature significantly decreasing trends over the whole central and eastern regions, except for the former over the eastern middle-mountain and hills (EM).

- Pre-monsoon precipitation features a significant positive trend over the central high mountain region (CH) and over all lowland regions. On the other hand, winter precipitation features a decreasing trend over most of Nepal; however, such a trend is significant only over the western middle mountains and hills (WM). Similarly, a significantly decreasing trend in the post-monsoonal precipitation has also been observed across Nepal, except over CH and WL.

- A coherent significant positive trend in the high-intensity-related extreme precipitation indices (RX1day, RX5day and R95) has been observed over the middle mountains and hills of the western region (WM) and central high mountains (CH), suggesting more intense precipitation therein. This is further supported by the evidence of significant positive trends in the monsoonal and annual precipitation, with a negative trend in the wet days over that region. In contrast, decreasing trends in the annual precipitation, wet days (WD) and high-intensity-related precipitation extremes (R95, RX1day), together with an increasing trend in the consecutive dry days (CDD) over the central and eastern middle mountains and hills reveals weakening of the intense precipitation extremes.

- Significant positive trend in consecutive dry days (CDD) but negative trend in wet days (WD) are observed across the country, suggesting the prolongation of the dry period.

Acknowledgments: The authors would like to thank the Department of Hydrology and Meteorology, Nepal, for the permission to use meteorological data. Ramchandra Karki's PhD scholarship was supported by Deutscher Akademischer Austauschdienst (DAAD) under the Research Grants—Doctoral Programmes in Germany, through University of Hamburg, Germany. Further, we acknowledge the TREELINE project funded by the German Research Foundation (SCHI 436/14-1, BO 1333/4-1, SCHO 739/14-1). Finally, we would like to thank the three anonymous reviewers for their valuable comments and suggestions which helped us on improving this paper.

Author Contributions: Ramchandra Karki initiated, designed the study, performed the data analysis and wrote the paper with critical input and contribution on analysis, interpretation of the results and writing from Shabeh ul Hasson and Jürgen Böhner while Udo Schickhoff and Thomas Scholten have contributed to the analysis and interpretation of the results.

Conflicts of Interest: The authors declare no conflict of interest.

References

1. Karl, T.R.; Knight, R.W. Secular trends of precipitation amount, frequency, and intensity in the United States. *Bull. Am. Meteorol. Soc.* **1998**, *79*, 231–241. [CrossRef]
2. Kunkel, K.E.; Easterling, D.R.; Redmond, K.; Hubbard, K. Temporal variations of extreme precipitation events in the United States: 1895–2000. *Geophys. Res. Lett.* **2003**, *30*, 51–54. [CrossRef]
3. Kunkel, K.E.; Easterling, D.R.; Kristovich, D.A.R.; Gleason, B.; Stoecker, L.; Smith, R. Recent increases in U.S. heavy precipitation associated with tropical cyclones. *Geophys. Res. Lett.* **2010**, *37*, 2–5. [CrossRef]
4. Alexander, L.V.; Zhang, X.; Peterson, T.C.; Caesar, J.; Gleason, B.; Klein Tank, A.M.G.; Haylock, M.; Collins, D.; Trewin, B.; Rahimzadeh, F.; et al. Global observed changes in daily climate extremes of temperature and precipitation. *J. Geophys. Res. Atmos.* **2006**, *111*, D05109. [CrossRef]
5. Klein Tank, A.M.G.; Peterson, T.C.; Quadir, D.A.; Dorji, S.; Zou, X.; Tang, H.; Santhosh, K.; Joshi, U.R.; Jaswal, A.K.; Kolli, R.K.; et al. Changes in daily temperature and precipitation extremes in central and south Asia. *J. Geophys. Res.* **2006**, *111*, D16105. [CrossRef]
6. Dai, A. Drought under global warming: A review. *WIREs Clim. Chang.* **2011**, *2*, 45–65. [CrossRef]
7. Coumou, D.; Rahmstorf, S. A decade of weather extremes. *Nat. Clim. Chang.* **2012**, *2*, 491–496. [CrossRef]
8. Groisman, P.Y.; Knight, R.W.; Karl, T.R. Changes in intense precipitation over the central United States. *J. Hydrometeorol.* **2011**, *13*, 47–66. [CrossRef]
9. Stocker, T.F.; Qin, D.; Platner, G.K. *IPCC Climate Change 2013: The Physical Science Basis*; Intergovernmental Panel on Climate Change Working Group I Contribution to the Fifth Assessment Report (AR5); Cambridge University Press: New York, NY, USA, 2013.
10. Westra, S.; Alexander, L.V.; Zwiers, F.W. Global increasing trends in annual maximum daily precipitation. *J. Clim.* **2013**, *26*, 3904–3918. [CrossRef]
11. Sheikh, M.M.; Manzoor, N.; Ashraf, J.; Adnan, M.; Collins, D.; Hameed, S.; Manton, M.J.; Ahmed, A.U.; Baidya, S.K.; Borgaonkar, H.P.; et al. Trends in extreme daily rainfall and temperature indices over South Asia. *Int. J. Climatol.* **2014**, *1637*, 1625–1637. [CrossRef]
12. United Nations Environment Programme (UNEP). *GEO Year Book 2003*; UNEP: Nairobi, Kenya, 2003.
13. Caesar, J.; Alexander, L.V.; Trewin, B.; Tse-ring, K.; Sorany, L.; Vuniyayawa, V.; Keosavang, N.; Shimana, A.; Htay, M.M.; Karmacharya, J.; et al. Changes in temperature and precipitation extremes over the Indo-Pacific region from 1971 to 2005. *Int. J. Climatol.* **2011**, *31*, 791–801. [CrossRef]
14. Donat, M.G.; Alexander, L.V.; Yang, H.; Durre, I.; Vose, R.; Dunn, R.J.H.; Willett, K.M.; Aguilar, E.; Brunet, M.; Caesar, J.; et al. Updated analyses of temperature and precipitation extreme indices since the beginning of the twentieth century: The HadEX2 dataset. *J. Geophys. Res. Atmos.* **2013**, *118*, 2098–2118. [CrossRef]
15. Goswami, B.N.; Venugopal, V.; Sengupta, D.; Madhusoodanan, M.S.; Xavier, P.K. Increasing trend of extreme rain events over India in a warming environment. *Science* **2006**, *314*, 1442–1445. [CrossRef] [PubMed]
16. Joshi, U.R.; Rajeevan, M. *Trends in Precipitation Extremes over India*; Research Report No.: 3/2006; National Climate Centre, India Meteorological Department: Pune, India, 2006.
17. Rajeevan, M.; Bhate, J.; Jaswal, A.K. Analysis of variability and trends of extreme rainfall events over India using 104 years of gridded daily rainfall data. *Geophys. Res. Lett.* **2008**, *35*, L18708.
18. Krishnamurthy, C.K.B.; Lall, U.; Kwon, H.H. Changing frequency and intensity of rainfall extremes over India from 1951 to 2003. *J. Clim.* **2009**, *22*, 4737–4746. [CrossRef]
19. Pattanaik, D.R.; Rajeevan, M. Variability of extreme rainfall events over India during southwest monsoon season. *Meteorol. Appl.* **2010**, *17*, 88–104. [CrossRef]
20. Hussain, M.S.; Lee, S. The regional and the seasonal variability of extreme precipitation trends in Pakistan. *Asia Pac. J. Atmos. Sci.* **2013**, *49*, 421–441. [CrossRef]
21. Roy, S.S.; Balling, R.C. Trends in extreme daily precipitation indices in India. *Int. J. Climatol.* **2004**, *24*, 457–466.
22. Revadekar, J.V.; Hameed, S.; Collins, D.; Manton, M.; Sheikh, M.; Borgaonkar, H.P.; Kothawale, D.R.; Adnan, M.; Ahmed, A.U.; Ashraf, J.; et al. Impact of altitude and latitude on changes in temperature extremes over South Asia during 1971–2000. *Int. J. Climatol.* **2013**, *33*, 199–209. [CrossRef]
23. 2013 North India floods. Available online: https://en.wikipedia.org/wiki/2013_North_India_floods (accessed on 25 October 2014).

24. Nepal: Landslides and Floods Information Bulletin. Available online: http://reliefweb.int/report/nepal/nepal-landslides-and-floods-information-bulletin-n-1-17-august-2014 (accessed on 25 October 2014).

25. Ministry of Agriculture and Cooperatives (MOAAC); World Food Programme (WFP); Food and Agriculture Organization (FAO). *2008/09 Winter Drought in Nepal-Crop and Food Security Assessment Joint Technical Report*; Ministry of Agriculture and Cooperatives, Government of Nepal, World Food Programme, Food and Agriculture Organization of the United Nations: Kathmandu, Nepal, 2009.

26. Revadekar, J.V.; Preethi, B. Statistical analysis of the relationship between summer monsoon precipitation extremes and foodgrain yield over India. *Int. J. Climatol.* **2012**, *32*, 419–429. [CrossRef]

27. Central Bureau of Statistics (CBS). *Statistical Information on Nepalese Agriculture*; Central Bureau of Statistics: Kathmandu, Nepal, 2013.

28. Baidya, S.K.; Shrestha, M.L.; Sheikh, M.M. Trends in daily climatic extremes of temperature and precipitation in Nepal. *J. Hydrol. Meteorol.* **2008**, *5*, 1.

29. Duncan, J.M.A.; Biggs, E.M.; Dash, J.; Atkinson, P.M. Spatio-temporal trends in precipitation and their implications for water resources management in climate-sensitive Nepal. *Appl. Geogr.* **2013**, *43*, 138–146. [CrossRef]

30. Hofstra, N.; Haylock, M.; New, M.; Jones, P.D. Testing E-OBS European high-resolution gridded data set of daily precipitation and surface temperature. *J. Geophys. Res. Atmos.* **2009**. [CrossRef]

31. Sharma, K.P. *Climate Change: Trends and Impacts on the Livelihoods of People*; Technical Report; Jalsrot Vikas Sanstha/Nepal Water Partnership: Kathmandu, Nepal, 2009.

32. Shrestha, A.B.; Bajracharya, S.R.; Sharma, A.R.; Duo, C.; Kulkarni, A. Observed trends and changes in daily temperature and precipitation extremes over the Koshi river basin 1975–2010. *Int. J. Climatol.* **2016**. [CrossRef]

33. Mann, H.B. Nonparametric tests against trend. *Econometrica* **1945**, *13*, 245–259. [CrossRef]

34. Kendall, M.G. *Rank Correlation Method*; Griffin: London, UK, 1975.

35. Kansakar, S.R.; Hannah, D.M.; Gerrard, J.; Rees, G. Spatial pattern in the precipitation regime in Nepal. *Int. J. Climatol.* **2004**, *24*, 1645–1659. [CrossRef]

36. Shrestha, A.B.; Aryal, R. Climate change in Nepal and its impact on Himalayan glaciers. *Reg. Environ. Chang.* **2010**, *11*, 65–77. [CrossRef]

37. Duncan, J.M.A.; Biggs, E.M. Assessing the accuracy and applied use of satellite-derived precipitation estimates over Nepal. *Appl. Geogr.* **2012**, *34*, 626–638. [CrossRef]

38. Karki, R.; Talchabhadel, R.; Aalto, J.; Baidya, S.K. New climatic classification of Nepal. *Theor. Appl. Climatol.* **2016**, *125*, 799–808. [CrossRef]

39. Nayava, J.L. Rainfall in Nepal. *Himal. Rev.* **1981**, *12*, 1–18.

40. Lang, T.J.; Barros, A.P. Winter storms in the central Himalayas. *J. Meteorol. Soc. Jpn.* **2004**, *82*, 829–844. [CrossRef]

41. Böhner, J. General climatic controls and topoclimatic variations in Central and High Asia. *Boreas* **2006**, *35*, 279–295. [CrossRef]

42. Böhner, J.; Miehe, G.; Miehe, S.; Nagy, L. Climate and weather variability: An introduction to the natural history, ecology, and human environment of the Himalayas, a companion volume to the flora of Nepal. *R. Bot. Gard. Edinb.* **2015**, *4*, 23–89.

43. Hasson, S.; Lucarini, V.; Pascale, S. Hydrological cycle over South and Southeast Asian river basins as simulated by PCMDI/CMIP3 experiments. *Earth Syst. Dyn.* **2013**, *4*, 199–217. [CrossRef]

44. Ichiyanagi, K.; Yamanaka, M.D.; Muraji, Y.; Vaidya, V.K. Precipitation in Nepal between 1987 and 1996. *Int. J. Climatol.* **2007**, *27*, 1753–1762. [CrossRef]

45. Barros, A.P.; Kim, G.; Williams, E.; Nesbitt, S.W. Probing orographic controls in the Himalayas during the monsoon using satellite imagery. *Nat. Hazards Earth Syst. Sci.* **2004**, *4*, 29–51. [CrossRef]

46. Dhar, O.N.; Nandargi, S. Areas of heavy precipitation in the Nepalese Himalayas. *Weather* **2005**, *12*, 354–356. [CrossRef]

47. Bookhagen, B.; Burbank, D.W. Topography, relief, and TRMM-derived rainfall variations along the Himalaya. *Geophys. Res. Lett.* **2006**, *33*, L08405.

48. Miehe, G. *Langtang Himal. A Prodormus of the Vegetation Ecology of the Himalayas. Mit Einer Kommentierten Flechtenliste von Josef Poelt*; Borntrager: Stuttgart, Germany, 1990.

49.   Putkonen, J. Continuous snow and rain data at 500 to 4400 m altitude near Annapurna, Nepal, 1999–2001. *Arct. Antarct. Alp. Res.* **2004**, *36*, 244–248. [CrossRef]

50.   Salerno, F.; Guyennon, N.; Thakuri, S.; Viviano, G.; Romano, E.; Vuillermoz, E.; Cristofanelli, P.; Stocchi, P.; Agrillo, G.; Ma, Y.; et al. Weak precipitation, warm winters and springs impact glaciers of south slopes of Mt. Everest (central Himalaya) in the last 2 decades (1994–2013). *Cryosphere* **2015**, *9*, 1229–1247. [CrossRef]

51.   Gerlitz, L.; Bechtel, B.; Böhner, J.; Bobrowski, M.; Bürzle, B.; Müller, M.; Scholten, T.; Schickhoff, U.; Schwab, N.; Weidinger, J. Analytic comparison of temperature lapse rates and precipitation gradients in a Himalayan treeline environment: Implications for statistical downscaling. In *Climate Change, Glacier Response, and Vegetation Dynamics in the Himalaya: Contributions toward Future Earth Initiatives*; Singh, R.B., Schickhoff, U., Mal, S., Eds.; Springer: Cham, Switzerland, 2016; pp. 49–64.

52.   Barros, A.P.; Lang, T.J. Monitoring the Monsoon in the Himalayas: Observations in Central Nepal, June 2001. *Mon. Weather Rev.* **2003**, *131*, 1408–1427. [CrossRef]

53.   Shrestha, A.B.; Awake, C.P.; Dibb, J.E.; Mayewski, P.A. Precipitation fluctuations in the Nepal Himalaya and its vicinity and relationship with some large scale climatological parameters. *Int. J. Climatol.* **2000**, *2*, 317–327. [CrossRef]

54.   Sharma, K.P.; Moore, B., III; Vorosmarty, C.J. Anthropogenic, climatic and hydrologic trends in the Koshi Basin, Himalaya. *Clim. Chang.* **2000**, *47*, 141–165. [CrossRef]

55.   Shrestha, M.L. Interannual variation of summer monsoon rainfall over Nepal and its relation to southern oscillation index. *Meteorol. Atmos. Phys.* **2000**, *75*, 21–28. [CrossRef]

56.   Gautam, D.K.; Regmi, S.K. Recent trends in the onset and withdrawal of summer monsoon over Nepal. *Ecopersia* **2013**, *1*, 353–367.

57.   Panthi, J.; Dahal, P.; Shrestha, M.; Aryal, S.; Krakauer, N.; Pradhanang, S.; Lakhankar, T.; Jha, A.; Sharma, M.; Karki, R. Spatial and temporal variability of rainfall in the Gandaki River Basin of Nepal Himalaya. *Climate* **2015**, *3*, 210–226. [CrossRef]

58.   Talchabhadel, R.; Karki, R.; Parajuli, B. Intercomparison of precipitation measured between automatic and manual precipitation gauge in Nepal. *Measurement* **2016**. [CrossRef]

59.   Zhang, X.; Yang, F. *User Manual; RClimDex 1.0*; Climate Research Branch Environment: Downsview, ON, Canada, 2004.

60.   Vincent, L.A.; Peterson, T.C.; Barros, V.R.; Marino, M.B.; Rusticucci, M.; Carrasco, G.; Ramirez, E.; Alves, L.M.; Ambrizzi, T.; Berlato, M.A.; et al. observed trends in indices of daily temperature extremes in South America 1960–2000. *J. Clim.* **2005**, *18*, 5011–5023. [CrossRef]

61.   Aguilar, E.; Auer, I.; Brunet, M.; Peterson, T.C.; Wieringa, J. *Guidelines on Climate Metadata and Homogenization*; World Meteorological Organization: Geneva, Switzerland, 2003.

62.   Wang, X.L. Penalized maximal F-test for detecting undocumented mean-shifts without trend-change. *J. Atmos. Ocean. Techol.* **2008**, *25*, 368–384. [CrossRef]

63.   Hasson, S.; Böhner, J.; Lucarini, V. Prevailing climatic trends and runoff response from Hindukush–Karakoram–Himalaya, upper Indus basin. *Earth Syst. Dyn.* **2015**, *6*, 579–653. [CrossRef]

64.   Hasson, S.U.; Gerlitz, L.; Schickhoff, U.; Scholten, T.; Böhner, J. Recent climate change over High Asia. In *Climate Change, Glacier Response, and Vegetation Dynamics in the Himalaya: Contributions toward Future Earth Initiatives*; Singh, R.B., Schickhoff, U., Mal, S., Eds.; Springer: Cham, Switzerland, 2016; pp. 29–48.

65.   Duan, W.; He, B.; Takara, K.; Luo, P.; Hu, M.; Alias, N.E.; Nover, D. Changes of precipitation amounts and extremes over Japan between 1901 and 2012 and their connection to climate indices. *Clim. Dyn.* **2015**, *45*, 2273–2292. [CrossRef]

66.   Zhang, X.; Alexander, L.; Hegerl, G.C.; Jones, P.; Tank, A.K.; Peterson, T.C.; Trewin, B.; Zwiers, F.W. Indices for monitoring changes in extremes based on daily temperature and precipitation data. *WIREs Clim. Chang.* **2011**, *2*, 851–870. [CrossRef]

67.   Bookhagen, B. Appearance of extreme monsoonal rainfall events and their impact on erosion in the Himalaya. *Geomat. Nat. Hazards Risk* **2010**, *1*, 37–50. [CrossRef]

68.   Casanueva, A.; Rodríguez-Puebla, C.; Frías, M.D.; González-Reviriego, N. Variability of extreme precipitation over Europe and its relationships with teleconnection patterns. *Hydrol. Earth Syst. Sci.* **2014**, *18*, 709–725. [CrossRef]

69.   Sen, P.K. Estimates of the regression coefficient based on Kendall's tau. *J. Am. Stat. Assoc.* **1968**, *63*, 1379–1389. [CrossRef]

70. Theil, H. A rank-invariant method of linear and polynomial regression analysis, I, II, III. In *Henri Theil's Contributions to Economics and Econometrics*; Springer: Amsterdam, The Netherlands, 1992; pp. 386–392, 512–525, 1397–1412.

71. Douglas, E.M.; Vogel, R.M.; Kroll, C.N. Trends in floods and low flows in the United States: Impact of spatial correlation. *J. Hydrol.* **2000**, *240*, 90–105. [CrossRef]

72. Zhang, X.; Vincent, L.A.; Hogg, W.D.; Niitsoo, A. Temperature and precipitation trends in Canada during the 20th century. *Atmos. Ocean* **2000**, *38*, 395–429. [CrossRef]

73. Yue, S.; Wang, C.Y. Regional streamflow trend detection with consideration of both temporal and spatial correlation. *Int. J. Climatol.* **2002**, *22*, 933–946. [CrossRef]

74. Yue, S.; Pilon, P.; Phinney, B. Canadian streamflow trend detection: Impacts of serial and cross-correlation. *Hydrol. Sci. J.* **2003**, *48*, 51–63. [CrossRef]

75. Yue, S.; Pilon, P.; Phinney, B.; Cavadias, G. The influence of autocorrelation on the ability to detect trend in hydrological series. *Hydrol. Process.* **2002**, *16*, 1807–1829. [CrossRef]

76. Von Storch, V.H. Misuses of statistical analysis in climate research. In *Analysis of Climate Variability: Applications of Statistical Techniques*; von Storch, H., Navarra, A., Eds.; Springer: Berlin, Germany, 1995; pp. 11–26.

77. Hamed, K.H.; Rao, A.R. A modified Mann-Kendall trend test for autocorrelated data. *J. Hydrol.* **1998**, *204*, 182–196. [CrossRef]

78. Burn, D.H.; Cunderlik, J.; Pietroniro, A. Climatic influences on streamflow timing in the headwaters of the Mackenzie River Basin. *J. Hydrol.* **2008**, *352*, 225–238. [CrossRef]

79. Shadmani, M.; Marofi, S.; Roknian, M. Trend analysis in reference evapotranspiration using Mann-Kendall and Spearman's Rho tests in arid regions of Iran. *Water Resour. Manag.* **2012**, *26*, 211–224. [CrossRef]

80. Wu, H.; Soh, L.; Samal, A. Trend analysis of streamflow drought events in Nebraska. *Water Resour. Manag.* **2008**, *22*, 145–164. [CrossRef]

81. Lacombe, G.; McCartney, M. Uncovering consistencies in Indian rainfall trends observed over the last half century. *Clim. Chang.* **2014**, *123*, 287–299. [CrossRef]

82. Vogel, R.M.; Kroll, C.N. Low-flow frequency analysis using probability plot correlation coefficients. *J. Water Resour. Plan. Manag.* **1989**, *115*, 338–357. [CrossRef]

83. Lettenmaier, D.P.; Wood, E.F.; Wallis, J.R. Hydro-climatological trends in the continental United-States, 1948–88. *J. Clim.* **1994**, *7*, 586–607. [CrossRef]

84. Livezey, R.E.; Chen, W.Y. Statistical field significance and its determination by Monte Carlo techniques. *Mon. Weather Rev.* **1983**, *111*, 46–59. [CrossRef]

85. Ventura, V.; Paciorek, C.J.; Risbey, J.S. Controlling the proportion of falsely rejected hypotheses when conducting multiple tests with climatological data. *J. Clim.* **2004**, *17*, 4343–4356. [CrossRef]

86. Renard, B.; Lang, M. Use of a Gaussian copula for multivariate extreme value analysis: Some case studies in hydrology. *Adv. Water Resour.* **2007**, *30*, 897–912. [CrossRef]

87. Efron, B. Bootstrap methods: Another look at the Jackknife. *Ann. Stat.* **1979**, *7*, 1–26. [CrossRef]

88. Weibull, W. A statistical theory of strength of materials. *Ing. Vetensk. Akad. Handl.* **1939**, *151*, 1–45.

89. Petrow, T.; Merz, B. Trends in flood magnitude, frequency and seasonality in Germany in the period 1951–2002. *J. Hydrol.* **2009**, *371*, 129–141. [CrossRef]

90. Department of Hydrology and Meteorology (DHM). *Study of Climate and Climatic Variation over Nepal*; Technical Report; Department of Hydrology and Meteorology: Kathmandu, Nepal, 2015. Available online: http://www.dhm.gov.np/climate/ (accessed on 1 October 2016).

91. Shrestha, D.; Singh, P.; Nakamura, K. Spatiotemporal variation of rainfall over the central Himalayan region revealed by TRMM precipitation radar. *J. Geophys. Res. Atmos.* **2012**, *117*, D22106. [CrossRef]

92. Dahal, R.K.; Hasegawa, S. Representative rainfall thresholds for landslides in the Nepal Himalaya. *Geomorphology* **2008**, *100*, 429–443. [CrossRef]

93. Practical Action Nepal Office. *Temporal and Spatial Variability of Climate Change over Nepal (1976–2005)*; Technical Report; Practical Action Nepal Office: Kathmandu, Nepal, 2009; Available online: http://practicalaction.org/file/region_nepal/ClimateChange1976--2005.pdf (accessed on 4 October 2014).

94. Mäkelä, A.; Shrestha, R.; Karki, R. Thunderstorm characteristics in Nepal during the pre-monsoon season 2012. *Atmos. Res.* **2014**, *137*, 91–99. [CrossRef]

95.   Hasson, S.U.; Pascale, S.; Lucarini, V.; Böhner, J. Seasonal cycle of precipitation over major river basins in South and Southeast Asia: A review of the CMIP5 climate models data for present climate and future climate projections. *Atmos. Res.* **2016**, *180*, 42–63. [CrossRef]

96.   Hasson, S.; Lucarini, V.; Pascale, S.; Böhner, J. Seasonality of the hydrological cycle in major South and Southeast Asian river basins as simulated by PCMDI/CMIP3 experiments. *Earth Syst. Dyn.* **2014**, *5*, 67–87. [CrossRef]

97.   Nayava, J.L. Variations of rice yield with rainfall in Nepal during 1971–2000. *J. Hydrol. Meteorol.* **2008**, *5*, 93–102.

98.   Khatiwada, K.R.; Panthi, J.; Shrestha, M.L. Hydro-climatic variability in the Karnali River Basin. *Climate* **2016**. [CrossRef]

99.   Dahal, P.; Shrestha, N.S.; Shrestha, M.L.; Krakauer, N.Y.; Panthi, J.; Pradhanang, S.M.; Jha, A.; Lakhankar, T. Drought risk assessment in central Nepal: Temporal and spatial analysis. *Nat. Hazards* **2016**, *80*, 1913–1932. [CrossRef]

100. Aryal, S. Rainfall and water requirement of rice during growing period. *J. Agric. Environ.* **2012**, *13*, 1–4. [CrossRef]

101. Wang, S.Y.; Yoon, J.H.; Gillies, R.R.; Cho, C. What caused the winter drought in western Nepal during recent years? *J. Clim.* **2013**, *26*, 8241–8256. [CrossRef]

102. Cannon, F.; Carvalho, L.M.V.; Jones, C.; Bookhagen, B. Multi-annual variations in winter westerly disturbance activity affecting the Himalaya. *Clim. Dyn.* **2015**, *44*, 441–455. [CrossRef]

103. Shekhar, M.S.; Chand, H.; Kumar, S.; Srinivasan, K.; Ganju, A. Climate-change studies in the western Himalaya. *Ann. Glaciol.* **2010**, *51*, 105–112. [CrossRef]

104. Sigdel, M.; Ma, Y. Variability and trends in daily precipitation extremes on the northern and southern slopes of the central Himalaya. *Theor. Appl. Climatol.* **2016**. [CrossRef]

105. Cho, C.; Li, R.; Wang, S.Y.; Ho, J.; Robert, Y. Anthropogenic footprint of climate change in the June 2013 Northern India flood. *Clim. Dyn.* **2016**, *46*, 797–805. [CrossRef]

106. Vellore, R.K.; Kaplan, M.L.; Krishnan, R.; Lewis, J.M.; Sabade, S.; Deshpande, N.; Singh, B.B.; Madhura, R.K.; Rama Rao, M.V.S. Monsoon-extratropical circulation interactions in Himalayan extreme rainfall. *Clim. Dyn.* **2016**, *46*, 3517–3546. [CrossRef]

107. Vellore, R.K.; Krishnan, R.; Pendharkar, J.; Choudhury, A.D.; Sabin, T.P. On the anomalous precipitation enhancement over the Himalayan foothills during monsoon breaks. *Clim. Dyn.* **2014**, *43*, 2009–2031. [CrossRef]

108. Song, X.; Song, S.; Sun, W.; Mu, X.; Wang, S.; Li, J.; Li, Y. Recent changes in extreme precipitation and drought over the Songhua River Basin, China, during 1960–2013. *Atmos. Res.* **2015**, *157*, 137–152. [CrossRef]

109. Shahid, S. Trends in extreme rainfall events of Bangladesh. *Theor. Appl. Climatol.* **2011**, *104*, 489–499. [CrossRef]

110. Sigdel, M.; Ikeda, M. Spatial and temporal analysis of drought in Nepal using standardized precipitation index and its relationship with climate indices. *J. Hydrol. Meteorol.* **2010**, *7*, 59–74. [CrossRef]

111. Kafle, H.K. Spatial and temporal variation of drought in far and mid-western regions of Nepal: Time series analysis (1982–2012). *Nepal J. Sci. Technol.* **2014**, *15*, 65–76. [CrossRef]

112. Miyan, M.A. Droughts in Asian least developed countries: Vulnerability and sustainability. *Weather Clim. Extremes* **2015**, *7*, 8–23. [CrossRef]

113. Yao, T.; Thompson, L.; Yang, W.; Yu, W.; Gao, Y.; Guo, X.; Yang, X.; Duan, K.; Zhao, H.; Xu, B.; et al. Different glacier status with atmospheric circulations in Tibetan Plateau and surroundings. *Nat. Clim. Chang.* **2012**, *2*, 663–667. [CrossRef]

114. Manton, M.J.; Haylock, M.R.; Hennessy, K.J.; Nicholls, N.; Chambers, L.E.; Collins, D.A.; Daw, G.; Finet, A.; Gunawan, D.; Inape, K.; et al. Trends in extreme daily rainfall and temperature in Southeast Asia and the South Pacific: 1961–1998. *Int. J. Climatol.* **2001**, *21*, 269–284. [CrossRef]

5

Linkage between Water Level Dynamics and Climate Variability: The Case of Lake Hawassa Hydrology and ENSO Phenomena

Mulugeta Dadi Belete [1,2,3,*], **Bernd Diekkrüger** [1] **and Jackson Roehrig** [2]

[1] Department of Geography, University of Bonn, 53115 Bonn, Germany; b.diekkrueger@uni-bonn.de

[2] Institute for Technology and Water Resources Management in the Tropics and Subtropics, Cologne University of Applied Sciences, 50679 Köln (Deutz), Germany; Jackson.roehrig@fh-koeln.de

[3] Institute of Technology, School of Water Resources Engineering, Hawassa University, Hawassa P.O. Box 005, Ethiopia

[*] Correspondence: mulugeta.belete9@gmail.com

Academic Editor: Yang Zhang

Abstract: Lake Hawassa is a topographically closed lake in the Central Main Ethiopian Rift Valley. The water level of this lake has been reported to dramatically rise without falling back to the original level. The cause of this rise is not yet sufficiently investigated and subjected to this study. This study argues that the general variability in the lake level and its resultant rise has significant linkage to the temperature variability at the Pacific Ocean. The linkage between water level dynamics and climate variability was analyzed through the application of diverse statistical techniques. It comprises the Mann-Kendall trend analysis to test monotonic variations over time; sequential regime shift index (RSI) to detect significant shifts in the mean values of time-series records of lake level; and coherence analysis to investigate the linear relationship between ENSO index and records of local hydrology. Despite the multiple rises and falls, the results of the trend analysis revealed that the lake level experienced a significant resultant upward trend with Mann-Kendall τ values of 0.558, 0.629, and 0.545 (at $\alpha = 0.05$ and $p < 0.01\%$) for monthly maximum, average and minimum values respectively. The sequential regime shift evidenced that most of the significant shifts coincide with the occurrences of ENSO events. Generally, the lake level tends to be high during El Niño and low during La Niña episodes. The typical examples are the coincidence of extreme historical maximum lake level to the strongest El Niño event of the century that occurred in 1997/98 and the lowest lake level record in the year 1975 with a strong La Niña year. The coincidence of climate regime shift in the Pacific Ocean in 1976/77 with an equivalent regime shift in the lake level is an additional confirmation for the possible climate-hydrology linkage. The likely involvement of anthropogenic factors (at least in modifying the effect of climate) is justified by the interplay between the non-trending rainfall and potential evapotranspiration and trending streamflow. The coherence analysis between 492 pairs of monthly step datasets of 3.4ENSO index and lake level changes is also found to have a significant linear relationship over frequencies ranging from 0.13 to 0.14 cycles/month or 1.56 to 1.68 cycles/year. This corresponds to a dominant average periodicity (coincident cycle) of about 7.4 months which is thought to be related to the time span of the two rainy season in the locality.

Keywords: Lake Hawassa; 3.4ENSO index; trend analysis; regime shift; coherence analysis; El Niño; La Niña

1. Introduction

One of the most significant and broadly impacting effects of climate variability on lakes is the changes in water level. Such changes reflect an alteration of the lake water balance, which can result from changes in precipitation, surface runoff, ground water flow, and evaporation from the lake surface [1]. The water in a lake is balanced by the basic hydrological relationship in which the change in water storage is governed by the water input and output to the system [2].

In the 1960s, lakes throughout East Africa rose [3], resulting from a series of remarkably wet years [4,5]. The spatial extent and the magnitude of fluctuations were considered as a signal to major global climate change [6]. According to Arnell et al. [6] and Bergonzini [7], African lakes are known to be very sensitive to climate variations with special sensitivity of closed lakes. The impact of non-climatic factors on water level variability in Ethiopia was also reported by Görner et al. [8] and Belay [9].

Lake Hawassa, a topographically closed lake in the Central Main Ethiopian Rift Valley, has been experiencing a progressive rise in water level during the past two decades (1981–1998) [10,11]. The concern of this rise achieved its peak in the aftermath of the extreme flooding of the surrounding area as a result of extreme rise in 1998/99. According to WRDB [12] and WWDSE [11], the lake level rise and the associated surface expansion affected about 162 urban and 2244 farmers' households, 13 different organizations, water supply schemes, 10 ha of sand quarry, roads, and forest land. In monetary terms, the total physical damage was estimated to be 43,490,524 Ethiopian birr (about €5.4 million).

Over the past few years, several researchers have studied the long-term water balance of Lake Hawassa, such as Gebreegiziabher [10], Ayenew [13], Deganovsky and Getahun [14], WWDSE [11], Ayenew and Gebreegiziabher [15], Gebremichael [16], and Shewangizaw [17]. Land use/cover changes have also been studied by Wagesho et al. [18], and WWDSE [11]. Despite the number of studies and their importance, the cause of lake level rise has not been concluded and not yet explicitly investigated.

The idea of "climate-hydrology link" was conceived in this study after the recognition of coincidence between the lowest lake level record in the year 1975 with a strong La Niña year and the maximum lake level in 1998 with the strongest El Niño year. La Niña and El Niño are anomalies in ocean surface water temperature. They are commonly termed as "teleconnections" [19].

Having the general objective to investigate the association between local hydro-climatic variables of Lake Hawassa hydro-system with the climate variability at Pacific Ocean, this study sets the following specific objectives:

- To analyze the long-term trends (variation over-time) and sequential regime shifts (variation across-time) for lake level, rainfall, streamflow, and potential evapotranspiration;
- To compare significant change points of the above hydro-climatic variables with the timing and intensity of North Pacific climate shifts/ El Niño/ La Niña occurrences; and
- To analyze the coherence between data series of Niño3.4 Index (N3.4) and Lake Hawassa water level.

2. Methodology

2.1. Description of the Study Area

Lake Hawassa watershed is located in the central North-East of the Ethiopian Rift Valley Basin (Figure 1) and covers an area of 1436.5 km^2. It contains five sub-watersheds: Dorebafena-Shamena, Wedesa-Kerama, Tikur Wuha, Lalima-Wendo Kosha and Shashemene-Toga. The geographical co-ordinates of the watershed are 6°45′ to 7°15′ North and 38°15′ to 38°45′ East latitude and longitude respectively. Hawassa city, named after the lake, is located at 275 km south of the capital city-Addis Ababa and is established in the very eastern shore of the lake [20].

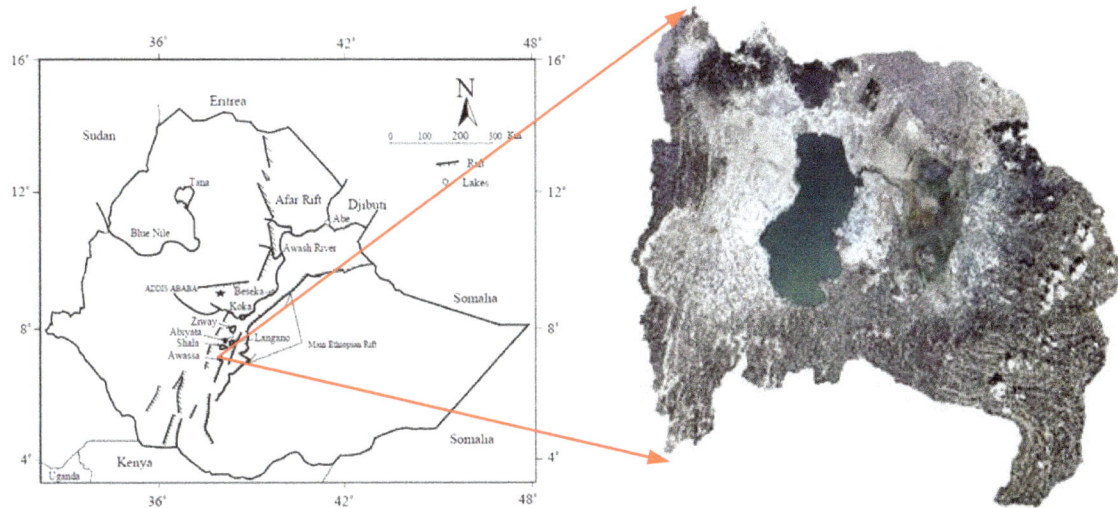

Figure 1. Location map of the study area (Source: Belete et al. [21])

According to MoWR [20], land use in the watershed is dominated by cultivation which occupies 66% of the land area with intensive cultivation. The major land cover splits into smallholder cultivation (95%) mechanized cultivation (5%) most of which is state owned. Although the lake water has been abstracted for supplementary irrigation, the total amount of water abstraction is negligible compared to all other water balance components.

According to Legesse et al. [22], the watershed is characterized by three main seasons. The long rainy season in the summer from June-September is known locally as Kiremt and is primarily controlled by the seasonal migration of the inter-tropical convergence zone (ITCZ), which lies to the north of Ethiopia at this period. The wet period represents 50%–70% of the mean annual total rainfall. The dry period (locally named as Baga) extends between October and February when the ITCZ lies to the south of Ethiopia [23]. During March and May, the "small rain" season (locally named as Belg) occurs when about 20%–30% of the annual rainfall falls. The climate in the area varies from dry to sub-humid according to the Thornthwaite's system of defining climate or moisture regions [24].

As computed from the long-term (1973–2010) rainfall record of Hawassa meteorological station, the annual average magnitude is computed to be 961mm and distributed as 50% for Kiremt (June–September); 20% for Baga (October–February) and 30% for Belg season (March–May). Figure 2 shows the long-term average monthly distribution of rainfall and temperature at Hawassa meteorological station.

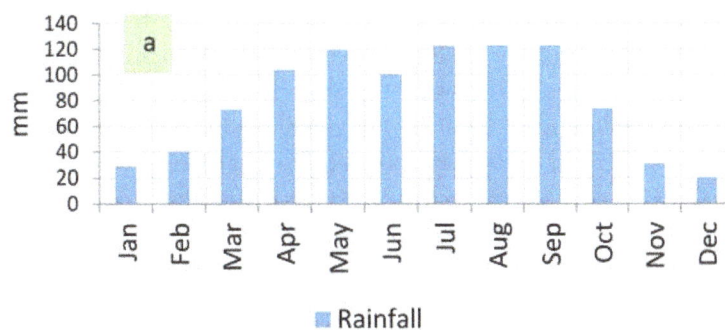

Figure 2. *Cont.*

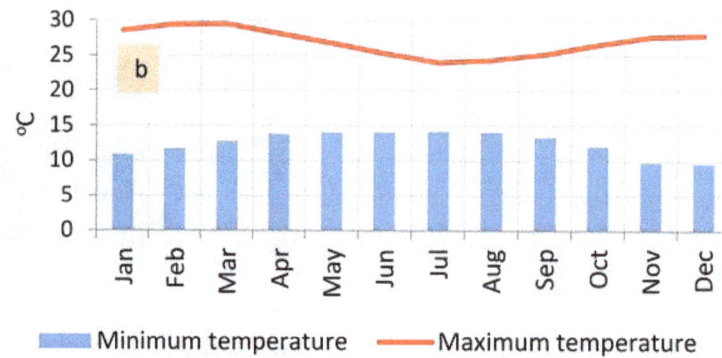

Figure 2. Distribution of monthly rainfall (**a**) and temperature (**b**) at Hawassa.

In term of topography, the majority of the watershed is flat to gently undulating but bounded by steep escarpments. The altitude ranges from 1680 m at Lake Hawassa to 2700 m on the Eastern escarpment: an altitude range of 1020 m. Most slopes (56%) are flat to gentle (0%–8%) with a further 33% moderately sloping (8%–30%) and only 5% steep to very steep (>30%) [24].

2.2. Data Availability

As shown in Table 1, there exists fairly long sequence of hydro-climatic data for Hawassa meteorological station which is the nearest station for the lake. Other meteorological stations in the watershed have limited data. Data gaps are filled by linear interpolation throughout this study.

Table 1. The core set of hydro-climatic data employed in the study.

Data Type	Temporal Scale	Period	Sources
Lake level records	Daily	1970–2010	Ministry of Water Resources
Streamflow	»	1980–2006	»
Rainfall	»	1972–2010	Meteorological Agency
Temperature	»	1986–2006	»
Wind speed	»	1986–2006	»
Relative humidity	»	1986–2006	»
Sun-shine hours	»	1986–2006	»

N.B. The water level records have been made from conventional local bench mark and time series data of the actual water level was derived by adjusting the records to the bathymetric map of the lake.

2.3. Detection of Long-Terms Trends Using Mann-Kendall Test

According to our knowledge, the statistical significance of long-term monotonic trend of Lake Hawassa water level had not been computed before. Statistical trend analysis is a hypothesis testing process in which the null hypothesis (H_0) states that there is no trend. Trend analysis enables to detect significant variations overtime. It is easily understood and communicated, and readily accepted due to its wide spread use [25]. In this study, the well-known Mann-Kendall (MK) statistical trend test [26,27] was employed to investigate trends in time series data. It is a kind of non-parametric test and compares the relative magnitudes of sample data rather than the data values themselves. The technique also allows us to investigate long-term trends of data without assuming any particular distribution. The other advantage is its low sensitivity to abrupt breaks due to inhomogeneous time series. In this study, the 5% level of significance was considered [28,29].

2.4. Sequential Regime Shift Detection Using Regime Shift Index (RSI)

The concept of "regime" in hydrology describes the temporal pattern of the variable under discussion over a period of time and "regime shift" was originally proposed in relation to oceanic ecosystem [30,31] to describe sudden drastic changes in temporal characteristics of a variable [11]. The definition of climatic regime shifts can be viewed as "differing average climatic levels over a multi-annual duration" [32]. Shifts in the mean are the most common type of shifts considered in literature [33,34].

A jump in a series that is detected by a regime shift test can imply changes in either climatic factors or watershed characteristics [35]. According to Breaker [36], change points occur where the changes are relatively abrupt. Formally, a change point exists at a time t_0, if all of the observations up to t_0 share a common statistical distribution, and those after t_0, share a different statistical distribution. Rodionov [33] introduced an algorithm for detecting sequential regime shifts in time series data based on sequential t-tests.

2.5. Estimation of Coherence between ENSO Index and Lake Level Variability

Time series data records of any two continuous variables suitable for computing a covariance, if of sufficient length for computing a stable fast Fourier transform (fft), can be transformed into the frequency domain for computation of a dimensionless squared spectral coherence. Transforming from the time to the frequency domain and computing the squared spectral coherence (CH) provides frequency-stratified results that can be tested for statistical significance using the F-distribution [37]. Coherence, also known as coherency spectrum, or magnitude-squared coherence, is a widely used measure for characterizing linear dependence between two time series and classical books on time series analysis present coherence as "the frequency domain analogue of the autocorrelation function" [38]. Further information on spectrum analysis can be referred from books such as [39,40].

The use of spectral coherence analysis is quite recent in the area of hydrology. The coherence analysis in this study was made following the idea of Jenkins and Watts [41] and Bloomfield [42] where more detailed explanation about the techniques can be referred. This technique was employed to analyze the relationship between Niño3.4 index and lake level data series. The significance of coherence resulting from this technique suggests that changes in one series are related to changes in the other. As subjected to this analysis, the two phenomena are assumed to be ergodic (*their statistical properties can be deduced from a single, sufficiently long, random sample of the processes*) and the system functions linearly. A total of 492 pairs of monthly step time series data were undergone through spectral analysis for the explicit estimation of "coherency" between these series. The Niño3.4 index (N3.4), which is the average SST anomaly within the region 5° S–5° N, 170°–120° W was used as a representative index for ENSO phenomena. This index is usually employed to predict rainfall in Ethiopia [43,44]. It is one of the most widely used ENSO index [45].

The presence of a trend in a time series data produces a spectral peak at zero frequency, and this peak can dominate the spectrum in that other important features are obscured [46]. Due to this, detrending should be part of the analysis. In this study, the time series were detrended using linear regression. The autocorrelations in the time series were also removed by differencing techniques with order 1.

For more reading about the Fourier transform, refer standard text books such as Jenkins and Watts [41]. Namdar-Ghanbari et al. [47] employed similar analysis to examine the relationships between ice, local climate and the teleconnections, Southern Ocean Oscillation (SOI), Pacific Decadal Oscillation (PDO), North Atlantic Oscillation (NAO), and Northern Pacific Index (NP).

As noted by Thomson and Emery [25], the final step in any coherence analysis is to specify the confidence limits for the coherence-square estimates. This step places the spectral results in a complete statistical context.

As presented by Namdar-Ghanbari et al. [47], the estimated coherencies are considered significant at the 99% and 95% level of confidence when they are larger than the critical value T derived from the upper 1% and 5% points of the F-distribution.

3. Results and Discussion

3.1. Result of Trend Analysis

The visual inspection of Figure 3 uncovers the underlying variability of the observed lake level by suggesting that, beside the annual cycle, the overall oscillation does not show periodicity that may be associated to the local situation such as the hydrologically-closed nature of the lake in that the occurrence of an extreme event may consecutively manifest itself (residual effect) in response to the imbalance between inflow and outflow water balance components. The highest peak was observed in November 1998 (22.54 m) followed by October and December of the same year (22.49 m each). The lowest level in this year (June) (21.8 m) was greater than 92.5% of historical records. This particular year was known for its peak records in many parts of the world. The cases of Lake Abaya (another Rift Valley lake in Ethiopia) [48] and other lakes in this basin [21]; Lake Nasser (Egypt), Lake Chad, Lake Turkana, Lake Tanganyika, Lake Victoria, and Lake Mwero [49] are among the examples.

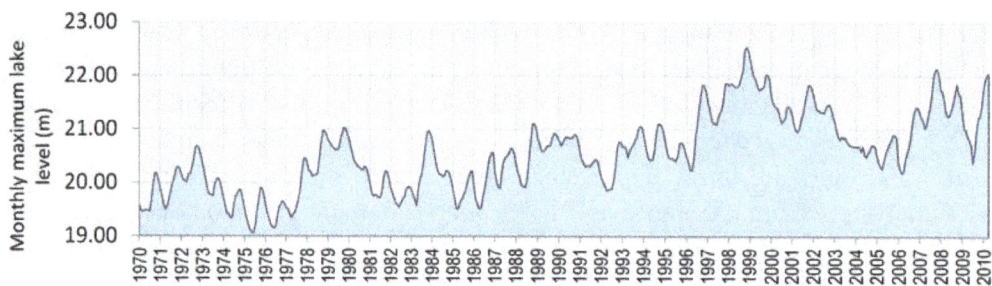

Figure 3. Hydrograph of monthly maximum lake level.

Despite the multiple rises and falls, the lake level experienced a significant resultant upward trend with Mann-Kendall τ values of 0.558, 0.629, and 0.545 (at $\alpha = 0.05$ and $p < 0.01\%$) for monthly maximum, average and minimum values respectively. The ultimate evolution of increasing trend is not gradual and consistent in direction (monotonic) rather sharp rises and falls have been frequently appearing and such variations are likely to bias the monotonic trend. Similar comment was given by [46] in that the use of trend analysis in climate change research depends greatly upon the time period studied, and results can be biased when an abrupt climate change is observed during the study period.

Regarding the connection of ENSO events to the extreme values of observed lake levels, the 1998 record (historical maximum) can easily be justified for its connection to the worst El Niño event of the twentieth century [50,51] as measured by changes in the Pacific [52]. Globally, this El Niño year caused loss of approximately 35–45 billion USD [53]. In contrast, the lowest lake level was observed in 1975 which is likely linked to the two consecutive strong La Niña events of 1973–1974 (the strongest in the period 1950–2012) and 1975–1976.

3.2. Results of Sequential Regime Shift Analysis

3.2.1. Lake Level Variability

Figure 4a–c demonstrate the observed annual average, maximum, and minimum lake levels have undergone a couple of sequential regime shifts reflecting the instability of the hydro-system. Synthesis of the RSI result for average lake level records is shown in Appendix A.

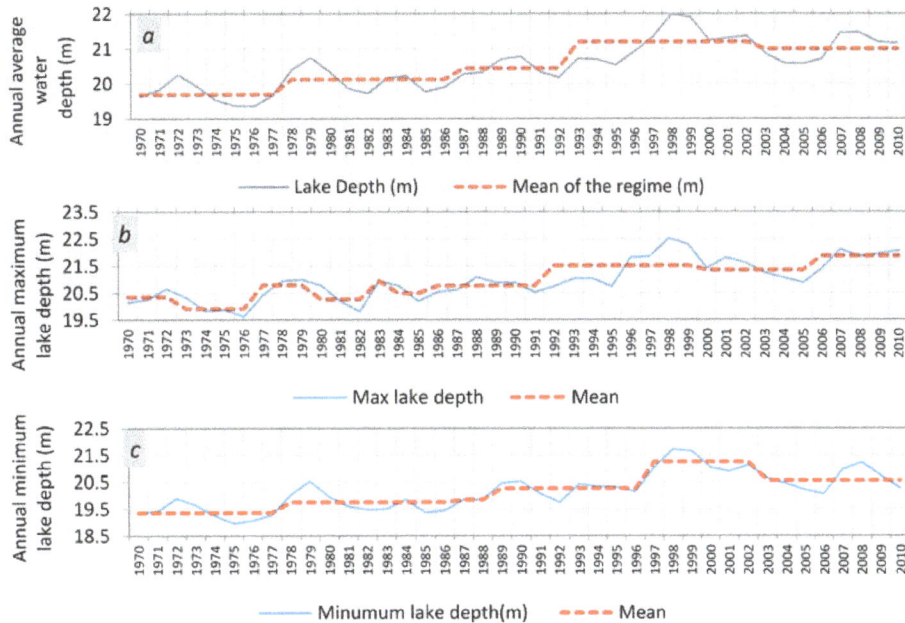

Figure 4. Sequential regime-shifts in annual lake levels (**a**) average; (**b**) maximum and (**c**) minimum.

The important aspect of the prevailing regime shifts lies on their occurrence in the year 1976–1978 (Figure 4a–c) which was known for the climatic regime shift period of the North Pacific [54,55]. The year 1977 also experienced the highest historical recorded annual total rainfall (1226 mm). The maximum lake level has undergone a regime shift in 1983 (Figure 4b), which is likely to be associated with the devastating El Niño of 1983. The other smaller shift in mean value of lake level occurred in 1986 which was likely caused by moderate but prolonged El Niño of 1986–1987. Another connection that is manifested by the overlap of regime shift of Lake Hawassa water level and the North Pacific climate regime shift was observed in 1989 (Figure 4c) and Yletyinen et al. [55] reported that in 1989, a new regime shift (in the climate of the Pacific) had also occurred but the changes were not as remarkable or pervasive as in the 1976–1977.

The highest regime shift was observed in 1992, which showed an upward shift in mean value of the lake level from 20.43 m to 21.2 m, implying a regime shift of 0.77 m. This regime was extended up to 2002 and known for its frequent El Niño years of 1991–1992 (strong), 1994–1995 (moderate), and the 1997–1998 El Niño (strong). Swanson and Tsonis [56] also noted that climate shift occurred around 2001/2002 too and Lake Hawassa also experienced water level regime shift in this year. The relatively sustained maximum lake level regime extended from 1992 up to 1999 (Figure 4b) signifies the occurrences of three El Niños (strong, moderate, strong consecutively) without the occurrence of La Niña in between.

The general up ward shifts between 1978 and 1998 are in agreement with the work of Peterson and Schwing [57]. They identified the PDO (Pacific Decadal Oscillation) index, another index to be negative for most years during 1948–1976 and positive during 1977–1998. In addition, Niebauer [58] observed that before the regime shift, the occurrence of El Niño and La Niña conditions was about even. Since the regime shift, El Niño conditions are about 3 times more prevalent and this further signifies the effect of climate.

3.2.2. Rainfall Variability

Figure 5 shows the sequential regime shifts in annual rainfall at Hawassa meteorological station that represents the over-lake rainfall. As depicted by the figure, the rainfall time series shows high variability with nine distinct regimes over the study period. The relatively long and stable regime extended from 1986 to 1994 (upward shift) followed by regimes of 1999–2004 (downward) and

2005–2010 (upward). The remaining regimes are short lived and most of the breaking points coincided with the occurrences of ENSO phenomena (1976, 1983, and 1994). The climate regime shift of North Pacific Ocean that occurred in 1976/1977 seems to manifest itself by causing an upward shift in both years. The annual total rainfall record of 1977 was the highest of the records (1226 mm). The shift in 1998 was also most likely linked to the transition from strong (1997–1998) to the two consecutive strong La Niñas (1998–1999 and 1999–2000).

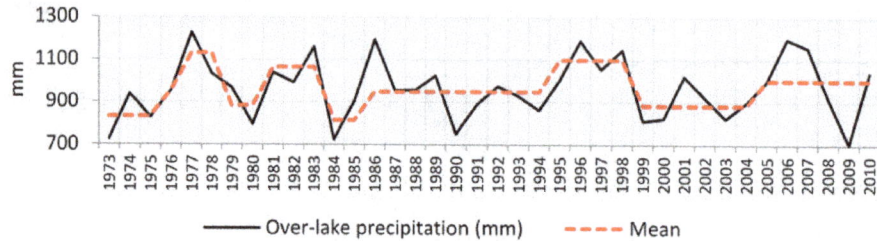

Figure 5. Regime shift of over-lake rainfall.

3.2.3. Variability in Streamflow

Figure 6 demonstrates the variability of stream flow of Tikur Wuha River (the only river draining into Lake Hawassa) across time. The Tikur Wuha River drains about 50% of the catchment while the rest of the catchment drains via ground water or ephemeral channel flow. The first breaking point occurred at 1986, which is known for its moderate El Niño. The years 1994 and 1997 are also another change points corresponding to the timing of ENSO events.

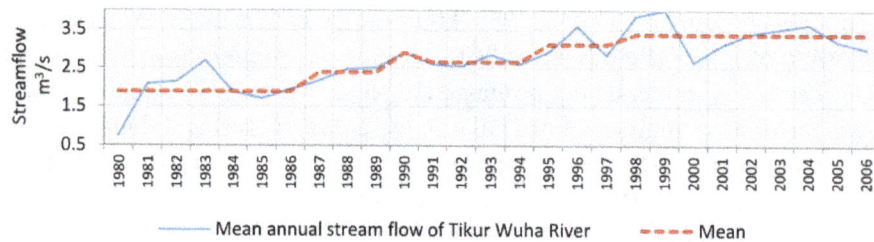

Figure 6. Regime shift in Tikur Wuha streamflow.

3.2.4. Variability in Potential Evapotranspiration (ET)

Time series of Evapotranspiration (ET) was derived by the Penman-Monteith model [59,60]. The model uses five climate variables (minimum and maximum temperature, relative humidity, wind speed, and sun-shine hours) to compute the potential evapotranspiration (ET). As shown in Figure 7, significant drops were exhibited in 1988 (strong La Niña year) and in 1995 (weak La Niña). An upward shift was also depicted in 1994 (Moderate El Niño year). Nevertheless, there was no significant trend in the potential evapotranspiration.

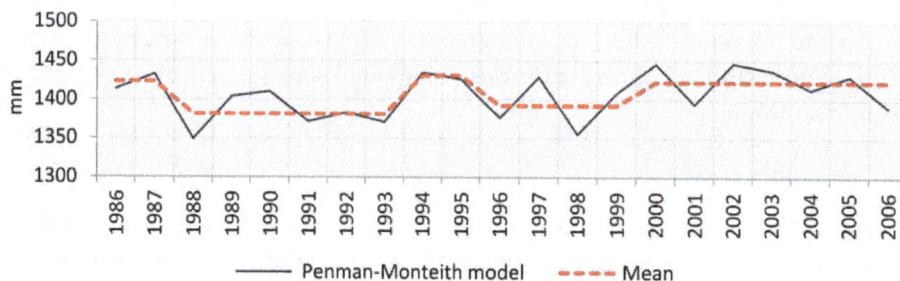

Figure 7. Regime shifts in potential evaporation as computed by the Penman-Monteith model.

3.2.5. Land Use/Cover Change as a Potential Anthropogenic Factor in Affecting the Lake Hydrology

Often, the effect of climate change and human activities on hydrological variables cannot be distinguished [61]. Climate and land use are key factors controlling the hydrological behavior of a catchment [62,63]. The assessment of the impacts of such diverse environmental changes is still one of the main challenges in recent hydrological research [64]. In terms of spatial scales, distinguishing the impact of land use changes on hydrology from the impact of climatic variability is more difficult at the catchment than at the plot or small catchment scale [65]. Many studies have considered these factors separately; however, they do not act in isolation, but rather interact to affect ecosystem structure and function [66].

A number of studies were performed in Lake Hawassa catchment in relation to the impact of land use/ cover of the local water cycle. Abrha [67] assessed the impact of land use/ cover changes on ground water recharge. The author found a significant change in ground water recharge is due to land use and land covers change (especially vegetation cover reduction) in the catchment. Gebreegziagher [10] also recognized the land-use change in the watershed of Lake Hawassa as the most likely cause for the increasing tendency of runoff overtime in combination with the effect of climate variability and climate change. The land use/ land cover dynamics of the watershed is presented in Table 2 below.

Table 2. Land use/ land cover changes in the catchment (units are in km^2) (After Abrha [67]).

	Agriculture	Grass Land	Bush Land	Shrubby Wood Land	Urban Area
1973	323.3	15.5	165.9	704.7	6
1986	466.2	59.5	180.3	548.6	8
2000	565.9	68.7	145.6	448.2	13

As shown in the above table, on average, 9.5 km^2/year of shrub wood land has been converted into other land use types mostly into agricultural lands (9 km^2/year).

3.2.6. Interaction of Geomorphological Processes with Lake Hydrology

The period 1996–1998 was extremely wet in the study area [10] and parallel to this, three consecutive surface cracks/gullies have developed in the years 1996, 1997 and 1998. The first gully developed in April 1996 with an extent of about 150 m length and 3 m depth after inundation of the area with massive flood following an above normal precipitation. The second gully developed across the central part of the catchment in April 1997 with the total length slightly more than 800 m and visible depth in the range of 3–8 m. The extent of this gully increased finally reaching a length of 2.4 km with depth ranges of 8–12 m [68]. Hydrologically, these gullies are important features that facilitate fast transfer of surface water into the aquifer and then to the lake. Field evidences also indicate the positive role of newly formed gullies that act as conduit to ground water inflow from the elevated areas to Lake Hawassa [69]. Unfortunately, these gullies developed in a region of the catchment which was not monitored using stream discharge gauges.

3.3. Results of Coherence Analysis

Figure 8 shows the result of coherence analysis. The values of coherency (y-axis) versus frequency (x-axis) between the Niño3.4 ENSO index and monthly mean lake level changes.

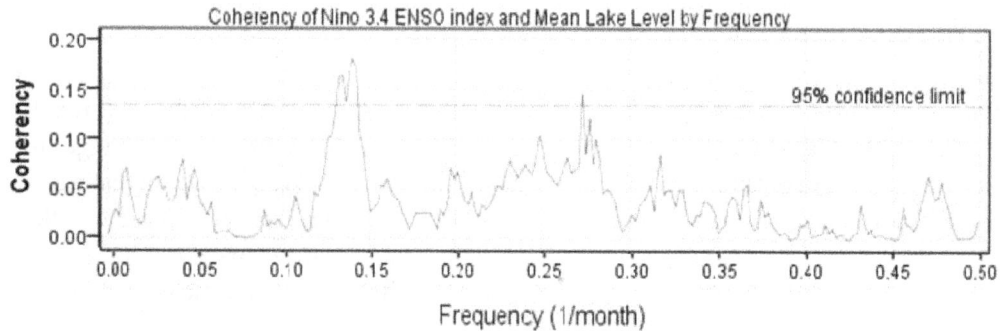

Figure 8. Coherence between ENSO index and lake level variability in frequency domain.

As evidenced by the result of coherence analysis (Figure 8), the cyclic nature of Lake Hawassa water level variability has significant linear relationship to the climate variability at some frequencies. It appears that there are two significant peaks at 95% confidence limit. Further probe to the prominent peak reveals that the peak occurred at a frequency between 0.13–0.14 cycle/month or 1.56–1.68 cycle/year. This corresponds to a period of about 7.14–7.69 months (=1/0.14–1/0.13) or a dominant average periodicity (coincident cycle) of about 7.4 months which is likely associated with the time span between the two wet periods that extend from June-September (main rainy season) and March and May (little rain). A relevant finding was reported by Namdar-Ghanbari and Bravo [70] in which the levels of Great Lakes and Trans-Niño Index (TNI) show significant coherence in the frequency range $(3–7)^{-1}$ cycles/year.

The vital importance of the above analyses is the detection of significant coherence at some specific frequency ranges and confirmed that significant portion of the lake level variability is caused by factors operating on a scale larger than processes in the watershed.

4. Conclusions

The diverse statistical analyses of this chapter provide a plausible explanation for the interaction between the local hydrology and climate anomalies at Pacific. More importantly, the evidences helped us to conclude about the effect of ENSO phenomena and climate shifts on the local hydrology of the lake. Generally, it is observed that high lake level tends to follow moderate to strong El Niño and the reverse is true for La Niña events.

The two prominent climate events which strongly influence the hydrology of Lake Hawassa are: the climate shift of North Pacific Ocean that occurred in 1976/1977 and the El Niño events that occurred in 1972–1973, 1982–1983, 1997–1998, and 2009–2010.

The general suggestions of this study supported the idea of Szestzay [71] in which water level fluctuations of closed lakes are considered as meaningful indicators of climatic changes. Nevertheless, there is no simple linear relationship between climate and lake level variability as shown in this study. The increasing water level of Lake Hawassa can be explained by the interaction of climate with a special emphasis on El Niño events, land use change and geomorphological processes which influence drainage density of the catchment and therefore residence time of water and total discharge. The occurrence of El Niño events alone cannot explain the increase in lake water level but is triggering processes which result in the observed dynamics.

The association of extreme lake level rises of Lake Hawassa to the occurrences of El Niño events (as in the case of 1998 flood) could have two management dimensions. On one hand, it would be difficult to mitigate the problem because of its dependence on macro-scale processes and on the other hand, those large El Niño events which are notorious for their extreme floods are acceptably predictable within period at lead times of up to two years [72]. Climate forecasts are also shown to be more accurate during El Niño and La Niña events and furthermore, stronger ENSO events lead to greater predictability of the climate [73]. These are opportunities to get alarms against the urgency of flood occurrences and it is recommended to mainstream the updated information regarding the probable occurrences of ENSO events and climate shifts in a regular emergency and preparedness actions to reduce the impact of potential flood risks.

Acknowledgments: The authors thanks the DAAD and Ethiopian government, who supported this study under the "engineering capacity building program-ecbp" project with a three-year scholarship provided to the first author. The anonymous reviewers are gratefully acknowledged for their constructive comments.

Author Contributions: All figures and tables and manuscript writing were done by Mulugeta Dadi Belete and Bernd Diekkrüger. The Ph.D. Thesis where the manuscript was derived from was supervised and examined by Bernd Diekkrüger and Jackson Roehrig.

Conflicts of Interest: The authors declare no conflict of interest.

Appendix A. SynThesis of RSI Output

Table A1. Synthesis of RSI output.

	Av.Lake Depth (m)	RSI	Mean	Weighed	Length	p	Outliers
1970	19.64	0.00	19.70	19.68	8		
1971	19.83	0.00	19.70	19.68	8		
1972	20.26	0.00	19.70	19.68	8		0.76
1973	19.92	0.00	19.70	19.68	8		
1974	19.53	0.00	19.70	19.68	8		
1975	19.37	0.00	19.70	19.68	8		
1976	19.36	0.00	19.70	19.68	8		
1977	19.67	0.00	19.70	19.68	8		
1978	20.40	0.63	20.12	20.10	9	0.02	
1979	20.74	0.00	20.12	20.10	9		0.68
1980	20.33	0.00	20.12	20.10	9		
1981	19.86	0.00	20.12	20.10	9		
1982	19.72	0.00	20.12	20.10	9		
1983	20.15	0.00	20.12	20.10	9		
1984	20.22	0.00	20.12	20.10	9		
1985	19.77	0.00	20.12	20.10	9		
1986	19.90	0.00	20.12	20.10	9		
1987	20.27	0.88	20.43	20.43	6	0.06	
1988	20.32	0.00	20.43	20.43	6		
1989	20.71	0.00	20.43	20.43	6		
1990	20.78	0.00	20.43	20.43	6		
1991	20.33	0.00	20.43	20.43	6		
1992	20.17	0.00	20.43	20.43	6		
1993	20.72	1.10	21.20	21.16	10	0.001	
1994	20.70	0.00	21.20	21.16	10		0.95
1995	20.53	0.00	21.20	21.16	10		0.69
1996	20.92	0.00	21.20	21.16	10		
1997	21.33	0.00	21.20	21.16	10		
1998	21.98	0.00	21.20	21.16	10		0.53
1999	21.90	0.00	21.20	21.16	10		0.59
2000	21.25	0.00	21.20	21.16	10		
2001	21.31	0.00	21.20	21.16	10		
2002	21.37	0.00	21.20	21.16	10		
2003	20.87	−0.29	21.00	20.99	8	0.34	
2004	20.58	0.00	21.00	20.99	8		
2005	20.56	0.00	21.00	20.99	8		
2006	20.72	0.00	21.00	20.99	8		
2007	21.46	0.00	21.00	20.99	8		0.93
2008	21.47	0.00	21.00	20.99	8		0.92
2009	21.19	0.00	21.00	20.99	8		
2010	21.16	0.00	21.00	20.99	8		

Where, RSI: Regime Shift Index; Mean: Equal-weighed arithmetic means of the regimes; Weighed: Weighed means of the regimes using the Huber's weight function with the parameter = 1; Length: Length of the regimes; P: Significance level of the difference between the mean values of the neighboring regimes based on the Student's two-tailed t-test with unequal variance; Outliers: Weight of the deviations from the weighed mean greater than 1 standard deviation(s).

References

1. Lenters, J.D.; Kratz, T.K.; Bowser, C.J. Effects of climate variability on lake evaporation: Results from a long-term energy budget study of Sparkling Lake, Northern Wisconsin (USA). *J. Hydrol.* **2005**, *308*, 168–195. [CrossRef]

2. Limgis. Winter School on Essentials in Limnology and Geographic Information System (GIS). 2001. Available online: http://wgbis.ces.iisc.ernet.in/energy/finance/New/limgis.html (accessed on 5 February 2013).

3. Lamb, H.H. Climate in the 1960s. *Geogr. J.* **1966**, *132*, 183–212. [CrossRef]

4. Flohn, H. East African rains of 1961/1962 and the abrupt change of the White Nile Discharge. *Palaeoecol. Afr.* **1987**, *18*, 3–18.

5. Nicholson, S.E. Environmental change within the historical period. In *The Physical Geography of Africa*; Goudie, A.S., Adams, W.M., Orme, A., Eds.; Oxford University Press: Oxford, UK, 1995.

6. Arnell, N.; Bates, B.; Lang, H.; Magnusson, J.J.; Mulholland, P. *Climate Change 1995: Impacts, Adaptations and Mitigations of Climate Change: Scientific Technical Analyses*; Cambridge Univ. Press: Cambridge, UK, 1996.

7. Bergonzini, L. Bilanshydriques de lacs (Kivu, Tanganyika, Rukwa and Nyassa) du riftest-africain Mus. *R. Afr. Centr. Tervuren Ann. Sc. Geol.* **1998**, *103*, 1–183.

8. Goerner, A.; Jolie, E.; Gloaguen, R. Non-climatic growth of the saline Lake Beseka, Main Ethiopian Rift. *J. Arid Environ.* **2009**, *73*, 287–295. [CrossRef]

9. Belay, E.A. Growing lake with growing problems: Integrated Hydrogeological Investigation on Lake Beseka, Ethiopia. Ph.D. Thesis, Rheinischen Friedrich-Wilhelms-Universität, Bonn, Germany, 2009.

10. Gebreegziabher, Y. Assessment of the Water Balance of Lake Awassa Catchment, Ethiopia. Master's Thesis, International Institute for Geo-information Science and Earth Observation, Enschede, The Netherlands, 2004.

11. WWDSE (Water Works Design and Supervision Enterprise). *The Study of Lake Awassa Level Rise*; Southern Nations Nationalities and Peoples Regional State; Water, Mines and Energy Resources Development Bureau: Addis Ababa, Ethiopia, 2001.

12. WRDB (Water Resources Development Bureau). *Study of Pollution of Lakes and Rivers*; AG Consulting Hydrogeologists and Engineers Plc: Addis Ababa, Ethiopia, 2007.

13. Ayenew, T. Environmental implications of changes in the levels of lakes in the Ethiopian Rift since 1970. *Reg. Environ. Chang.* **2004**, *4*, 192–204. [CrossRef]

14. Deganovsky, A.M.; Getahun, B.A. Water balance and level regime of Ethiopian lakes as integral indicators of climate change. In Proceedings of the 12th World Lake Conference (Taal2007), Jaipur, India, 28 October–2 November 2008.

15. Ayenew, T.; Gebreegziabher, Y. Application of a spreadsheet hydrological model for computing the long-term water balance of Lake Awassa, Ethiopia. *Hydrol. Sci. J.* **2006**, *51*, 418–431. [CrossRef]

16. Gebremichael, H. Modeling and Forcasting Hawassa Lake Level Fluctuation. Master's Thesis, Addis Ababa University, Addis Ababa, Ethiopia, 2007.

17. Shewangizaw, D. Assessing the Effect of Land Use Changes on the Hydraulic Regime of Lake Hawassa. Master's Thesis, Addis Ababa University, Addis Ababa, Ethiopia, 2010.

18. Wagesho, N.; Goel, N.K.; Jain, M.K. Investigation of non-stationarity in hydro-climatic variables at Rift Valley Lakes Basin of Ethiopia. *J. Hydrol.* **2012**, *444*, 113–133. [CrossRef]

19. Wallace, J.M.; Gutzler, D.S. Teleconnections in the geopotential height field during the Northern Hemisphere winter. *Mon. Weather Rev.* **1981**, *109*, 784–812. [CrossRef]

20. MoWR. *The Federal Democratic Republic of Ethiopia-Ministry of Water Resources: Rift Valley Lakes Basin Integrated Resources Development Master Plan Study Project. Phase 3 Report: Lake Hawassa Sub-basin Integrated Watershed Management Feasibility Study. Part 1 and 2*; Halcrow Group Limited and Generation Integrated Rural Development (GIRD) Consultants: Addis Ababa, Ethiopia, 2010.

21. Belete, M.D.; Diekkrüger, B.; Roehrig, J. Characterization of the water level variability of the main Ethiopian Rift Valley Lakes. *Hydrology* **2015**, *3*, 1. [CrossRef]

22. Legesse, D.; Vallet-Coulomb, C.; Gasse, F. Hydrological response of a catchment to climate and land use changes in Tropical Africa: Case study South Central Ethiopia. *J. Hydrol.* **2003**, *275*, 67–85. [CrossRef]

23. Legesse, D.; Vallet-Coulomb, C.; Gasse, F. Analysis of the hydrological response of a tropical terminal lake, Lake Abiyata (Main Ethiopian Rift Valley) to changes in climate and human activities. *Hydrol. Process* **2004**, *18*, 487–504. [CrossRef]

24. Dessie, N. Hydrogeological Investigation of Lake Hawassa Catchment. Master's Thesis, Addis Ababa University, Addis Ababa, Ethiopia, 1995.

25. Thomson, R.E.; Emery, W.J. *Time-Series Analysis Methods: Data Analysis Methods in Physical Oceanography*, 2nd ed.; Elsevier Science: Amsterdam, The Netherlands, 2001.

26. Mann, H.B. Non-parametric tests against trend. *Econometrica* **1945**, *13*, 245–259. [CrossRef]

27. Kendall, M.G. *Rank Correlation Measures*; Charles Griffin: London, UK, 1975.

28. Tabari, H.; Marofi, S. Changes of pan evaporation in the West of Iran. *Water Resour. Manag.* **2011**, *25*, 97–111. [CrossRef]

29. Tabari, H.; Somee, B.S.; Zadeh, M.R. Testing for long-term trends in climatic variables in Iran. *Atmos. Res.* **2011**, *100*, 132–140. [CrossRef]

30. Steele, J.H. Regime shifts in fisheries management. *Fish Res.* **1996**, *25*, 19–23. [CrossRef]

31. Hare, S.R.; Mantua, N.J. Empirical evidence for North Pacific regime shifts in 1977 and 1989. *Prog. Ocean.* **2000**, *47*, 103–145. [CrossRef]

32. Overland, J.E.; Percival, D.B.; Mofjeld, H.O. Regime shifts and red noise in the North Pacific. Deep Sea Research. Part I. *Oceanogr. Res. Pap.* **2006**, *53*, 582–588. [CrossRef]

33. Rodionov, S.N. A sequential algorithm for testing climate regime shifts. *Geophys. Res. Lett.* **2004**, *31*, L09204. [CrossRef]

34. Rodionov, S.N. A brief overview of the regime shift detection methods. In *Large-Scale Disturbances (Regime Shifts) and Recovery in Aquatic Ecosystems: Challenges for Management toward Sustainability*; Velikova, V., Chipev, N., Eds.; UNESCO-ROSTE/BAS Workshop on Regime Shifts: Varna, Bulgaria, 2005.

35. TSOA (Texas State Auditor's Office). *Data Analysis: Analyzing Data—Trend Analysis-2*; Methodology Manual rev. 5/95; Texas State Auditor's Office: Austin, TX, USA, 1995.

36. Breaker, L.C. A closer look at regime shifts based on coastal observations along the eastern boundary of the North Pacific. *Cont. Shelf Res.* **2007**, *27*, 2250–2277. [CrossRef]

37. Biltoft, C.A.; Eric, R.P. Spectral coherence and the statistical significance of turbulent flux computations. *J. Atmos. Ocean. Technol.* **2009**, *26*, 403–409. [CrossRef]

38. Hernando, O.; Bellegem, S.V. *Coherence Analysis of Non-Stationary Time Series: A Linear Filtering Point of View*; UCL—EUEN/CORE—Center for Operations Research and Econometrics: Louvain, Belgium, 2006.

39. Koopmans, L.H. *The Spectral Analysis of Time Series*; Academic Press: New York, NY, USA, 1974.

40. Bendant, J.S.; Piersol, A.G. *Random Data Analysis and Measurement Procedures*, 2nd ed.; Wiley: New York, NY, USA, 1986.

41. Jenkins, G.M.; Watts, D.G. *Spectral Analysis and Its Applications*; Holden Day: San Francisco, CA, USA, 1968.

42. Bloomfield, P. *Fourier Analysis of Time Series: An Introduction*; Wiley: New York, NY, USA, 1976.

43. Korecha, D.; Barnston, A.G. Predictability of June-September rainfall in Ethiopia. *Am. Meteorol. Soc. Mon. Weather Rev.* **2007**, *135*, 628–650. [CrossRef]

44. Babu, A. The impact of Pacific sea surface temperature on the Ethiopian rainfall. In Presented at Workshop on High-Impact Weather Predictability Information System for Africa and AMMA-THORPEX Forecasters' Handbook, Trieste, Italy, 5–8 October 2009.

45. Barnston, A.G.; Chelliah, M.; Goldenberg, S.B. Documentation of a highly ENSO-related SST region in the equatorial Pacific. *Atmos.-Ocean* **1997**, *35*, 367–383.

46. GEOS. *Notes on Time Series Analysis*; University of Arizona: Tucson, AZ, USA, 2013.

47. Namdar-Ghanbari, R.; Bravo, H.R.; Magnuson, J.J.; Hyzer, W.G.; Benson, B.J. Coherence between lake ice cover, local climate and teleconnections (Lake Mendota, Wisconsin). *J. Hydrol.* **2009**, *374*, 282–293. [CrossRef]

48. Belete, A. Climate Change Impact on Lake Abaya Water Level. Master's Thesis, Addis Ababa University, Addis Ababa, Ethiopia, 2009.

49. Mercier, F.; Cazenave, A.; Maheu, C. Inter-annual lake level fluctuations (1993–1999) in Africa from Topex/Poseidon: Connections with ocean-atmosphere interactions over the Indian Ocean. *Glob. Planet. Chang.* **2002**, *32*, 141–163. [CrossRef]

50. Tereshchenko, I.; Filonov, A.; Gallegos, A.; Monzo´n, C.; Rodrı´guez, R. 1997–98 and the hydrometeorological variability of Chapala, a shallow tropical lake in Mexico. *J. Hydrol.* **2002**, *264*, 133–146. [CrossRef]

51. Strub, P.T.; James, C. Altimeter-derived surface circulation in the large-scale NE Pacific Gyres.: Part 2: 1997–1998 anomalies. *Prog. Ocean.* **2002**, *53*, 185–214. [CrossRef]

52. Marucci, S.D. The Impact of Southern Oscillation Phenomenon on the Panama Canal and Its Markets. In Proceedings of the International Association of Maritime Economists Annual Conference, Panama City, Panama, 13–15 November 2002.

53. Sponberg, K. *Navigating the Numbers of Climatological Impact. Compendium of Climatological Impacts*; University Corporation for Atmospheric Research 1, National Oceanic and Atmospheric Administration, Office of Global Programs: Boulder, CO, USA, 1999.

54. Miller, A.J.; Cayan, D.R.; Barnett, T.P.; Graham, N.E.; Oberhuber, J.M. The 1976–77 climate shift of the Pacific Ocean. *Oceanography* **1994**, *7*, 21–26. [CrossRef]

55. Yletyinen, J.; Blenckner, T.; Biggs, R. North Pacific Ocean. In *Regime Shifts Database*; Stockholm Resilience Centre Stockholm University: Stockholm, Sweden, 2012. Available online: http://www.regimeshifts.org/about/item/406-north-pacific-ocean (accessed on 10 January 2013).

56. Swanson, K.L.; Tsonis, A.A. Has the climate recently shifted? *Geophys. Res. Lett.* **2009**, *36*, L06711. [CrossRef]

57. Peterson, W.T.; Schwing, F.B. A new climate regime in Northeast Pacific Ecosystems. *Geophys. Res. Lett.* **2003**, *30*, 1–4. [CrossRef]

58. Niebauer, H.J. Variability in Bering Sea ice cover as affected by a "regime shift" in the North Pacific in the period 1947–96. *J. Geophys. Res.* **1998**, *103*, 27717–27737. [CrossRef]

59. Penman, H.L. Natural evaporation from open water, bare soil, and grass. *Proc. R. Soc. Lond. A* **1948**, *193*, 120–146. [CrossRef]

60. Monteith, J.L. Evaporation and environment. In *Symposium of the Society for Experimental Biology, the State and Movement of Water in Living Organisms*; Fogg, G.E., Ed.; Academic Press Inc.: New York, NY, USA, 1965; pp. 205–234.

61. Uhlenbrook, S. Climate and man-made changes and their impacts on catchments. In Proceedings of the Joint Conference of APLU and ICA, Prague, Czech, 23–26 June 2009.

62. Hörmann, G.; Horn, A.; Fohrer, N. The evaluation of land-use options in mesoscale catchments–prospects and limitations of eco-hydrological models. *Ecol. Model.* **2005**, *187*, 3–14. [CrossRef]

63. Li, Z.; Liu, W.Z.; Zhang, X.C.; Zheng, F.L. Impacts of land use change and climate variability on hydrology in an agricultural catchment on the Loess Plateau of China. *J. Hydrol.* **2009**, *377*, 35–42. [CrossRef]

64. Elfert, S.S.; Bormann, H. Simulated impact of past and possible future land use changes on the hydrological response of the Northern German lowland 'Hunte' catchment. *J. Hydrol.* **2010**, *383*, 245–255. [CrossRef]

65. Archer, D. Scale effects on the hydrological impact of upland afforestation and drainage using indices of flow variability: The River Irthing, England. *Hydrol. Earth Syst. Sci.* **2003**, *7*, 325–338. [CrossRef]

66. Kulakowski, D.; Bebi, P.; Rixen, C. The interacting effects of land use change, climate change and suppression of natural disturbances on landscape forest structure in the Swiss Alps. *Oikos* **2011**, *120*, 216–225. [CrossRef]

67. Abrha, L. Assessing the Impact of Land Use and Land Cover Change on Ground Water Recharge Using RS and GIS: A Case of Awassa Catchment, South Ethiopia. Master's Thesis, Addis Ababa University, Addis Ababa, Ethiopia, 2007.

68. Yirgu, G.; Paola, D.; Kebede, G.M.; Hagos, F.; Teferi, M.A.; Tilahun, G.; Gemetesa, N. *Study Report on Ground Fracturing in the Muleti Area, Awasa Zuria Woreda*; Addis Ababa University: Addis Ababa, Ethiopia; SNNPR Professional Paper: Hawassa, Ethiopia, 1997.

69. Ayenew, T. Water management problems in the Ethiopian rift: Challenges for development. *J. Afr. Earth Sci.* **2007**, *48*, 222–236. [CrossRef]

70. Namdar-Ghanbari, R.; Bravo, H.R. Coherence between atmospheric teleconnections, Great Lakes water levels and regional climate. *Adv. Water Resour.* **2008**, *31*, 1284–1298. [CrossRef]

71. Szesztay, K. Water balance and water level fluctuation of lakes. *Hydrol. Sci.-Bull.* **1974**, *19*, 73–84. [CrossRef]

72. Chen, D.; Mark, A.C.; Alexey, K.; Stephen, E.Z.; Daji, H. Predictability of Niño over the past 148 years. *Nature* **2004**, *428*, 733–736. [CrossRef] [PubMed]

73. Goddard, L.; Dilley, M. Catastrophe or opportunity. *J. Clim.* **2005**, *18*, 651–665. [CrossRef]

6

Spatial and Temporal Responses of Soil Erosion to Climate Change Impacts in a Transnational Watershed in Southeast Asia

Pham Quy Giang [1,*], Le Thi Giang [1] and Kosuke Toshiki [2]

[1] Faculty of Land Management, Vietnam National University of Agriculture, Trau Quy, Gia Lam, Hanoi, Vietnam

[2] Faculty of Regional Innovation, University of Miyazaki. 1-1, Gakuenkibanadainishi, Miyazaki 8892192, Japan; toshiki.k@cc.miyazaki-u.ac.jp

* Correspondence: quygiang1010@vnua.edu.vn

Academic Editor: Yang Zhang

Abstract: It has been widely predicted that Southeast Asia is among the regions facing the most severe climate change impacts. Despite this forecast, little research has been published on the potential impacts of climate change on soil erosion in this region. This study focused on the impact of climate change on spatial and temporal patterns of soil erosion in the Laos–Vietnam transnational Upper Ca River Watershed. The Soil and Water Assessment Tool (SWAT) coupled with downscaled global climate models (GCMs) was employed for simulation. Soil erosion in the watershed was mostly found as "hill-slope erosion", which occurred seriously in the upstream area where topography is dominated by numerous steep hills with sparse vegetation cover. However, under the impact of climate change, it is very likely that soil erosion rate in the downstream area will increase at a higher rate than in its upstream area due to a greater increase in precipitation. Seasonally, soil erosion is predicted to increase significantly in the warmer and wetter climate of the wet season, when higher erosive power of an increased amount and intensity of rainfall is accompanied by higher sediment transport capacity. The results of this study provide useful information for decision makers to plan where and when soil conservation practice should be focused.

Keywords: climate change; modelling; soil erosion; spatial; temporal

1. Introduction

As more evidence on human-induced climate change has accumulated, rapidly increasing attention of researchers around the world has been paid to its potential impacts on the environment. Soil erosion is one of the environmental phenomena most affected by climate change. There is a general consensus in previous research that increases in the amount and intensity of rainfall will lead to greater rates of soil erosion unless protection measures are taken. According to Parry et al. [1], soil erosion rates can be affected by changes in climate for a wide variety of reasons: changes in plant canopy due to shifts in plant biomass production caused by changes in moisture regime; changes in litter cover on the ground due to changes in plant residue decomposition rates, moisture-dependent soil microbial activity, and plant biomass production rates; changes in surface runoff ratios due to changes in soil moisture caused by shifting precipitation regimes and evapotranspiration rates; changes in soil erodibility caused by a decrease in soil organic matter concentrations; a shift from non-erosive snow to erosive rainfall due to temperature rise; melting of permafrost, which induces an erodible soil state from a previously non-erodible one; and changes in land use to adapt to new climatic regimes.

A number of studies have been conducted to examine the change of soil erosion in response to changing climate conditions, especially change in rainfall under different global climate models

(GCMs). For example, Nearing [2] used the output from two GCMs (HadCM3 and CGCM1) and relationships between monthly precipitation and rainfall erosivity to assess potential changes in rainfall erosivity in the USA. The study found that the changes were significant, but results from the two models differed both in magnitude and regional distributions. Zhang et al. [3] used HadCM3 to assess potential changes in rainfall erosivity in the Huanghe River Basin of China, projecting increases in rainfall erosivity by as much as 11% to 22% by 2050. Pruski and Nearing [4] used HadCM3 to simulate erosion under maize and wheat cropping systems at eight locations in the USA for the 21st century, with consideration of the primary physical and biological mechanisms affecting erosion. The study found that where precipitation was projected to increase, estimated erosion increased by 15% to 100%. However, the results were more complex where precipitation was projected to decrease, due largely to interactions between plant biomass, runoff, and erosion, and either increases or decreases in overall erosion could occur. Zhang and Nearing [5] predicted the potential impacts of climate change on soil erosion in central Oklahoma State of the USA, while the HadCM3-predicted mean annual precipitation during 2070–2099 decreased by 6.2%, 13.6%, and 7.2%, for GGa1, A2, and B2, respectively, the predicted erosion increased by 67%–82% for GGa1, increased by 18%–30% for A2, and remained similar for B2. The larger increase in erosion in the GGa1 scenario was attributed to larger variability in monthly precipitation and an increased frequency of large storms in the model simulation. According to Parry et al. [1], extreme rainfall events contribute a disproportionate amount of soil erosion relative to the total rainfall contribution, and this effect will be exacerbated if the frequency of such events increases in the future.

In Southeast Asia, climate change is widely projected to severely hit the region in coming decades. Despite this forecast, little research has been published on the potential impacts of climate change on soil erosion in this region. Research focusing on transnational watersheds is especially scarce due to the lack of collaboration and data sharing among riparian states. This study aimed to predict the impact of climate change on spatial and temporal patterns of soil erosion in the Laos–Vietnam transnational Upper Ca River Watershed (UCRW). The UCRW is a large agricultural watershed playing an important role in the economic development of both countries, but it is severely affected by soil erosion due to its hilly-slope topography, high intensity of rainfall, and other climate and land cover characteristics. Identifying spatial and temporal patterns of soil erosion under the impact of climate change could therefore provide useful information about where and when soil conservation practice should be focused in order to protect the soil for long-term agricultural cultivation.

2. Materials and Methods

2.1. The Study Area and Data

The Ca River Watershed is situated in north-central Vietnam and north-east Laos. In Laos, the watershed covers an area of more than 9000 km^2 of the Provinces of Houaphanh and Xiangkhouang. In Vietnam, it covers the entire area of the Provinces of Nghe An and Ha Tinh and a part of Thanh Hoa Province. The Upper Ca River Watershed defined in this study is the upper part of the watershed which has its outlet at Yen Thuong Hydrological Station of Vietnam (105°23′ E, 18°41′ N), draining an area of approximately 22,800 km^2. Geographic location and detailed information of the UCRW are shown in Figure 1.

In the UCRW, precipitation is abundant, but is spatially and seasonally uneven in its distribution. Average annual precipitation observed varies from 1200 mm to 2100 mm depending on the weather station, with an average of approximately 1700 mm, of which the rainy season accounts for more than 80%. The UCRW is dominated by rugged terrain with approximately 70% of the watershed area having a slope of greater than 15 degrees and 20% having a slope of greater than 30 degrees. Highly steep areas are mostly located in the middle part of the watershed, where high mountains are alternated by low and narrow strips along streams. Soils in the UCRW include mostly Humic Acrisols (strongly weathered acid soil, with generally low inherent soil fertility) and Plinthic Ferralsols (deeply weathered red or yellow soil, dominated by low activity clays including mainly kaolinite and sesquioxides). Except for the alluvial soils in the low valleys, soils in the area are generally acidic, poor in nutrients, and highly susceptible to erosion. The natural land cover is mostly evergreen and semi-deciduous tropical moist forest, although mixed forest can be found throughout the watershed. Anthropogenic land-cover in the UCRW includes annual rain-fed crops (e.g., corn, groundnuts, beans, and upland rice), irrigated rice, perennial crops (e.g., banana, sugarcane), orchards, pastures, bare soil, and residential areas. Maps presenting climate, topography, and ground cover of the UCRW are shown in Figure 2.

In this study, daily weather data including precipitation, temperature, potential evapotranspiration (PET), and relative humidity was collected from the Hydro-Meteorological Data Center of Vietnam. In addition, five points where weather data was available from the National Centers for Environmental Prediction (NCEP) Climate Forecast System Reanalysis (CFSR) were used for simulation by the SWAT model. A digital elevation model (DEM) with 3 arc-second resolution covering the entire Ca River Watershed was obtained from the NASA Shuttle Radar Topographic Mission. Land use data and soil data were collected from multiple sources; and were then reclassified in accordance with the SWAT model input requirement.

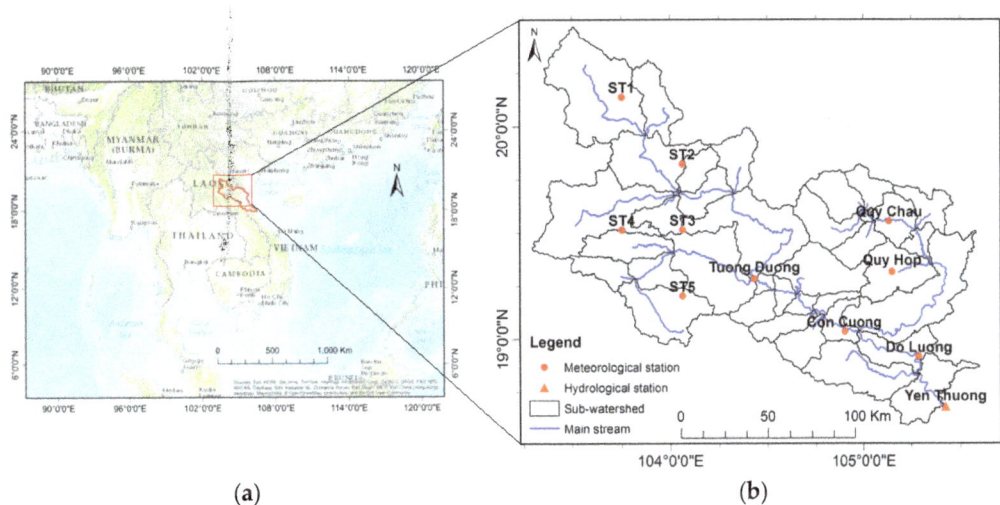

(a)   (b)

Figure 1. Geographic location of the entire Ca River Watershed in (**a**) Southeast Asiaand; (**b**) the Upper Ca River Watershed (UCRW).

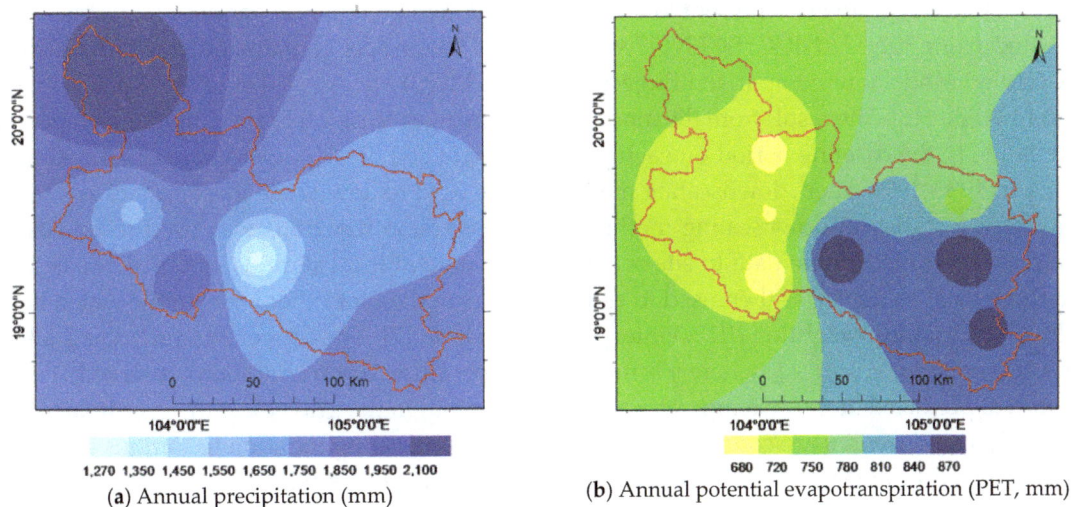

(**a**) Annual precipitation (mm)   (**b**) Annual potential evapotranspiration (PET, mm)

Figure 2. *Cont.*

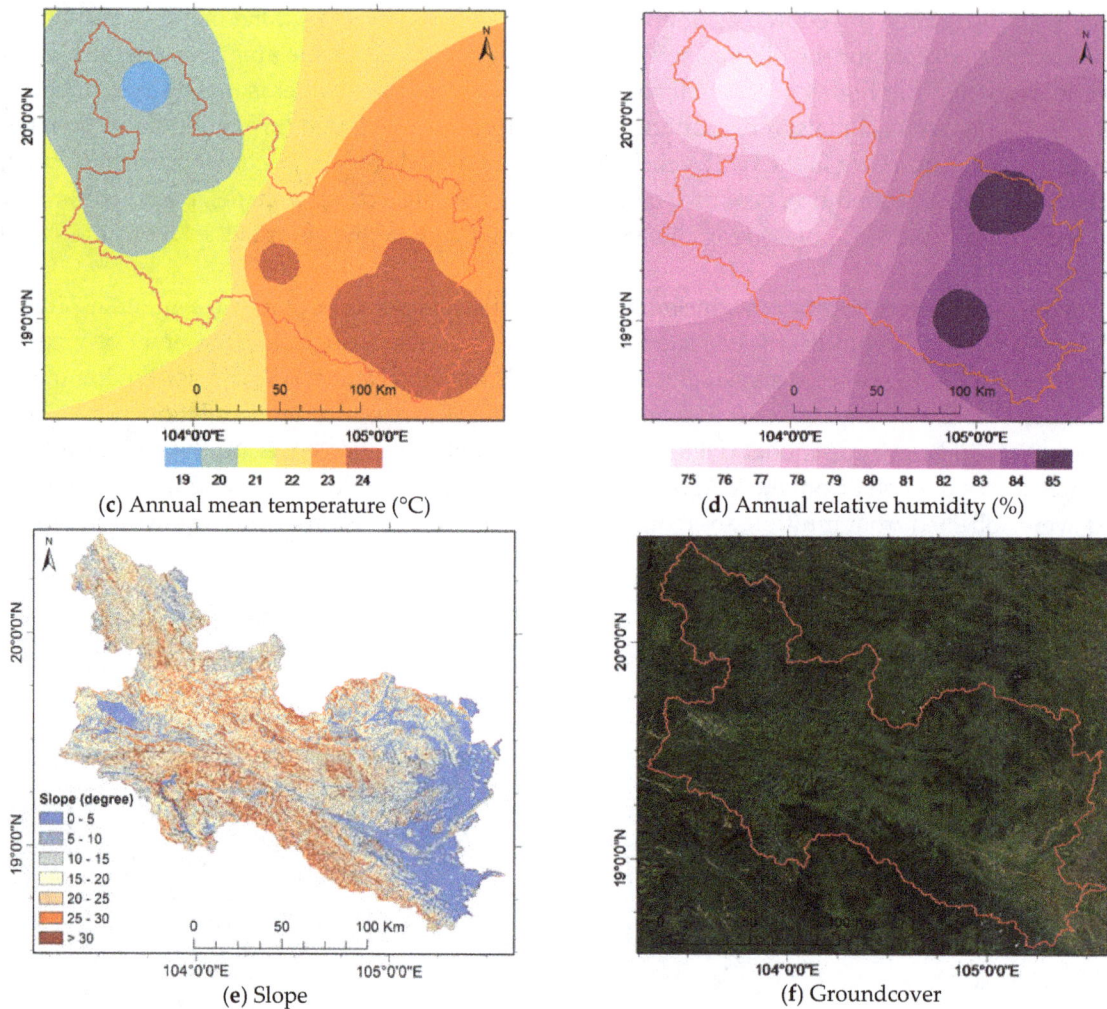

(c) Annual mean temperature (°C)

(d) Annual relative humidity (%)

(e) Slope

(f) Groundcover

Figure 2. Spatial pattern of climate data, slope, and groundcover of the UCRW.

2.2. Prediction of Future Climate Change

In the present study, future climate projections under three emission scenarios B1, B2, and A2 from the Fourth Assessment Report (AR4) of the Intergovernmental Panel on Climate Change (IPCC)—which respectively represent low, medium, and high greenhouse gas emission levels—were generated using the MAGICC/SCENGEN model [6]. The model contains output of 20 GCMs of the Coupled Model Intercomparison Project Phase 3 (CMIP3) archive. A list of the 20 GCMs can be found in Wigley [6]. The result from the model display changes in monthly precipitation, temperature, etc. relative to the baseline period on each prediction grid cell. The baseline period defined in this study was 1980–1999. Then, downscaling methods were used to downscale these data to at-site daily data. First, a statistical downscaling method with conversion functions was used to transfer the large-scale monthly climate data to site-scale monthly data at local stations, and downscaled data from these local stations was then used to estimate average data for other areas in the watershed by the Inverse Distance Weighting (IDW) interpolation method [7]. The conversion function is a linear regression equation $y = ax + b$; where y is monthly temperature (or precipitation) observed at a local station, x is predicted monthly temperature (or precipitation) in the grid cell at the coordinates of the local station, and a and b are constants. The statistical downscaling for climate stations in Vietnam can be referred to in IMHEN [8]. Spatial interpolation is a process which can estimate or predict values at unsampled points based on values of sampled points. It is widely used to predict unknown values for various climatic data such as precipitation, temperature, humidity, and evaporation, etc.

IDW is an interpolation method that estimates cell values by averaging the values of measured points surrounding each processing cell. The assumption of IDW is that the variable being mapped decreases in influence with distance from its sampled locations. In other words, the closer a point is to the center of the cell being estimated, the more weight, or influence [9–11]. The general equation of IDW is as follows:

$$\hat{Z}(s_0) = \sum_{i=1}^{N} \lambda_i Z(s_i) \tag{1}$$

where:

$\hat{Z}(s_0)$ is the value of temperature (or precipitation) being predicted for location s_0;

N is the number of measured sample points around the prediction location that will be employed in the prediction;

λ_i is the weight assigned to measured point i. The weight will decrease when distances increase.

$Z(s_i)$ is the the observed value at the location s_i.

The weights are determined by the following equation:

$$\lambda_i = d_{i0}^{-p} / \sum_{i=1}^{N} d_{i0}^{-p} \tag{2}$$

where:

$$\sum_{i=1}^{N} \lambda_i = 1$$

As the distance becomes larger, the weight is reduced by a factor of p.

The quantity d_{i0} are the distance between the prediction location (s_0) and each of the measured locations (s_i).

The power parameter p influences the weighting of the measured location's value on the value of prediction location; that is, as the distance between the measured sample locations and the prediction location increases, the influence (or weight) that the measured point will have on the prediction will exponentially decrease. The weights for the measured sample locations that will be employed in the prediction are scaled, their sum is therefore equal to 1 [11,12]. In this study, to consider the effect of topography on temperature/precipitation, observed baseline temperature/precipitation was used as the "Weight Field", and optimal p values were identified based on root mean square error (RMSE) of the prediction. The optimal p value is the value that produces the lowest RMSE.

Next, the downscaled monthly data at the local stations were downscaled again to daily data using the MODAWEC weather generator model [13]. The daily data for precipitation and temperature at the local stations were then employed for the simulation by the SWAT model. More detailed methodology for the generation of future climate change scenarios in this study can be referred to Pham et al. [14]

2.3. Simulation of Soil Erosion

In the present study, a soil erosion simulation was performed using the SWAT model [15]. SWAT was developed for the computation of runoff and sediment and agricultural chemical yields in large complex watersheds with varying soils and land use management conditions over long time periods [15–17]. In this model, a watershed is divided into multiple sub-watersheds that are then further subdivided into unique soil/land-use characteristics called hydrologic response units (HRUs).

SWAT calculates soil erosion within each HRU using the Modified Universal Soil Loss Equation (MUSLE) [18], which is shown in Equation (3).

$$Sed = 11.8 \times \left(Q_{surf} \times q_{peak} \times area_{hru} \right)^{0.56} \times K \times C \times P \times LS \times CFRG \tag{3}$$

where:

Sed is the sediment yield in a given day (metric tons);

Q_{surf} is the surface runoff volume (mm/ha);

q_{peak} is the peak surface runoff rate (m^3/s);

$area_{hru}$ is the area of the HRU (ha);

K is the Universal Soil Loss Equation (USLE) soil erodibility factor, which is available from the Soil Survey Geographic (SSURGO) data;

C is the USLE cover and management factor and can be derived from land cover data;

P is the USLE support practice factor, which is a field specific value;

$CFRG$ is the coarse fragment factor.

LS is the topographic factor. It is a function of the land slope length (L_{hill}), the angle of slope (α_{hill}), and the exponential term m in the equation below:

$$LS = (L_{hill}/22.1)^m \times \left(65.41 \times sin^2(\alpha_{hill}) + 4.56 \times sin\,\alpha_{hill} + 0.065\right) \tag{4}$$

Parameters including K, C, P, LS, and $CFRG$ are to be adjusted in the calibration process. The use of a physically-based hydrology model such as SWAT when coupled with the MUSLE may reduce uncertainty in soil erosion prediction when compared to the original USLE-based calculations of long-term sediment yields that are highly sensitive to topographic factors [19].

The peak runoff rate (q_{peak}) is the maximum runoff flow rate occurring with a given rainfall event. It is an indicator of the erosive power of the rainfall event, and in SWAT, it is calculated as follows:

$$q_{peak} = \frac{\alpha_{tc} \times Q_{surf} \times Area}{3.6 \times t_{conc}} \tag{5}$$

where:

α_{tc} is the fraction of daily rainfall that occurs during the time of concentration (time of concentration is the amount of time from the beginning of a rainfall event until the entire sub-basin area is contributing to flow at the sub-basin outlet) [17];

Q_{surf} is the surface runoff (mm H_2O);

$Area$ is the area of sub-basin (km^2);

3.6 is a unit conversion factor;

t_{conc} is the time of concentration for the sub-basin (h).

2.4. SWAT Model Calibration and Validation

Model calibration and validation are necessary and critical steps in any model application. Calibration is an iterative procedure of parameter adjustment and refinement, as a result of comparing simulated and observed values of interest. Validation is a process to ensure that the calibrated model properly assesses all the variables and conditions which can affect model results and demonstrate the ability of the calibrated model to predict field observations for periods separate from the calibration effort [20,21]. SWAT input parameters must be held within a realistic range of uncertainty. The first step in the calibration in SWAT is to determine the most sensitive parameters for the investigated watershed. The method of sensitivity analysis in the SWAT model combines Latin Hypercube (LH) [22] and one-factor-at-a-time (OAT) sampling [23]. During sensitivity analysis, SWAT runs ($p + 1$) $\times m$ times, where p is the number of parameters being evaluated and m is the number of LH loops. For each loop, a set of parameter values is selected such that a unique area of the parameter space is sampled. That set of parameter values is used to run a baseline simulation for that unique area. Then, using OAT, a parameter is randomly selected, and its value is changed from the previous simulation by a user-defined percentage. SWAT is run on the new parameter set, and then a different parameter is randomly selected and varied. After all the parameters have been varied, the LH algorithm locates a new sampling area by changing all the parameters [24,25].

Two types of analysis are available in the sensitivity analysis tool in SWAT. The first type of analysis uses only simulated data to identify the impact of adjusting a parameter value on some specific model outputs, such as average streamflow. The second type of analysis uses observed data to provide an overall "goodness of fit" estimation between the simulated and the observed time series. The first type of analysis may help to identify parameters that improve a particular process or characteristic of the model, while the second type of analysis identifies the parameters that are affected by the characteristics of the study watershed [24,25]. In this study, the selection of parameters for stream discharge and sediment calibration was carried out based on the sensitivity analysis, following the guidelines reported by Veith and Ghebremichael [24] and Van Liew and Veith [25].

SWAT model calibration and validation were performed using observed data of sediment yield recorded at the watershed outlet during 40 years (1971–2010), in which 25 years (1971–1995) were used for calibration and 15 years (1996–2010) were used for validation. To evaluate the model predictions for both time periods, four statistical indicators, including the coefficient of determination (R^2), Nash–Sutcliffe simulation efficiency (*Nash*), and Percent Bias (*PBIAS*), and root mean square error—observation standard deviation ratio (*RSR*) were used. R^2 was used as an indicator to measure the strength of the linear relationship between simulated data and observed data, meanwhile the *Nash* indicator was used to indicate how well the plot of observed versus simulated data fits the 1:1 line—or in other words, how close the simulated data was to the observed data. *PBIAS* was used to determine if the average tendency of the simulated data to be larger or smaller than their observed counterparts, which points out whether the model tends to underestimate or overestimate the simulated variable. In addition to R^2, *Nash*, and *PBIAS*, *RSR* was used to incorporate the benefits of error index statistics (i.e., root mean square error) and a scaling/normalization factor (i.e., standard deviation—STDEV), so that the resulting statistic and reported values can be applied to various constituents. A comprehensive description of these indicators can be found in Pham et al. [14]. These indicators are calculated as in Equations (6)–(9), and their values were then compared with the guideline values introduced by Moriasi et al. [26], which recommended that for monthly time-step simulations, sediment prediction can be judged as "satisfactory" if $Nash > 0.50$, $RSR \leq 0.70$, and $PBIAS \leq \pm55\%$, as "good" if $0.65 < NSE \leq 0.75$, $0.50 < RSR \leq 0.60$, and $\pm15\% \leq PBIAS \leq \pm30\%$, and as "very good" if $0.75 < NSE \leq 1.00$, $PBIAS < \pm15\%$, and $0.00 \leq RSR \leq 0.50$.

$$R^2 = \left[\frac{\sum_{i=1}^{n}\left(X_i^o - X^{omean}\right)\left(X_i^s - X^{smean}\right)}{\sqrt{\sum_{i=1}^{n}\left(X_i^o - X^{omean}\right)^2}\sqrt{\sum_{i=1}^{n}\left(X_i^s - X^{smean}\right)^2}} \right]^2 \tag{6}$$

$$Nash = 1 - \frac{\sum_{i=1}^{n}\left(X_i^o - X_i^s\right)^2}{\sum_{i=1}^{n}\left(X_i^o - X^{omean}\right)^2} \tag{7}$$

$$PBIAS = \frac{\sum_{i=1}^{n}\left(X_i^o - X_i^s\right)}{\sum_{i=1}^{n} X_i^o} \times 100\% \tag{8}$$

$$RSR = \frac{RMSE}{STDEV_o} = \frac{\sqrt{\sum_{i=1}^{n}\left(X_i^o - X_i^s\right)^2}}{\sqrt{\sum_{i=1}^{n}\left(X_i^o - X^{omean}\right)^2}} \tag{9}$$

where:

X_i^o and X_i^s are the observed and simulated values of the variable X, respectively;

X^{omean} and X^{smean} are the mean of the observed values and the mean of the simulated values of the variable X, respectively;

n is the total number of observations.

In this study, X represents sediment yield.

A graphical comparison of time series observed versus simulated sediment yield in the present study can be referred to Pham et al. [27]. The results of evaluation statistics calculation show that there was a high agreement between simulated data and observed data in both calibration and validation periods, as all evaluation statistics computed for both periods were strong. Specifically, the R^2 was 0.88

and 0.87 and *Nash* was 0.89 and 0.87 for calibration and validation periods, respectively. Meanwhile, the values of *RSR* were 0.34 and 0.35 and the values of *PBIAS* were 0.9% and 4.14%. Positive values of *PBIAS* indicator found in this study (0.90% for calibration and 4.14% for validation) indicate that there was an underestimation bias for both time periods, although the bias was insignificant. Comparing with the guideline of Moriasi et al. [26] described previously, the performance of the SWAT model in this study can be assessed as "very good". This confirms that the model is applicable to the present study.

3. Results and Discussion

3.1. Projected Warming

Figure 3 presents projected monthly average temperature for the basin average. Temperature is predicted to increase in all months, with warming rates ranging from 0.6 °C to 1.2 °C by the 2030s, from 1 °C to 2.5 °C by the 2060s, and from 1.4 °C to 3.9 °C by the 2090s, depending on the emission scenario and month. The highest warming rates are expected for February and April, with 3.6 °C and 3.9 °C by the 2090s according to scenario A2. Other months with a high warming rate are June and September, which are expected to warm by as much as 3.5 °C by the 2090s compared to the baseline average. In contrast, the smallest warming rates are predicted for March and August, with 2.6 °C and 2.3 °C, respectively, by the 2090s according to the high emission scenario (A2). From a seasonal point of view, temperature is likely to rise faster in the dry season (winter and spring) than in the wet season (summer and autumn). A faster warming rate in winter and spring may cause destruction of the seasonal cycle, gradually leading summer to come earlier than normal and winter and spring to become shorter. This issue has been discussed in several recent studies [28,29].

Considering the temperature prediction behaviors of the three selected scenarios B1, B2, and A2, it should be noted that the difference in behavior of the three scenarios changes with time. The three scenarios give similar predictions until the near future period (2030s), and then the predictions begin to diverge: temperature increases fastest under A2, followed by B2, and then B1. This is in line with the characteristics of the emission scenarios, which evolve similarly until the middle of the 21st century, when A2 becomes more negative due to the continuous increase in greenhouse gas emissions as a result of the increase in population growth. In contrast, B1 becomes less negative due to the slowing of population growth, and therefore a reduction in greenhouse gas emissions [1,30]. Similar findings have also been reported in several recent studies using IPCC TAR or IPCC AR4 models on a regional scale [31–33].

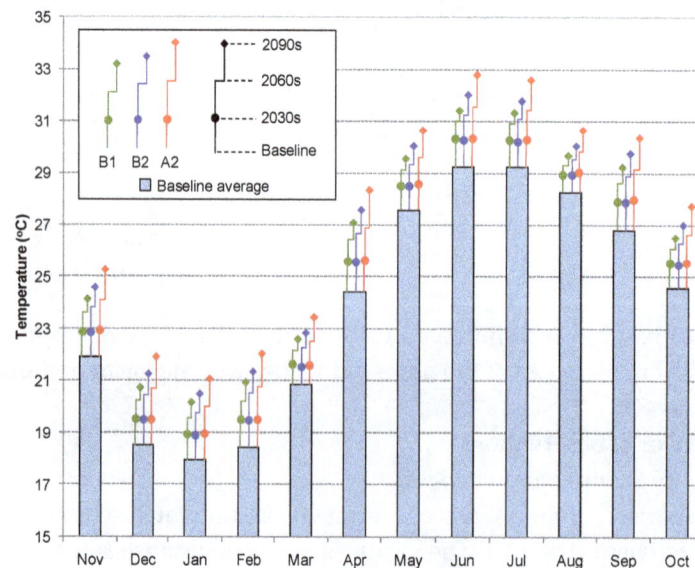

Figure 3. Projected monthly temperature for the future periods under three emission scenarios.

3.2. Projected Change in Precipitation

The results of climate projection show that precipitation in the UCRW is likely to change, varying by months. Figure 4 shows the projected precipitation in future periods. It can be easily seen that precipitation is predicted to increase for February, and for all months from June to December, and to decrease for January, March, April, and May. Significant decreases are likely to occur in January and April, while significant increases are likely to occur from July to October. Comparing the three scenarios, the order scenarios with respect to the magnitude of precipitation change until the 2030s is A2 > B1 > B2. From the 2030s onwards, this order shifts to A2 > B2 > B1, with the difference among the scenarios growing larger throughout the century. Maps showing spatial patterns of changes in mean annual precipitation for the 2090s are presented in Figure 5.

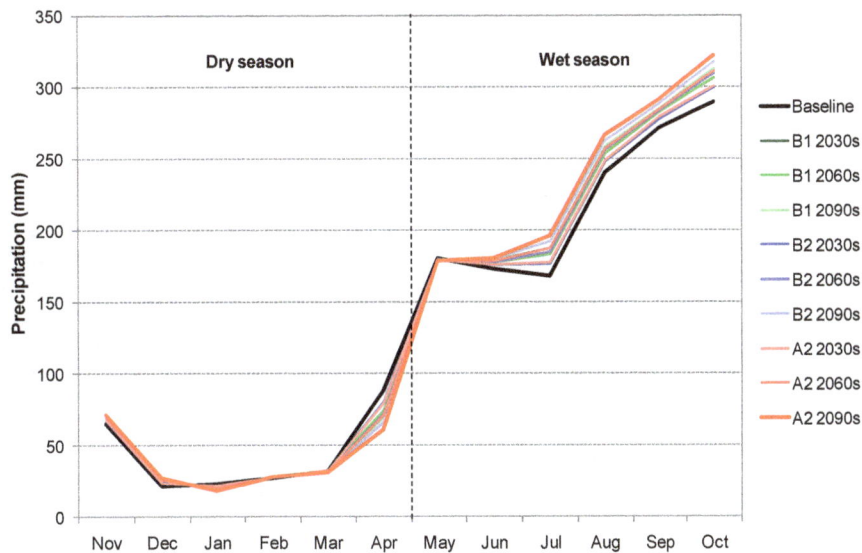

Figure 4. Projected mean monthly precipitation. The bold black line presents data for the baseline period, whereas the bold redline presents data with the largest change compared to the baseline period.

3.3. Spatial Pattern of Changes in Erosion Rate

Figure 6 presents spatial patterns of erosion for the UCRW during the baseline period. The erosion rate ranges from 2.7 tons/ha/year to 51.6 tons/ha/year with a mean value of 21.3 tons/ha/year. In general, the upstream areas were found to have more severe erosion than the downstream areas. Among the 39 sub-watersheds in the UCRW, upstream sub-watersheds 4, 7, 8, and 9 were identified as having the highest erosion rates (more than 40 tons/ha/year), while downstream sub-watersheds 20, 36, and 37 were found to have the lowest erosion rates (less than 5 tons/ha/year). One possible reason for this is that upstream sub-watersheds have steep slopes and receive a large amount of rainfall (Figure 2a), and are subjected to shifting cultivation, while downstream sub-watersheds are mostly flat (Figure 2e). Interestingly, although the UCRW receives a large amount of rainfall in both upstream and downstream areas and less rainfall in its middle part, erosion was found to occur severely in the middle stream sub-watersheds (for example, sub-watersheds 24 and 25). This indicates that not only surface runoff plays an important role in the movement of soil particles, but the erosion process is substantially affected by other factors. We refer to Equation (3) to examine factors causing a high erosion rate for the identified areas. Equation (3) calculates erosion rate as a result of combined effects of rainfall (through surface runoff), soil erodibility factor (K), cover and management factor (C), support practice factor (P), topographic factor (LS), and coarse fragment factor ($CFRG$). Since soil in the watershed is mostly Ferralsols (deeply weathered red or yellow soil, dominated by low activity clays including mainly kaolinite and sesquioxides), soil-related factors (K and $CFRG$) could almost have the same values for the entire watershed. The P factor is related to soil conservation practice; it reflects the effects of practices that will reduce the amount and rate of the water runoff and thus reduce the amount of erosion. In this study, we assumed that no supporting practice was applied in the watershed. The C factor is related to land surface cover. C varies from zero to one, with high values for barren land and crop land (e.g., row crops, rice), and low values for forest land and pasture. Although most of the crop land (C was set at 0.3) is located downstream, the area of crop land is small. In addition, the upstream area—and especially middle stream area—are subjected to shifting cultivation and deforestation, resulting in large areas of barren land, which is highly sensitive to erosion. The topographic factor LS is a function of the land slope length and the angle of slope Thus, the larger the slope length, and the larger the slope angle are, the larger the topographic factor will be. In other words, the longer and steeper the slope of surface, the higher the risk for erosion. Therefore, although the middle stream area receives less rainfall than other parts (Figure 2a), due to the rugged terrain with long and steep hill slopes (Figure 2e) soil erosion was found to occur severely in this area. In short, the physical meaning of the two key factors (i.e., C and especially LS) discussed above can explain the spatial patterns of soil erosion in the UCRW.

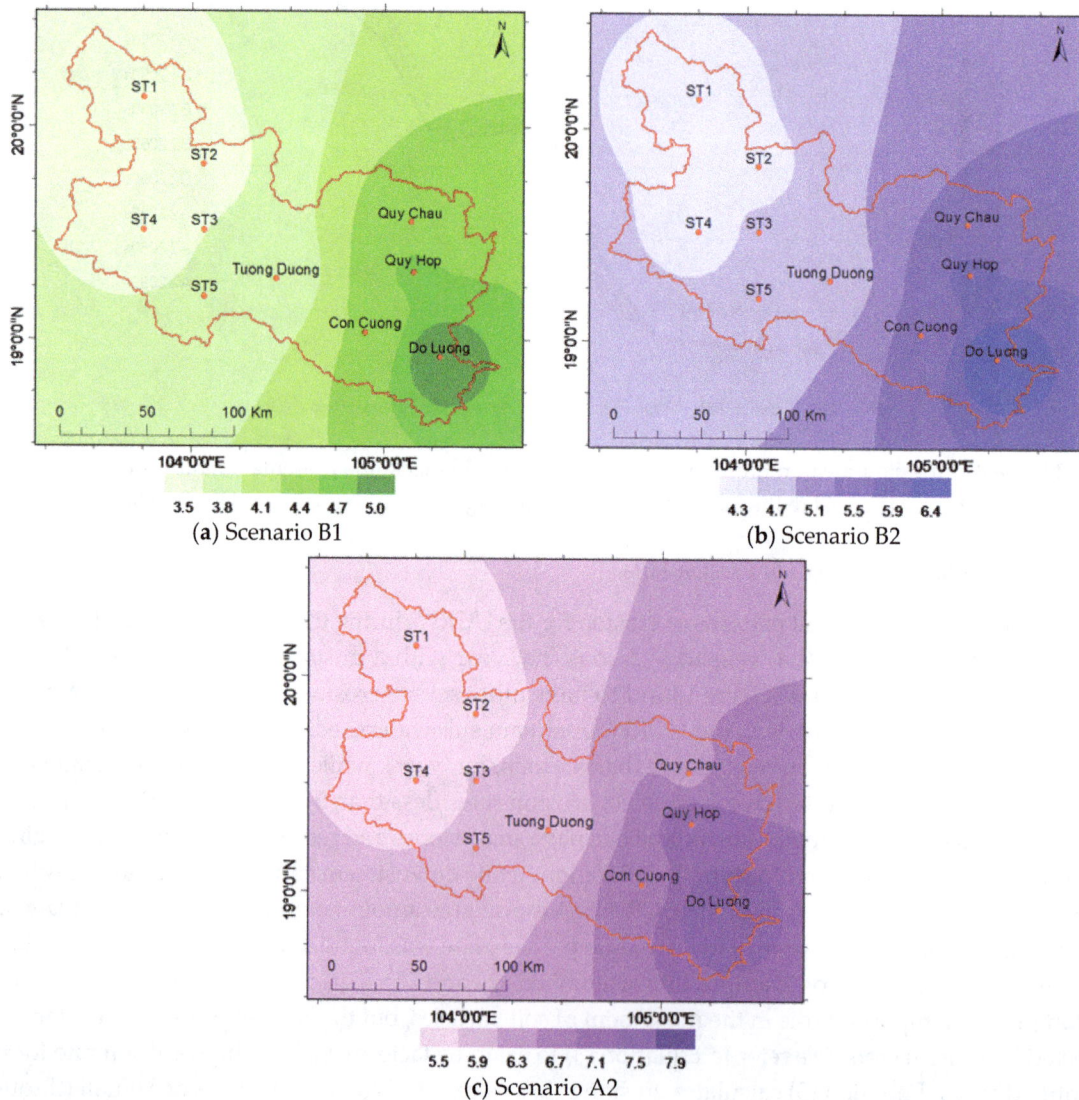

Figure 5. Maps present increases in average annual precipitation in the 2090s compared to the baseline period, units in percent.

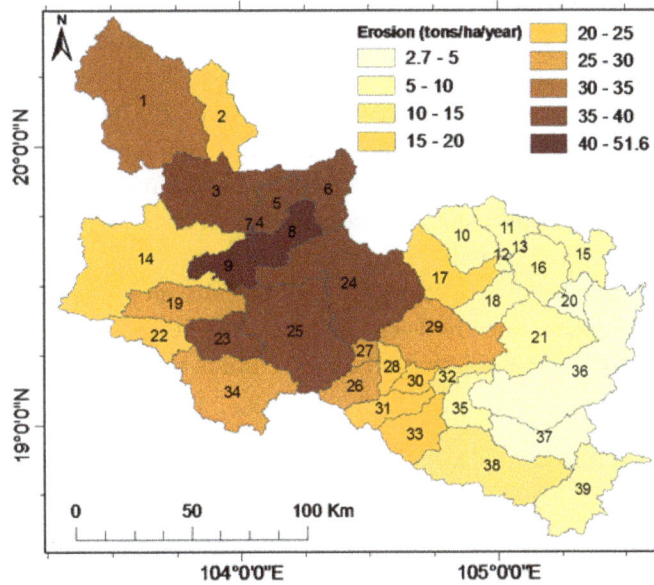

Figure 6. Average annual erosion rate in each sub-watershed in the baseline period.

Figure 7 shows the variation in erosion rates by sub-watersheds in the future time stages under the three emission scenarios. Although erosion is predicted to change differently depending on the sub-watershed, time stage, and emission scenario, an increasing trend throughout the 21st century can be seen in all three scenarios. Erosion is likely to increase the most with scenario A2, followed by scenario B2, and the least with scenario B1. For the near future time stage (2030s) the predictions of erosion under the three emission scenarios are very similar. This is consistent with the characteristic of the increase in temperature (Figure 3) and precipitation (Figure 4). On a basin-wide average, annual erosion rate is projected to be approximately 22.5 tons/ha/year by the 2030s, 23.5–24.0 tons/ha/year by the 2060s, and 24.1–25.6 tons/ha/year by the 2090s.

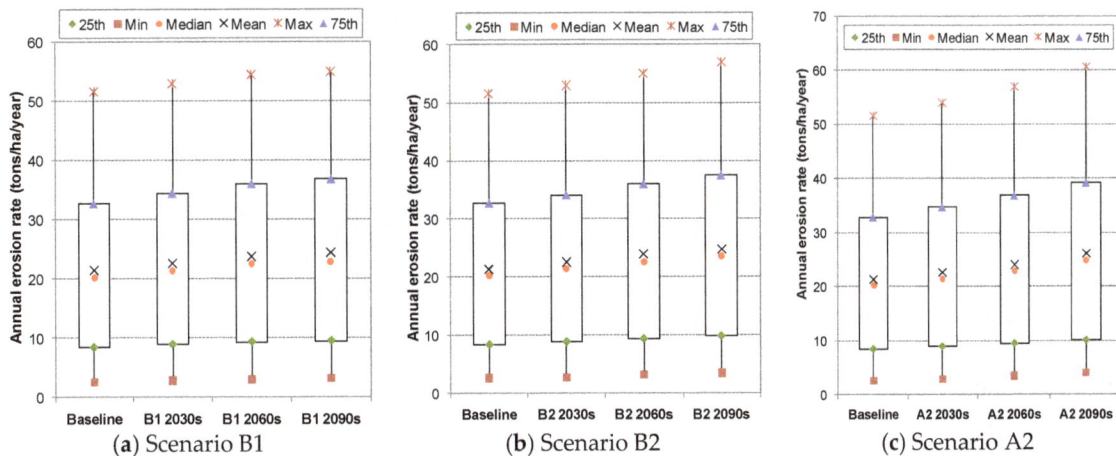

Figure 7. Box and whisker plots showing variation of annual erosion rates in sub-watersheds of the UCRW in the future time stages.

A map showing the spatial pattern of increase in erosion rate in the UCRW by the 2090s according to the high emission scenario A2 is shown in Figure 8. All sub-watersheds are projected to experience an increase in erosion rates ranging from approximately 13.3% to approximately 32.6%. It is noticeable that middle stream and downstream areas are projected to be more sensitive to climate change than upstream areas in terms of soil erosion. This may be due to the projected precipitation and temperature, both of which are projected to increase more in the middle stream and downstream areas than in the upstream areas. In general, a larger increase in precipitation will produce a greater increase in surface runoff, which is a direct factor causing more erosion (Equation 3). Increases in temperature may to a certain extent facilitate plant growth, but temperature in the UCRW is already high, and an increase in temperature in the area may cause stress to plants, consequently leading to a reduction in plant biomass [1]. On the other hand, an increase in temperature and a resulting increase in moisture will likely cause a faster residue decomposition process due to an increase in microbial activity [1,4,34]. These would lead to a reduction in biological ground cover, making the area more exposed to the projected increasing precipitation, and ultimately lead to more erosion. Large increases in erosion rate in some middle stream sub-watersheds can also be explained by the steep topography of these sub-watersheds. The effect of increased rainfall due to climate change on erosion in steep areas has been reported previously [19].

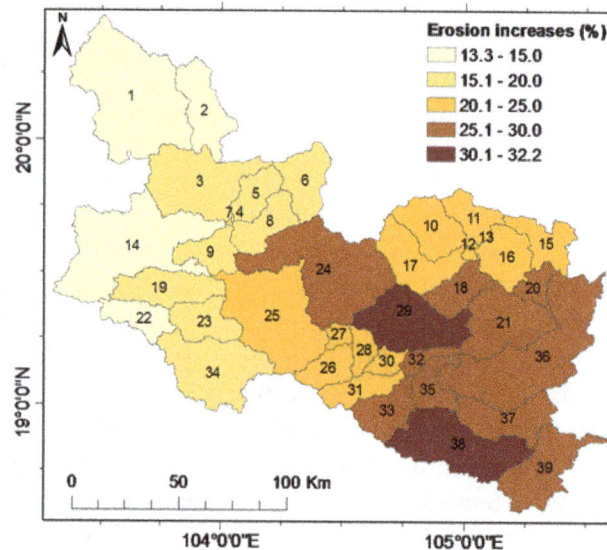

Figure 8. Increases in soil erosion in the 2090s compared to the baseline period according to the high emission scenario A2.

3.4. Temporal Pattern of Changes in Erosion Rate

The temporal pattern of climate change impacts on erosion rate in the UCRW was investigated. The column chart in Figure 9 presents average monthly erosion rate for the baseline period, while the line chart presents projected change in erosion rate in the future time stages relative to the baseline period. Average intra-annual (monthly) erosion rate of the baseline period varies from 0.15 tons/ha (February) to 5.3 tons/ha (September), with an annual total of 21.3 tons/ha. In general, erosion rates in the wet season months are much higher than in the dry season months. The seasonal erosion rate for the dry season is 2.2 tons/ha, and for the wet season is 19.1 tons/ha. Predicted monthly change in the erosion rate ranges from a 39.5% decrease to a 33.4% increase depending on the specific month, emission scenario, and time period. From a seasonal point of view, the erosion rate is predicted to increase from mid wet season to early dry season (July to December), and to decrease from mid dry season to early wet season (January to June). Increases in the erosion rate in the months from July to December can be explained by the projected increases in precipitation in these months. When precipitation increases, its erosive power and sediment transport capacity increase, resulting in an increase of erosion. However, it should be noted that the magnitudes and patterns of the change in erosion rate are not consistent with that of the change in precipitation. This is because besides precipitation (and as a result, surface runoff), erosion rate depends on other factors such as evaporation, topography, soil, land cover, and plant growth, etc. For example, a decrease in the erosion rate during the months from January to June may be related to changes in antecedent soil water content during precipitation events under future climate conditions. Even when an increase in precipitation was predicted, an increase in evaporation (due to the warmer climate) appeared to cause a reduction in soil water content which may lead to an increase in saturation deficit. This means that more precipitation is needed to bring the soil to saturation and generate the same amount of runoff as under the current climate conditions. The role of antecedent soil water content on soil erosion in saturation excess dominated areas has been reported [19,34]. In addition, increased temperature and increased precipitation in the dry-cold months (spring) may also accelerate plant growth, which results in increased biological cover, and less erosion as a result. However, it should be noted that the decrease in erosion rate in February and June is very small, and the pattern of the decrease is complicated. While in the other months, the magnitude of changes (increase or decrease) becomes larger with time, the magnitude of decrease in these two months decreases gradually toward zero. This means that if a larger increase in precipitation was predicted, the change in erosion rate in these two months would be positive (turn to an increase, following the increasing trend of rainfall).

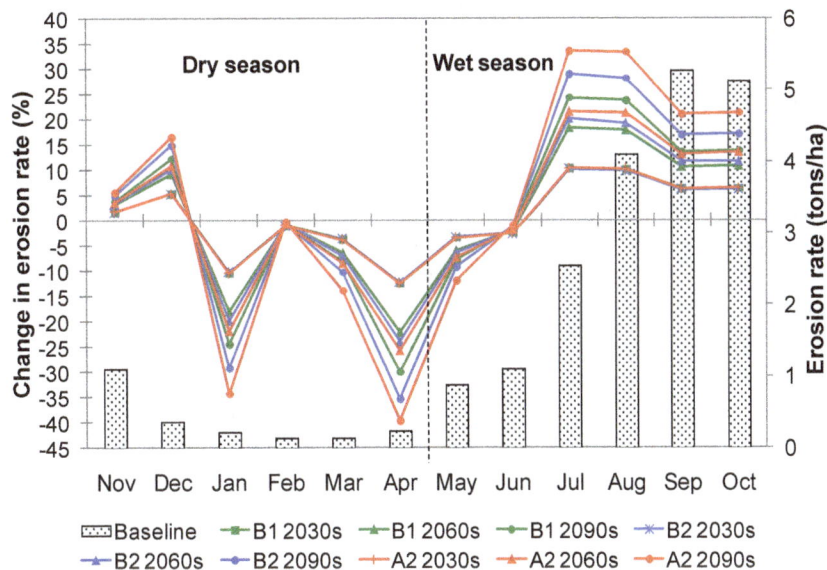

Figure 9. Projected changes in soil erosion relative to the baseline period.

4. Conclusions

This study investigated spatial and temporal responses of soil erosion to climate change impacts in the Upper Ca River Watershed, which is shared upstream by Laos and downstream by Vietnam. Soil erosion in the study watershed was mostly found as "hill-slope erosion", which occurred seriously in the upstream and middle stream areas, where topography is dominated by numerous steep hills with sparse vegetation cover. However, under the impact of climate change, it is likely that soil erosion rate in the downstream area will increase at a higher rate than in the upstream and middle stream areas. From a seasonal point of view, soil erosion is predicted to increase significantly in the warmer and wetter climate of the wet season, when higher sediment transport capacity is accompanied by the higher erosive power of an increased rainfall amount and intensity. These results provide useful information for decision makers and local people to plan where and when soil conservation practice should be focused. To protect the soil from erosion, countermeasures should focus on the peak months of the wet season, such as August, September and October, when erosion occurs the most seriously and is likely to increase with a high rate. For the steep upstream and middle stream areas where groundcover is mostly dominated by degraded or exploited forests and shifting cultivation land, countermeasures aiming to increase groundcover and to prevent hill-slope erosion such as reforestation, afforestation, terrace farming, and construction of check dams and retaining walls should be planned and implemented by the local governments. For the downstream area where the topography is less steep and agricultural fields are located, measures by local farmers such as crop rotation, cover cropping, and mulching should be encouraged.

Acknowledgments: The authors are grateful to the agencies of the Government of Vietnam for providing the data necessary for this study. We would like to thank the anonymous reviewers of the manuscript for their thorough review and helpful comments and suggestions.

Author Contributions: The design of study, data collection and analysis, model simulation, and manuscript writing were done by Pham Quy Giang. All authors contributed to the discussion of the study results and editing the manuscript.

Conflicts of Interest: The authors declare no conflict of interest.

References

1.   Parry, M.L.; Canziani, O.F.; Palutikof, J.P.; van der Linden, P.J.; Hanson, C.E. *Climate Change 2007: Working Group II: Impacts, Adaptation and Vulnerability*; Cambridge University Press: Cambridge, UK, 2007.

2.   Nearing, M.A. Potential changes in rainfall erosivity in the United States with climate change during the 21st century. *J. Soil Water Conserv.* **2001**, *56*, 229–232.

3.   Zhang, G.H.; Nearing, M.A.; Liu, B.Y. Potential effects of climate change on rainfall erosivity in the Yellow River basin of China. *Trans. ASAE* **2005**, *48*, 511–517. [CrossRef]

4.   Pruski, F.F.; Nearing, M.A. Climate-induced changes in erosion during the 21st century for eight U.S. locations. *Water Resour. Res.* **2002**, *38*, 1298. [CrossRef]

5.   Zhang, X.C.; Nearing, M.A. Impact of climate change on soil erosion, runoff, and wheat productivity in Central Oklahoma. *Catena* **2005**, *61*, 185–195. [CrossRef]

6.   Wigley, T.M.L. *MAGICC/SCENGEN 5.3: User Manual Version 2*; National Center for Atmospheric Research: Boulder, CO, USA, 2008; p. 81.

7.   Shepard, D. A two-dimensional interpolation function for irregularly-spaced data. In Proceedings of the 1968 ACM National Conference, Las Vegas, NV, USA, 27–29 August 1968.

8.   Vietnam Institute of Meteorology, Hydrology, and Climate Change (IMHEN). *Impacts of Climate Change on Water Resources and Adaptation Measure: Final Report*; IMHEN: Hanoi, Vietnam, 2010; p. 120.

9.   Burrough, P.A. *Principles of Geographical Information Systems for Land Resources Assessment*; Oxford University Press: Oxford, UK, 1986; p. 194.

10. Watson, D.F. *Contouring: A Guide to the Analysis and Display of Spatial Data*; Pergamon Press: Oxford, UK, 1992.

11. Environmental Systems Research Institute (ESRI). *ArcGIS 9, Using Arc Geostatistical Analyst*; Environmental Systems Research Institute Inc.: Redlands, CA, USA, 2003; p. 300.

12. Pham, Q.G. Effectiveness of different spatial interpolators in estimating heavy metal contamination in shallow groundwater: A case study of arsenic contamination in Hanoi, Vietnam. *Environ. Nat. Resour. J.* **2011**, *9*, 31–37.

13. Liu, J.; Williams, J.R.; Wang, X.; Yang, H. Using MODAWEC to generate daily weather data for the EPIC model. *Environ. Model. Softw.* **2009**, *24*, 655–664. [CrossRef]

14. Pham, Q.G.; Toshiki, K.; Sakata, M.; Kunikane, S.; Tran, Q.V. Modelling Climate Change Impacts on the Seasonality of Water Resources in the Upper Ca River Watershed in Southeast Asia. *Sci. World J.* **2014**. [CrossRef]

15. Arnold, J.G.; Srinivasan, R.; Muttiah, R.S. Large area hydrologic modeling and assessment part I: Model development. *J. Am. Water Resour. Assoc.* **1998**, *34*, 73–89. [CrossRef]

16. Williams, J.; Arnold, J.G. *A System of Hydrologic Models. U.S.Geological Survey. Water Resources Investigations Report*; U.S. Geological Survey: Reston, VA, USA, 1993; pp. 93–4018.

17. Neitsch, S.L.; Arnold, J.G.; Kiniry, J.R.; Williams, J.R. *Soil and Water Assessment Tool Theoretical Documentation*; Version 2009; Blackland Research Center: Temple, TX, Texas, USA, 2011; p. 618.

18. Williams, J.R. Sediment-yield prediction with universal equation using runoff energy factor. Present and Prospective Technology for Predicting Sediment Yield and Sources. In Proceedings of the Sediment Yield Workshop, USDA Sedimentation Lab., Oxford, MS, USA, 28–30 November 1972; pp. 244–252.

19. Mukudan, R.; Soni, M.P.; Elliot, M.S.; Donald, C.P.; Aavudai, A.; Mark, S.Z.; Adão, H.M.; David, G.L.; Tammo, S.S. Suspended sediment source areas and future climate impact on soil erosion and sediment yield in a New York City water supply watershed, USA. *Geomorphology* **2013**, *183*, 110–119. [CrossRef]

20. Donigian, S. Watershed model calibration and validation: The HSPF experience. In Proceedings of the Water Environment Federation, St. George, UT, USA, 10–12 October 2001.

21. Arnold, J.G.; Moriasi, D.N.; Gassman, P.W.; Abbaspour, K.C.; White, M.J.; Srinivasan, R.; Santhi, C.; Harmel, R.D.; van Griensven, A.; van Liew, M.W.; et al. SWAT: Model use, calibration and validation. *Trans. ASABE* **2012**, *55*, 1491–1508. [CrossRef]

22. McKay, M.D.; Beckman, R.J.; Conover, W.J. A Comparison of Three Methods for Selecting Values of Input Variables in the Analysis of Output from a Computer Code. *Technometrics* **1979**, *21*, 239–245. [CrossRef]

23. Van Griensven, A. Sensitivity, Auto-Calibration, Uncertainty and Model Evaluation in SWAT 2005. Available online: http://biomath.ugent.be/~ann/swat_manuals/SWAT2005_manual_sens_cal_unc.pdf (accessed on 6 March 2017).

24. Veith, T.L.; Ghebremichael, L.T. How to: Applying and interpreting the SWAT Auto-calibration tools. In Proceedings of the Fifth International SWAT Conference Proceedings, Boulder, CO, USA, 5–7 August 2009.

25. Van Liew, M.W.; Veith, T.L. Guidelines for Using the Sensitivity Analysis and Auto-Calibration Tools for Multi-Gage or Multi-Step Calibration in SWAT. Available online: http://www.academia.edu/21601923/Guidelines_for_Using_the_Sensitivity_Analysis_and_Auto-calibration_Tools_for_Multi-gage_or_Multi-step_Calibration_in_SWAT (accessed on 6 March 2017).

26. Moriasi, D.N.; Arnold, J.G.; Van Liew, M.W.; Binger, R.L.; Harmel, R.D.; Veith, T. Model evaluation guidelines for systematic quantification of accuracy in watershed simulations. *Trans. ASABE* **2006**, *50*, 885–900. [CrossRef]

27. Giang, P.Q.; Toshiki, K.; Sakata, M.; Kunikane, S. Modelling the seasonal response of sediment yield to climate change in the Laos-Vietnam Transnational Upper Ca River Watershed. *EnvironmentAsia* **2014**, *7*, 152–162.

28. Stine, A.R.; Huybers, P.; Fung, I.Y. Changes in the temperature in the phase of the annual cycle of surface temperature. *Nature* **2009**, *457*, 435–440. [CrossRef] [PubMed]

29. Jacques, F.M.B.; Shi, G.; Li, H.; Wang, W. An early-middle Eocene Antarctic summer monsoon: Evidence of 'fossil climates'. *Gondwana Res.* **2014**, *25*, 1422–1428. [CrossRef]

30. Intergovernmental Panel on Climate Change (IPCC). *Special Report on Emissions Scenarios: A Special Report of Working Group III of the Intergovernmental Panel on Climate Change*; Cambridge University Press: Cambridge, UK, 2000; p. 570.

31. Ribalaygua, J.; Pino, M.R.; Pórtoles, J.; Roldán, E.; Gaitán, E.; Chinarro, D.; Torres, L. Climate change scenarios for temperature and precipitation in Aragón (Spain). *Sci. Total Environ.* **2013**, *463*, 1015–1030. [CrossRef] [PubMed]

32. Liu, L.; Hong, Y.; Hocker, J.E.; Shafer, M.A.; Carter, L.M.; Gourley, J.J.; Bednarczyk, C.N.; Yong, B.; Adhikari, P. Analyzing projected changes and trends of temperature and precipitation in the southern USA from 16 downscaled global climate models. *Theor. Appl. Climatol.* **2012**, *109*, 345–360. [CrossRef]

33. Jiang, X.; Yang, Z.L. Projected changes of temperature and precipitation in Texas from downscaled global climate models. *Clim. Res.* **2012**, *53*, 229–244. [CrossRef]

34. Williams, J.R.; Nearing, M.A.; Nicks, A.; Skidmore, E.; Valentine, C.; King, K.; Savabi, R. Using soil erosion models for global change studies. *J. Soil Water Conserv.* **1996**, *51*, 381–385.

7

The Uncertain Role of Biogenic VOC for Boundary-Layer Ozone Concentration: Example Investigation of Emissions from Two Forest Types with a Box Model

Boris Bonn [1,*], Jürgen Kreuzwieser [1], Felicitas Sander [2], Rasoul Yousefpour [2], Tommaso Baggio [3] and Oladeinde Adewale [4]

[1] Chair of Tree Physiology, Albert Ludwig University, Georges-Koehler-Allee 053, D-79110 Freiburg i.Br., Germany; juergen.kreuzwieser@ctp.uni-freiburg.de

[2] Chair of Forestry Economics and Forest Planning, Albert Ludwig University, Tennenbacher Str. 4, D-79106 Freiburg i. Br., Germany; Felcitas.Sander@posteo.de (F.S.); rasoul.yousefpour@ife.uni-freiburg.de (R.Y.)

[3] Department of Land, Environment, Agriculture and Forestry, University of Padova, Agripolis, Viale dell'Università 16, I-35020 Legnaro (PD), Italy; tbaggio93@gmail.com

[4] UMR LERFoB, AgroParisTech, INRA, 54000 Nancy, France; alabifavour365@gmail.com

* Correspondence: Boris.Bonn@ctp.uni-freiburg.de

Abstract: High levels of air pollution including ground level ozone significantly reduce humans' life expectancy and cause forest damage and decreased tree growth. The French Vosges and the German Black Forest are regions well-known for having the highest tropospheric ozone concentrations at remote forested sites in Central Europe. This box model study investigates the sensitivity of atmospheric chemistry calculations of derived ozone on differently resolved forest tree composition and volatile organic compound emissions. Representative conditions were chosen for the Upper Rhine area including the Alsatian Vosges/France and the Black Forest/Germany during summer. This study aims to answer the following question: What level of input detail for Alsace and Black Forest tree mixtures is required to accurately simulate ozone formation? While the French forest in Alsace—e.g., in the Vosges—emits isoprene to a substantially higher extent than the forest at the German site, total monoterpene emissions at the two sites are rather similar. However, the individual monoterpene structures, and therefore their reactivity, differs. This causes a higher ozone production rate for Vosges forest mixture conditions than for Black Forest tree mixtures at identical NO_x levels, with the difference increasing with temperature. The difference in ozone formation is analyzed in detail and the short-comings of reduced descriptions are discussed. The outcome serves as a to-do-list to allow accurate future ozone predictions influenced by the climate adaptation of forests and the change in forest species composition.

Keywords: tropospheric ozone pollution; influence of tree mixture; forest; BVOC emissions

1. Introduction

Tropospheric air pollution is known to affect human health [1] and plant growth substantially [2–4]. Studies on the impact of ground level ozone on plants have demonstrated serious damage to the surface and cell structure of species, causing plant injury and death and, consequently, reduced economic bargain [5–7]. Because of several stresses herbivorous plants have adopted several defense mechanisms, which enable them to tolerate different stresses at different intensities, for example different temperatures or ozone levels [8]. One of these mechanisms seems to work by using the reactions of semi-volatile organic compounds with ozone [9,10]. It has been demonstrated for example that ozone forces the release of highly reactive sesquiterpenes [10], which has the effect of reducing ozone mixing ratios near the plant's surface [10]. As well as having a detoxification effect, biogenic volatile organic compounds (BVOCs) can act as a tool for plant communication, e.g., allowing other neighbouring plants to get prepared for a threat, for instance against herbivory attack or ozone [11].

Tropospheric ozone—in this context boundary layer or ground level ozone—is formed in a photolytic production cycle that includes nitrogen oxides ($NO_x = NO + NO_2$), volatile organic compounds (VOCs), and radiation. Based on the photosynthesis of ozone, hydroxyl radicals (OH) are formed that partially react with VOCs, yielding organic peroxy radicals (RO_2). These RO_2 radicals partially react with NO, forming NO_2 and an alkoxy radical producing hydrogen peroxide (HO_2), which regains OH by reacting with NO, yielding a second NO_2 molecule. This cycle can occur 40–60 times until being terminated. Each NO_2 molecule can be photolysed resulting in ozone production. Depending on the conditions, this tropospheric ozone production cycle can be limited by the availability of VOCs—i.e., typical urban conditions with high NO_x emissions—or by NO_x—i.e., typical background conditions that are found in forests with a substantial amount of biogenic VOCs (BVOCs). The area of production addresses a horizontal range of several hundred metres to 30 km depending on wind speed and radiation, and is most intense at points of transition from VOC- to NO_x-limited regions.

In this area, sufficient NO_x is still present and the BVOC level increases due to emissions, for example by the forest. The list of compounds that contribute to BVOCs is extensive, with isoprene (C_5H_8) and monoterpenes ($C_{10}H_{15}$) being the highest contributors by amount released and reactivity. The group of sesquiterpenes ($C_{15}H_{24}$) is a much more reactive BVOC group, for which even smaller ambient concentrations are sufficient to cause similar effects as those caused by isoprene or the monoterpenes. While isoprene is a single compound with approximately 1900 oxidation reactions and 600 products [12] until it is oxidized to carbon dioxide (CO_2), the group of monoterpenes abbreviates a set of more than 900 differently structured species of an identical molar mass, but with different amounts of carbon-carbon double bonds and chemical reactivity ranging by four orders of magnitude for ozone and by two orders of magnitude for the hydroxyl radical (OH). Including a single monoterpene in a detailed chemical scheme may add 1550 further reactions and 520 species [13], thus substantially increasing the number of calculations. OH is involved in the tropospheric ozone formation cycle [14,15] and altering its concentration, as well as the amount of total VOCs, potentially causes a change in tropospheric ozone. Therefore, although they protect the emitting plant from surface ozone, the released BVOCs are expected to affect the tropospheric ozone production cycle [16,17] and its strength for the areas downwind the site of emission [17]. As a consequence, both ozone precursor groups—i.e., NO_x and VOCs—need to be considered in order to understand the effects and feedback processes for different climate conditions. This is being addressed by recent studies of trends in ozone in North America and Europe [18–20]. They indicate an opposite trend in ground level ozone concentrations in cities and remote locations. While ozone tends to increase in urban areas due to a reduction in NO_x—i.e., the major sink of ozone in these areas—ozone was found to decrease moderately in rural areas, where ozone formation is NO_x limited. The role of BVOCs is more complex because of the large number of compounds with different reactivities and chemical degradation [21].

Despite these important features, atmospheric chemical reaction schemes currently implemented in different model types and scales—i.e., box, regional, global or climate models—are minimized because of simulation capacities and limitations. Moreover, not all the chemical degradation reactions from oxidation initiation until increasing CO_2 levels are known or agreed upon. Detailed chemical schemes, such as the Master Chemical Mechanism (MCM) version 3.3.1 of Leeds University [12,22], include isoprene, aromatic [23] and non-aromatic compounds [24], three exemplary monoterpenes (d-limonene, α- and β-pinene) and a single sesquiterpene (β-caryophyllene) [25] with about 12,500 individual reactions in total including further organic and inorganic reactions. This is drastically reduced to 105–237 reactions and 46–80 species (50 stable and 30 intermediate species) for chemical algorithms such as CB05, MM5, RACM or RADM2 [26,27], which are commonly used not only by regional, global and climate models but also for air quality and pollution forecasts [28–30]. A reliable match between simulation and observation is therefore key for predicting health issues and ecosystem damage.

In this study, we focus on ground level—i.e., boundary layer—ozone as an atmospheric pollution marker and the influence of forest tree mixture—i.e., tree emissions of BVOCs. The Southwestern part of Germany—i.e., Black Forest—is well known for its topmost hourly averaged ozone pollution levels during the summer, which have exceeded 200 µg/m^3—i.e., approx. 100 parts per billion by volume (ppb$_v$) (measurement height: ca. 4 m a.s.l.)—several times. It is interesting that, despite temperature, humidity and radiation being quite similar, summertime ozone concentrations nearby in the Alsatian Vosges (France)—about 50–60 km west of Black Forest—are on average 28% lower (Figure 1). Significant differences between the two locations are evident throughout the year. The high pollution level in the Black Forest is linked to nitrogen oxides advected from the major transport routes in the Upper Rhine valley—separating Alsace on the West and Baden-Württemberg on the East—and high concentrations of biogenic VOCs released by the forest ecosystem (primarily trees). Ozone favouring conditions are listed in Table 1, which provides the Pearson correlation coefficients of different parameters, with the amount of ozone observed during summer by a background reference station of the local Agency for Air Quality and Environment (LUBW) in the Southern part of the Black Forest ("Schwarzwald-Sued", Muenstertal, 47.8099° N, 7.7645° E—see Section 2.3 for more details). Nitrogen monoxide (NO) and dioxide (NO_2) were separated into two ranges—i.e., NO_x limited and VOC limited ozone production. For the condition of interest, the limitation of summertime tropospheric ozone production changes at NO_2 volume mixing ratios of about 3.5 ppb$_v$. High global radiation and temperatures cause elevated ozone as they enhance biogenic VOC emissions. Additionally, semi-volatile species—such as VOCs previously bound to organic aerosol particles—are released to the gas-phase and contribute to the amount of total VOCs that support ozone formation. Ozone production in the area of interest is nitrogen dioxide (NO_2) limited and the more that is available the more ozone will be formed. NO acts differently, as notable concentrations will decrease ozone by way of a chemical reaction—especially during the night—but will also enhance production of NO_2, which is used in ozone production during the day [15]. On the contrary, wind speed acts as a reductive, as less dilution of precursor gases occurs and so more time is available to locally produce ozone.

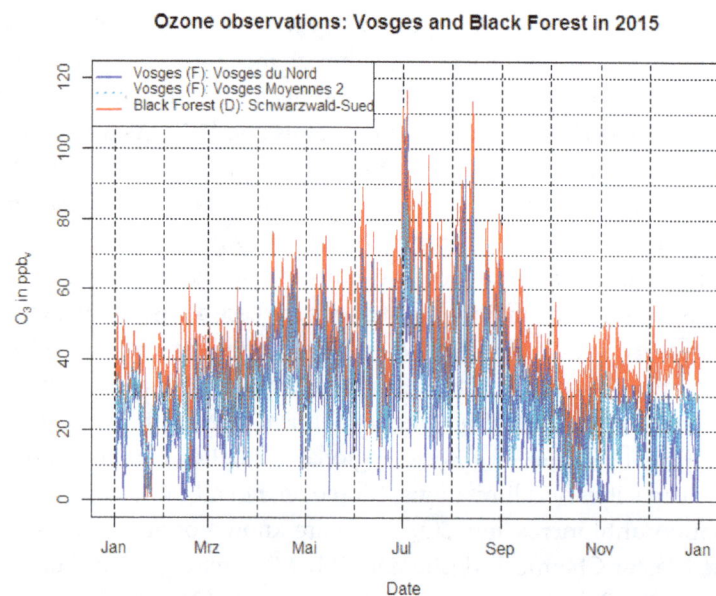

Figure 1. Hourly observations of background boundary layer atmospheric ozone volume mixing ratio values in the Alsatian Vosges ('Vosges du Nord' and 'Vosges Moyennes 2', courtesy of Pierre Robellet, l'Association pour la Surveillance et l'Etude de la Pollution Atmosphérique en Alsace (ALPA)) and the Black Forest ('Schwarzwald-Sued'), resolved by the local agencies for air quality.

Because the focus of this study is on the influence of vegetation, we use NO and NO_2 values from present observations but alter forest conditions. We consider two nearby differently composed ecosystems to investigate and analyse the resulting effects of ecosystem composition on BVOC emissions, resulting ozone mixing ratios, and formed secondary organic aerosols (SOA) in mass and number. Calculations are done using a box model, which considers forest properties and atmospheric chemistry: (a) at different temperatures; (b) at different ecosystem compositions (Alsace and the Black Forest); and (c) with and without considering SOA formation. For checking the sensitivity of the ozone calculations and predictions to the accurate description of individual processes, a comparison between observations (hourly resolved) and simulations during a period with low and high boundary layer ozone levels will be carried out in Section 3.1 (Figure 2). This will be used for demonstrating the interaction of local forests with air pollution and as the basis of a discussion about the need for further input to accurately predict ozone in the context of changing forest mixtures and climate change.

Table 1. Correlation coefficients r (Pearson), *p* values and number of data pairs N for different meteorological and atmospheric chemistry parameters with ground level ozone volume mixing ratios observed at Schwarzwald-Sued/Black Forest between June and August 2015.

Parameter	r	*p*	N
T	0.78	$<2.2 \times 10^{-16}$	2053
NO$_2$ *	0.19 *	9.8×10^{-13} *	1952 *
	0.01 **	0.66 **	2036
global radiation	0.11	2.1×10^{-8}	2053
NO	0.01 *	5.8×10^{-9}*	1952 *
	−0.05 **	0.03 **	2036 **
wind speed	−0.09	1.6×10^{-4}	1756
wind direction	−0.16	1.4×10^{-8}	1756

* denotes NO_2 mixing ratios below 3.5 ppb$_v$; ** all NO_2 measurements (0–10 ppb$_v$). Values within the detection limits of the instruments are excluded.

2. Materials and Methods

2.1. Box Model

In this study, the CAABA box model [31], extended by biogenic VOC emissions and SOA formation [32], was used to cover the full range of gas-phase chemical reactions and phase changes in detail. More information about those reactions and processes can be found elsewhere [12,13,22–25]. While the organic chemistry scheme is known for the initial reactions and products [33], some degradation reactions have been considered differently by different groups and schemes, resulting in a small level of uncertainty. Here we concentrate on the Master Chemical Mechanism by Leeds University in its latest version [12], which was obtained via the website http://mcm.leeds.ac.uk/MCM, for process studies. This includes time development of selected trace gas observations—such as NO, NO_2, CO, non-methane hydrocarbons—and meteorology—temperature T, relative humidity (rH), global radiation (GR) and mixing layer height (MLH)—as input. It considers about 6700 primary, secondary and radical chemical species and about 17,000 chemical reactions in the gas-phase, of which 2237 are traced by the output. However, only a small number of VOC species are provided as input as far as they are available from earlier studies. With respect to biogenic VOCs, the emission rates of isoprene, monoterpenes and sesquiterpenes are calculated (see Section 2.3) and mixed in each time-step into the entire box volume. The description of SOA processes—formation and destruction—considers 174 relevant VOC species based on volatility and substantial production, and has been described elsewhere in more detail [32]. For each time-step, the individual partitioning coefficients $K_{org,i}$ [34] have been estimated and the equilibrium partitioning between both phases was calculated, readjusting the amount of corresponding species between absorbed and volatile phases. In this context, the present ambient organic aerosol mass needed was derived by the total particulate mass observations with a cut-off diameter of 10 micrometres (PM_{10}) at Schauinsland [35]. The total PM_{10} was considered to consist of 40% organic material, which represents a typical monthly mean value for PM_{10} analysis for Schauinsland (in the Black Forest, Germany) [36]. The contribution of different forest species (see Section 2.2) was read in and the emissions of each species were treated individually and summed up before mixing into the box.

2.2. Forest Inventories

Forest inventories were considered for (a) the regions of Alsace [37] and (b) Baden-Wuerttemberg, based on the BWI inventory 2012 [38] (Table 2). To obtain the dry weight biomass per square metre, the stock volume of the tree species was multiplied by the density factor (basic specific gravity, 0.55 Mg/m^3 for broadleaves and 0.45 Mg/m^3 for conifers [39]). The values were then multiplied by the expansion factor (1.612 for broadleaves and 1.3 for conifers [40]) to obtain the above-ground biomass.

An overview of the species, the mean dry weight biomass per square metre, and the contribution of (a) Alsace including the Alsatian Vosges (France) and (b) the Black Forest (Baden-Wuerttemberg, Germany), is provided in Tables 2 and 3. While Norway spruce (*Picea abies*) represents the dominant tree type in the Black Forest, followed by silver fir (*Abies alba*) and European beech (Fagus sylvatica), the tree mixture is notably different in the Vosges. There, beech makes the highest contribution by area. Silver fir, Norway spruce, oak (*Quercus spp.*) and Scots pine (*Pinus sylvestris*) follow next, all with similar contributions between 6% and 22%. In summary, the French Alsatian forest consists 67%:33%, the French Alsatian Vosges of about 50% coniferous and 50% deciduous trees and the Black Forest type tree mixture has a ratio of about 75%:25%, i.e., a notably higher coniferous contribution.

Table 2. Contribution of different tree species to regional forests in Alsace (F, larger area), Vosges (F, smaller area) and the Black Forest (D), related to forest area based on the latest inventories [37,38].

Tree Species (lat.)	Tree Species (engl.)	Alsace [%]	Vosges [%]	Black Forest [%]
Alnus spp.	Alder	2.2 $^\Delta$	1.3 $^\Delta$	1.2 *
Fraxinus excelsior	European ash	4.4 $^\Delta$	2.5 $^\Delta$	0 *
Fagus sylvatica	European beech	23.5	22.7	15.3
Betula spp.	Birch	3.3 $^\Delta$	2.5 $^\Delta$	1.9 *
Robinia pseudoacacia	Black locust	2.2 $^\Delta$	1.3 $^\Delta$	0
Pseudotsuga menziesii	Douglas fir	-	-	5.1
Carpinus betulus	Hornbeam	4.4 $^\Delta$	1.9 $^\Delta$	1.2 *
Larix decidua	European larch	-	-	1
Tilia spp.	Lime	3.3 $^\Delta$	1.9 $^\Delta$	1 *
Acer spp	Maple	4.4 $^\Delta$	2.5 $^\Delta$	1.2 *
Quercus spp.	Oak	11.1	10.8	3.1
Pinus sylvestris	Scots pine	9.2	5.9	5
Populus spp.	Poplar	2.2 $^\Delta$	1.3 $^\Delta$	1.2 *
Abies alba	Silver fir	10.2	22.4	18.5
Picea abies	Norway spruce	9.2	15.8	42.8
	Other broadleaves	6.6 $^\Delta$	3.8 $^\Delta$	
	Other conifers	n.s.	n.s.	
	Total	96.2	96.4	90.8

* equally split due to a lumped group within the corresponding inventory; $^\Delta$ based on a lumped group within the corresponding inventory and split taking expert knowledge into account (lumping done differently for different countries). n.s. = statistically not significant within the corresponding inventory.

2.3. Emissions

The hourly emissions E_X per gram of dry weight biomass were considered for three different biogenic VOC species and classes, i.e., isoprene (C_5H_8), monoterpenes ($C_{10}H_{16}$) and sesquiterpenes

($C_{15}H_{24}$). Isoprene and monoterpenes affect tropospheric ozone production and destruction [41–43] because of their notable atmospheric mixing ratios (>100 ppt_v), while sesquiterpenes are less abundant (lowest ppt_v range) and only act reductive on surface ozone predominantly in the direct vicinity of emission sites. Isoprene and monoterpenes were therefore included in the tropospheric chemistry simulations, while sesquiterpenes were considered for new particle formation calculations only. Each individual BVOC emission rate E_X was described in terms of its dependency on leaf or needle temperature [44] (γ_T) and photosynthetic active radiation (γ_P) in accordance with references [45,46], as far as direct measurement data or literature were available.

$$E_X = E_{Xi} \times \gamma_P \times \gamma_T \times \gamma_{dr} \times \gamma_a \tag{1}$$

The potential interference of drought and ageing factors had to be ignored ($\gamma_{dr} = \gamma_a = 1$) because of insufficient information on individual species and these therefore require future investigation. The details of tree species' specific basal emission rates E_{Xi} at 30 °C, temperature dependencies and dry weight biomasses are summarized in Tables 3 and S1 (sesquiterpenes). Because of strong isoprene emitters such as oak among the Alsatian Vosges forest trees, the basal isoprene emission rate $E_{0,isop}$ at 30 °C and 1000 $\mu mol/m^2/s$ is about ten times more than that of the German Black Forest (18.3 to 1.9 $\mu g/g(dw)/h$, respectively). Thus, under identical meteorological conditions, isoprene emissions are remarkably higher in the Vosges (Alsace).

Monoterpene emissions were calculated as a mixture of de novo and pool emissions of the 16 different tree types considered. The sum of emitted monoterpenes was split according to their structure—i.e., endo- and exo-cyclic for each tree species according to the literature—and were treated as α- (endocyclic) and β-pinene (exocyclic) in the following.

At first glance, the monoterpene emissions of the Black Forest under standard conditions are less different to those of the Vosges than are the isoprene emissions. The forests in the Alsatian Vosges release only about 30% more than the Black Forest species under standard conditions. As a consequence, the French forest represents a substantially higher BVOC source than the nearby German forest, with resulting effects on tropospheric ozone production.

The estimate of total emission rates $\Sigma E_X(i)$ and the transfer to ambient concentrations of compound X [X] included several steps (Equation (1)): (i) summing the individual contributions by tree species weighted by relative contribution to forest area; (ii) multiplication with escaping fraction and (iii) distribution into the meteorological boundary layer height (MLH) (see Section 2.5 below).

$$[X] = [X]_0 + \Sigma(E_X(i) \times g(dw)_i) \times survival/MLH/M_{wx} \times \Delta t/3600 \text{ s} \times 10^{-12} \times N_A \tag{2}$$

$[X]_0$ is the resulting concentration of the final time-step before adding the emission; $g(dw)_i$ is the individual dry weight biomass of a tree species; M_{wx} is the molar weight of the BVOC (such as 68 g/mole for isoprene); Δt is the calculated model time-step; and N_A represents the Avogadro constant of 6.022×10^{23} molecules/mole. Survival comprises the amount of BVOC escaping the canopy into the open atmosphere that is accessible to photochemistry:

$$Survival = 1 - CH/(0.3 \times u \times /(k_{O3}{}^{BVOC} \times [O_3] + k_{OH}{}^{BVOC} \times [OH] + k_{NO3}{}^{BVOC} \times [NO_3]) + CH \tag{3}$$

As a first estimate, the crown height (CH) was set to 10 m for all the species included, because of the lack of further detailed information. The mean tree height was assumed to be 25 m.

Table 3. Assumed mean dry weight biomass (DWB), basal emission factors E_0 for isoprene and monoterpenes (MT) at 30 °C and temperature coefficients β_{MT} of the individual tree species considered ([46] and ref. therein). *DWB values are based on the Black Forest inventory (BWI 2012) [38].

Tree Species (lat.)	Tree Species (engl.)	DWB * [g/m^2]	$E_{0,isop}$ [µg/g(dw)/h]	$E_{0,MT}$ [µg/g(dw)/h]	β_{MT} [K^{-1}]
Alnus spp.	Alder	109 ± 96	0.018	0.13	0.09
Fraxinus excelsior	European ash	139 ± 116	0.012	0.012	0.09
Fagus sylvatica	European beech	180 ± 140	0	43.5	0.31
Betula spp.	Birch	105 ± 31	0	6.7	0.09
Robinia pseudoacacia	Black locust	85 ± 75	11.9	3.34	0.09
Pseudotsuga menziesii	Douglas fir	141 ± 45	0.008	0.064	0.08
Carpinus betulus	Hornbeam	132 ± 110	0.1	0.0093	0.09
Larix decidua	European larch	125 ± 91	0.4	13.1	0.07
Tilia spp.	Lime	110 ± 92	5.5	0	0.09
Acer spp	Maple	132 ± 110	3.9	0	0.09
Quercus spp.	Oak	179 ± 83	20.4	13.1	0.121
Pinus sylvestris	Scots pine	77 ± 39	0	7.48	0.09
Populus spp.	Poplar	85 ± 75	76.3	3.45	0.09
Abies alba	Silver fir	174 ± 54	0.038	28.8	0.135
Picea abies	Norway spruce	177 ± 31	0.05	0.886	0.11

2.4. Secondary Organic Aerosol Treatment

Based on the saturation vapour pressure and estimated ambient mixing ratios, the partitioning of 174 organic oxidation products of the most important anthropogenic and biogenic VOCs between gas- and particle-phases were treated explicitly. Saturation vapour pressures of the SOA species were calculated based on group contribution methods, splitting the corresponding molecules into functional groups for which a parameterization was derived from a series of similar structured compounds [47,48]. Corresponding saturation vapour pressures were used to obtain the individual partitioning coefficients as a function of temperature [34]. The essential amount of total organic particular mass M_{org} was derived from ambient PM_{10} measurements at Schauinsland (Black Forest, Germany, EMEP station DE0003R, 47.915 °N, 7.909 °E, 1205 m a.s.l.). From former studies, it can be assumed that the observed PM_{10} consist of approximately 40% organic compounds [35]. Subsequent iteration achieved the realistic partitioning equilibrium of individual species, depending on the individual saturation concentrations, the amount present, and the total mass of organic aerosols. Further details on the basics and the procedure can be found elsewhere [32,33].

2.5. Simulations

Simulations are focused on two areas close by—i.e., the Alsatian Vosges in Eastern France and the Black Forest in the Southwestern part of Germany. Both areas are covered by substantial amounts of forest, that are each different in their tree species composition. While the Upper Rhine valley is known as a major traffic and transportation route with substantial emissions of nitrogen oxide (NO_x), observations within the Black Forest display Germany's uppermost ground level ozone pollution levels. This is suspected to be caused by the transport of NO_x from the Upper Rhine valley towards the Black Forest that then mixes with BVOC emissions from the forest. Thus, the effect of forest tree species' composition on tropospheric ozone production and mixing ratio is expected to be both most sensitive and most likely.

In order to investigate this, exemplary box model simulations were performed for observed meteorological and atmospheric chemistry conditions for June, July and August 2015. The focus was set on the first two weeks in August, which covered the highest annual temperature (34.0 ± 0.1 °C) and GR (920 ± 1 W/m^2), with maximum surface ozone (113.6 ± 0.2 ppb$_v$) and notable PM_{10} (18 ± 1 µg/m^3) values, especially for remote areas. The observation site "Schwarzwald-Sued" of LUBW at Muenstertal (47.8099° N, 7.7645° E, 902 m a.s.l.) was picked as the reference station. It is located in the upper part of the Black Forest in Southwestern Germany at 902 m a.s.l., is classified as remote, and is surrounded by mixed forest. Surface ozone observations display the highest ozone level of this area and are assumed to be representative for the Black forest area. The following parameters were monitored at this site, provided in 30-minute time resolution, and were used for the simulation: Temperature, dew point, atmospheric pressure, GR, NO, and NO_2. Surface ozone measurements were used only for comparison and as a starting value at time 0. Several parameters were set to be constant due to no direct measurements and thus no information about a daily pattern. Those parameters and their corresponding values were derived from earlier studies [49–51] or they were assumed based on current mean observations for remote locations: methane (CH_4) = 1900 ppb_v and CO = 100 ppb_v.

The meteorological boundary (mixing) layer height MLH in metres was treated as:

$$MLH = \max(10., (60. + 40. \times \sin((\text{hour-days}) \times 24 - 10)/24 \times \pi \times 2) + 1000 \times$$
$$\exp(-0.035 \times ((\text{hour-days}) \times 24 - 17))^2) \times (-270 + 75.7 \times (T\text{-}273.15))/2400)) \times 2. \tag{4}$$

With a maximum of 1600 m at 2 p.m., which is in line with the heights observed by Kalthoff et al. [52], i.e., 1400 and 1900 m at 1:48 p.m. in June 1996. Photosynthetic active radiation (PAR), required for BVOC emission calculations, was scaled by global radiation measurements and assumed as:

$$PAR = 700 \ \mu mol \ m^{-2} \ s^{-1}/0.4 \times GR(\text{measured})/GR(\text{daily maximum in summer}). \tag{5}$$

3. Results

3.1. Match with Observations

In order to check the reproducibility of observed ground level ozone mixing ratios by the box model, both observed and simulated ozone values were compared for June, July and August—i.e., the time of highest annual ozone mixing ratios in the meteorological boundary layer, at the reference site for the Black forest in Muenstertal. The results of an exemplary week at the beginning of August are displayed in Figure 2. Different sensitivity runs were conducted to find the best deposition velocity and minimum NO_2 mixing ratios for the best agreement between observations and simulation based on the range of published values.

With respect to ozone deposition velocity (v_{dep}), the value is influenced by the structure of the section of forest that the air mass had crossed before arrival. Predominantly coniferous forests result in a very small deposition rate, while deciduous forests cause about twice to three times the value (Figure 2, left). A mixed-to-coniferous forest structure was found to match best and a v_{dep} value of 0.1 cm/s was chosen in the following. With respect to the minimum value of nitrogen dioxide (limiting factor), different simulations between one and five ppb were performed during the daytime (Figure 2, right). This concluded in an overall match between observed and simulated ground level ozone with a correlation coefficient of 0.81 for a daytime minimum NO_2 mixing ratio of 3.5 ppb_v and a night time minimum 0.5 ppb_v. Both different minimum values of NO_2 are crucial as NO_2 displays two different effects, i.e., a substantial sink for ozone during the night and an essential source for tropospheric ozone production during day (NO_x limited regime). The values are partially higher than observed, as the detected ozone is not necessarily formed on-site but during the air mass transport before arrival. This transport commonly originates in regions of elevated NO_x (transport routes in the Upper Rhine valley and Freiburg) and during the crossing if the NO_x in the Black Forest region gets vertically diluted and chemically converted during its travelling time. The described set-up was considered to be the best match and is treated as the reference for the following.

As Figure 2 indicates, the air chemistry box model results display generally a stronger daily variation than the observations. Differences between observation and simulations are apparent, especially during night. This is caused by changes in vertical mixing and changing slope currents for different times of the day, which is challenging to reproduce with a 0D-model. A highly resolved 3D model with sufficient boundary values and parameter distributions would be needed, which are currently lacking. Furthermore, different NO_x mixing ratios will prevail during both day and night—i.e., smaller values during the night, where advection from anthropogenically-affected areas like the Upper Rhine valley is negligible, and higher values during the day, when vertical mixing and transport of NO_x is notable. The effect is critical for ozone sink and production, as production in this remote area is evidently NO_x limited (see e.g., correlation coefficient r of surface ozone with nitrogen dioxide in Table 1) and a lack of NO_x will result in reduced ozone production. This is supported by a moderate anti-correlation of daytime ground level ozone with monoterpenes (r = −0.16) and isoprene (r = −0.09) for Black Forest simulations during the highest ozone periods in July and August 2015. As the box model is unable to simulate regional transport, the focus was on reproducing the daily and nightly maximum ground level ozone mixing ratios. NO_x was therefore overestimated. SOA contributions to tropospheric ozone from advection—i.e., the evaporation of previously formed organic aerosol mass—are smaller if considered compared to the neglected +(36 ± 25)%.

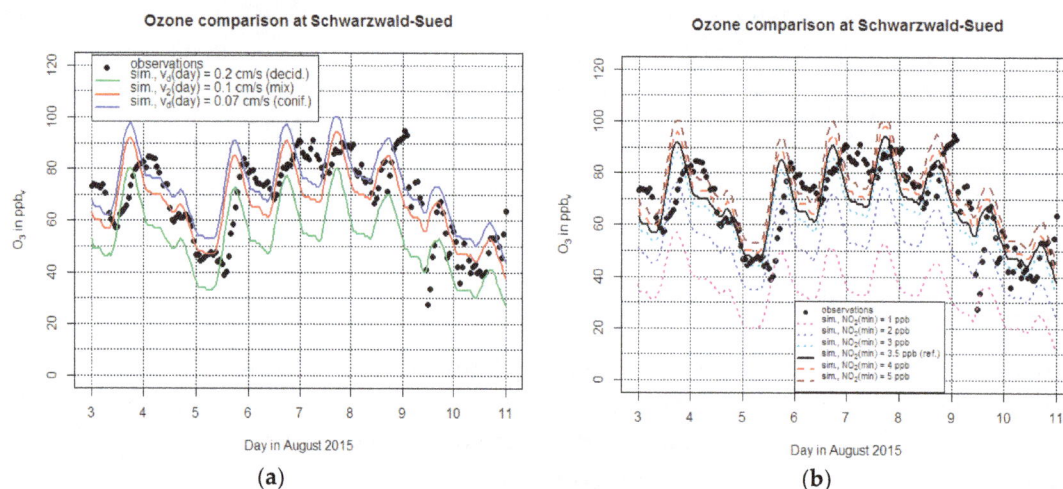

Figure 2. Comparison of observed and simulated ground level ozone volume mixing ratios at the remote reference site in the Black Forest. Dots display observed ozone values and lines display simulated ones. Different lines show the NO_2 effect on ozone, relying on air mass history, which is unknown for the box model. (**a**) Sensitivity runs for different ozone deposition velocities; (**b**) Sensitivity for different minimum NO_2 mixing ratios during the formation process.

3.2. Effect of BVOCs

While the link between tropospheric ozone formation and NO_x in a NO_x limited area is evident—i.e., the more NO_x the more ozone is formed—the situation with the VOCs depends on the amount present and the VOCs' reactivity with ozone, OH and NO_3. In mid-latitude forests, ambient VOCs are predominantly of biogenic origin [14]. The most reactive ones are isoprene and terpenes (mono- and sesquiterpenes), with atmospheric lifetimes between seconds and several hours. Isoprene (C_5H_8) and its reaction products contribute nearly exclusively to the gas-phase and gently force ozone production, but isoprene destroys ozone by direct reaction, resulting in a minor reduction (see right-hand column in Table 4). Monoterpenes act in a similar way, but react to a larger extent with ozone, bind hydroperoxy radicals (HO_2) into larger molecular structures. They slow down the ozone production cycle and form SOA mass yielding a stronger negative correlation.

Table 4. Correlation coefficients r of different atmospheric chemistry parameters with ozone simulated for both conditions, i.e., Vosges (Alsace) forest and Black Forest between June and August 2015.

Parameter		r(Vosges)	p(Vosges)	r(Black Forest)	p(Black F.)
HO_2	O_3	0.39	$<2.2 \times 10^{-16}$	0.39	$<2.2 \times 10^{-16}$
OH	O_3	0.35	$<2.2 \times 10^{-16}$	0.35	$<2.2 \times 10^{-16}$
Isoprene	O_3	-0.07	5.5×10^{-8}	-0.07	4.1×10^{-8}
Monoterpenes	O_3	-0.26	$<2.2 \times 10^{-16}$	-0.24	$<2.2 \times 10^{-16}$
SOA	O_3	-0.28	$<2.2 \times 10^{-16}$	-0.29	$<2.2 \times 10^{-16}$
Isoprene	SOA	-0.15	1.7×10^{-11}	-0.11	1.5×10^{-9}
Monoterpenes	SOA	-0.11	5.4×10^{-6}	-0.09	4.1×10^{-8}
T	SOA	-0.46	$<2.2 \times 10^{-16}$	-0.46	$<2.2 \times 10^{-16}$

3.3. Alsatian Vosges Forest vs. Black Forest: Effects of BVOC Mixtures

Next, we investigate the impact of different vegetation structures—i.e., forest tree species—on surface ozone, and SOA levels at otherwise identical conditions—i.e., meteorology and inorganic compound concentrations. In order to do so, two neighbouring forested areas with similar meteorological conditions but different species relevance—i.e., the Vosges and the Black Forest—are used (see Table 2, Section 2.2). Because of the different emission rates, the Alsace type forests emit significantly more isoprene, while monoterpenes are released to a similar extent by both forest mixtures (Figure 3). The exact ratio of monoterpenes provided by the Alsatian Vosges forest type to Black forest type mixtures depends on temperature and PAR.

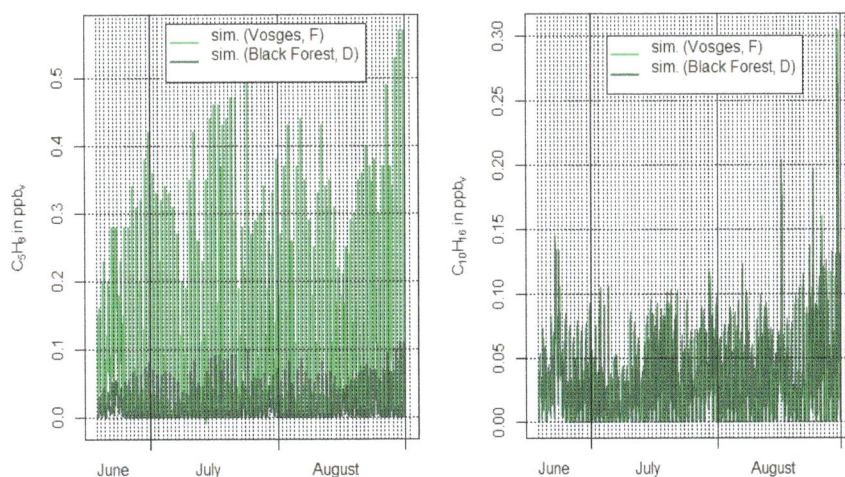

Figure 3. Estimated average summertime isoprene (**left**) and monoterpene (**right**) mixing ratios above the Vosges forest type and Black Forest type forest. Light green represents Vosges forest type conditions and dark green Black Forest type ones.

3.3.1. Effect of Isoprene and Monoterpenes

The different BVOC emission strengths and corresponding ambient mixing ratios result in ground level ozone pollution changes between -3 and $+35\%$ above the forests of elevated isoprene emissions (Alsatian Vosges mixture forest) or several ppb_v (Figure 4). The larger the isoprene emission increase, the higher the ozone increase. On the contrary, monoterpenes have two opposite effects: (i) the increase of non-methane hydrocarbons (NMHC) yielding elevated ozone production; and (ii) the formation of SOA—i.e., reduction of gaseous VOCs by absorption to the particulate phase. Especially lhe latter onecan display notable strength: -19 to $+2\%$ for Vosges mixture type and -9 to $+9\%$ for Black Forest type forests. The mean effect is negative for both types of forest and non-negligible.

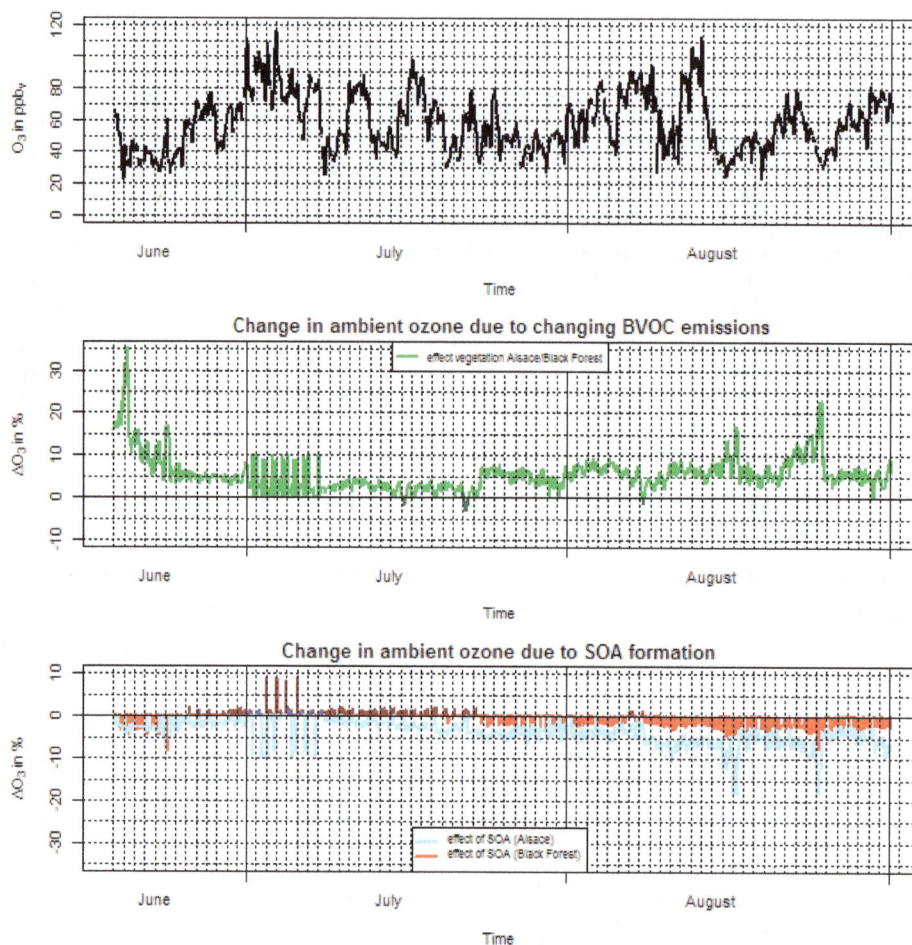

Figure 4. Summertime tropospheric (ground level) ozone mixing ratio: Observations in the Black Forest (**top**), effect of changing tree species contributions (**middle**) and of secondary organic aerosol (**bottom**).

3.3.2. Implications for Organic Aerosol Particles

The total implications for ambient particulate matter concentrations tends to be minor and can change direction. If external SOA masses are advected and repartitioned to the gas-phase to a certain extent, they will intensify ozone production, while BVOCs emitted at the site of interest will either contribute to ozone formation or partition to the particle phase and will be therefore unable to join ozone production. The first effect is positive—e.g., for the Black Forest, a median of +31% is derived—and for the second effect a negative impact was calculated (e.g., the Black Forest median August 2015: −14%) with different intensities throughout the day. Therefore, although the total SOA effect tends to be minor (Figure 5), opposing effects may be major and show up if the local situation is disturbed by local wind. With respect to the estimated total SOA mass concentration changes due to different emissions and oxidation between both forest types are about 0.5 μg/m³, which is less than 3% (Figure 5, on the left). The magnitude of the effect is related to different temperatures and the competition between higher emissions and organic aerosol mass precursor formation and the increased volatility of semi-volatile aerosol species.

The effect of different forest types becomes more evident for the estimated biogenic new particle formation rates [53] (Figure 6), which are relevant for total particle number concentration and indirect climate feedback processes such as cloud properties [14,15]. Despite a notable daily scattering in the ratio of the formation rate at D_p = 3 nm of the Black Forest type to Alsatian Vosges type conditions, the change of different weather systems and temperature is clearly visible. A daily mean enhancement of formation rates under the Black Forest conditions of between 109% ± 25% compared to the Alsatian

Vosges conditions can be seen. This can be explained by different emission characteristics that are influenced by daily mean temperature (84.4%, Pearson's correlation coefficient). In this context, a major aspect of uncertainty is the emission of individual sesquiterpenes and thus mean reactivity, which are affected by different tree species, tree injuries, sesquiterpene specification and the local variation of needle/leaf temperatures [44]. In our case, the mean reactivity of sesquiterpenes with ozone was assumed to be $k(sqt, O_3) = 4 \times 10^{-16}$ cm^3 molec^{-1} s^{-1} based on individual sesquiterpene emission contributions for different species [54–56] and the corresponding reaction rate coefficients [33]. Please note that our calculations provide an average response for a mean forest tree type mixture. Heterogeneity of vegetation and variability of atmospheric trace gases will cause a remarkable scattering, even in suburban areas not directly affected by traffic emissions of particles, as seen in Berlin [57].

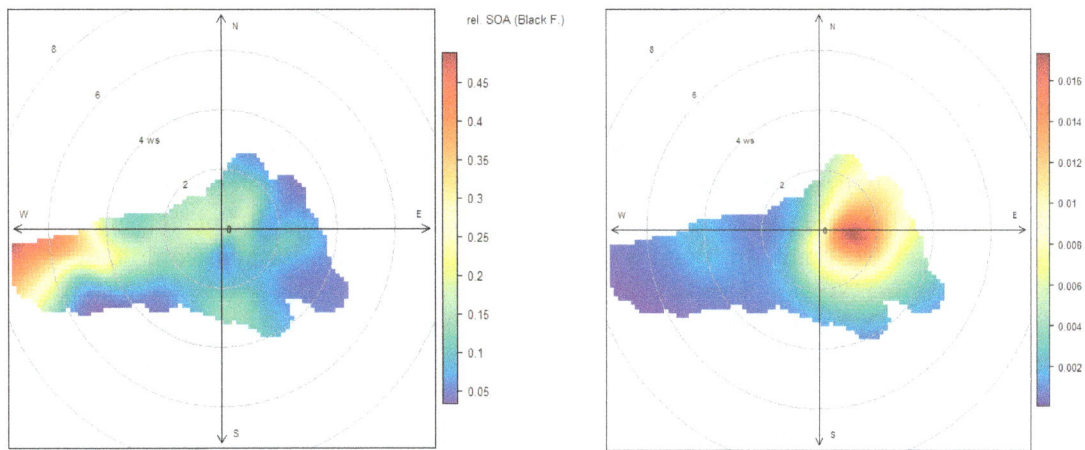

Figure 5. Changing forest ecosystem composition (Alsatian Vosges forest to Black Forest) and its consequences for ambient SOA mass concentrations. **Left**: intensities of calculated SOA as a function of wind speed and direction at Muenstertal; **Right**: absolute calculated SOA difference in μg/m^3 as a function of wind speed and direction at Muenstertal.

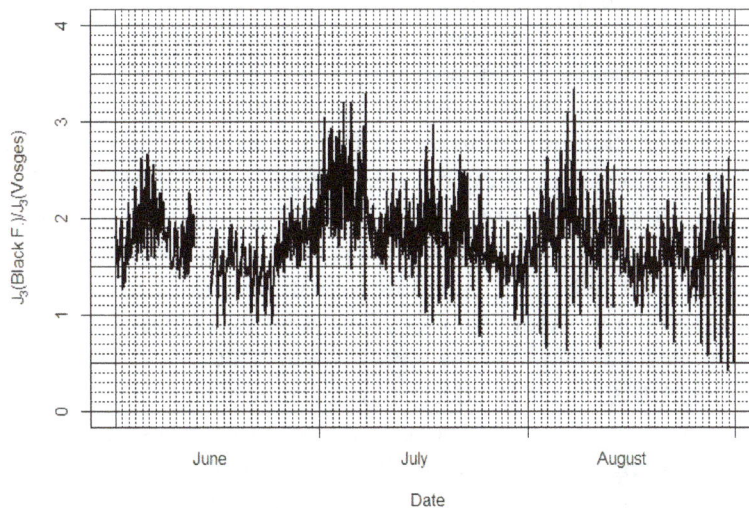

Figure 6. Forest effect on newly formed aerosol particles from natural sources [40] during the summer. Ratio of particle formation rates in Black Forest and the Vosges area.

3.3.3. Implications for OH, HO_2 and Nitrogen Species

The changing emissions of BVOCs influences the cleansing capacity of the atmosphere and the NO lifetime. In the Black Forest, OH concentration was calculated to be approximately 5% (25th percentile: 10%, 75th percentile: 2%) smaller than in the Alsatian Vosges type forest. An even stronger effect was found for HO_2: -17 (25th perc.: -25%, 75th perc.: -12%). While isoprene intensifies the tropospheric ozone production cycle [58], monoterpenes tend to bind the hydroperoxy radical as organic hydroperoxide and to slow down the reformation of OH. Therefore, the lifetime of nitrogen monoxide is increased by exactly 5% and moderately increases its reaction rate with ozone yielding NO_2. In the presence of sunlight, ozone is formed back (null cycle). In the presence of organic peroxy radicals, organic nitrates are formed to a higher extent, which have longer residence times either in the gas– or particle-phases [59] and reduce the cycling intensity between NO and NO_2 with respect to ozone production.

3.4. Effects of Temperature

Several important aspects require further investigation: (i) temperature; (ii) soil water content; and (iii) ecosystem interactions, of which only the first one is accessible for the box model approach used here based on publications. While a general parameterization for soil water content (SWC) that is impact independent of tree species is available [45]—i.e., including reduction of emissions down to 40% SWC at the wilting point, this point is expected to be species-specific and would influence the results depending on SWC significantly. Consideration of these feedback processes within ecosystems requires a more detailed knowledge of drought tolerance and interactions of the individual species significantly contributing to the forest. This is currently lacking and was therefore not included in this study.

In changing climate conditions, PAR will not change as GR will stay put. But infrared radiation will enhance and will be trapped within the Earth's climate system [60]. The mean temperature will increase and so will enzyme activity, until its upper limits has been reached and until the storage pools of terpenes, such as resin ducts, have been emptied. Because of that, we kept all conditions—i.e., NO_x, anthropogenic VOCs, global and photosynthetical radiation—fixed, and varied temperature in the range of +0 up to +8 K compared to the present conditions as described by the IPCC report [60]. Therefore, the speed of the chemical reaction will be enhanced because of enhanced collision rates, biogenic emissions will increase as they are linked to increased temperatures, and SOA will evaporate or form to a smaller extent according to changing saturation vapour pressures and partitioning coefficients.

Figures 7 and 8 present the simulation results for August 2015. As temperature and biogenic emissions increase, ozone is formed more rapidly and intensely, which is dampened by BVOC reactions with ozone and by formation of larger organic hydroperoxides and nitrates, gently reducing NO_x availability and tropospheric ozone production. This is reflected in the development of daily ground level ozone maxima (Figure 8), which increase continuously as temperatures increase.

Similar observations can be made for SOA production differences between both forest types. Although being insignificant at temperatures smaller than 30 °C, the situation changes above this temperature. At 30.1 ± 1.1 °C the SOA effect becomes more intense for Black Forest type forests compared to both Alsatian—i.e., Alsace in general and Alsatian Vosges—type forests and intensifies at higher temperatures. This originates from the higher contribution f of stored monoterpenes at Black Forest type compared to Vosges mixture type forests for standard conditions at T = 30 °C (f(E_{MT},stored, Black Forest) = 0.44, f(E_{MT}, stored, Alsatian Vosges forest) = 0.38). As temperature increases the mean stored monoterpene emission increases by 0.139/K for both forest types, while the de novo monoterpene emission (PAR and T dependent) only increases by about 0.04/K. Thus, the relative importance of the stored monoterpene resources grows with ascending temperatures.

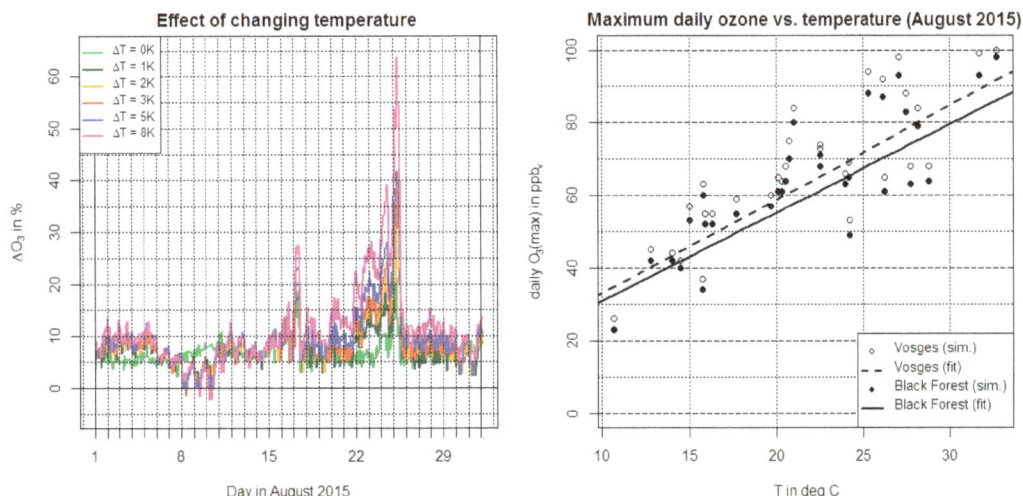

Figure 7. Change in boundary layer ozone mixing ratio because of changing air temperature for August 2015 as timeline (**left**) and as daily max vs. temperature (**right**).

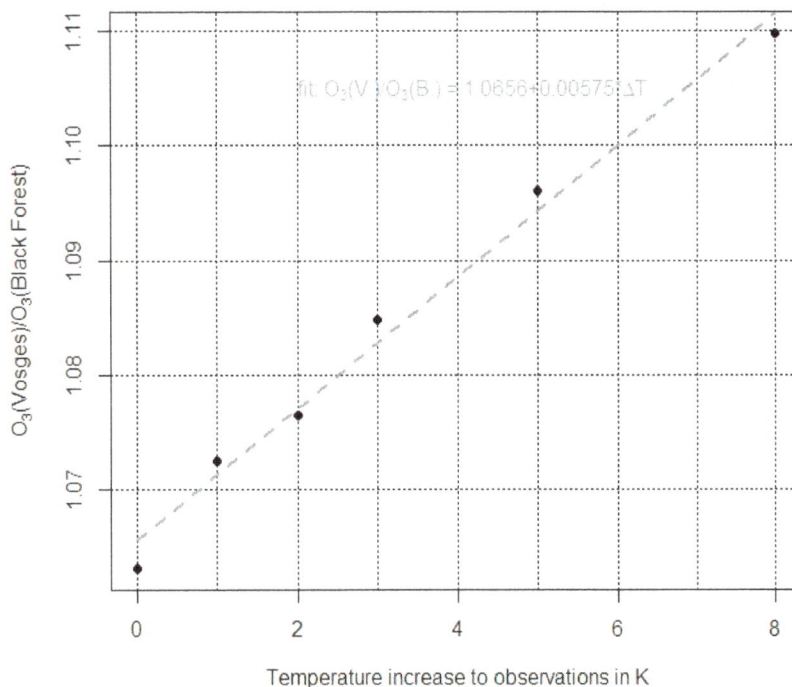

Figure 8. Effect of temperature on calculated ozone differences for different forest compositions.

This becomes more evident when checking the wind direction and speed. Apparently, the maximum intensity of calculated SOA is linked to westerly wind directions (max. temperatures) and the highest wind speeds (precursors from regional emissions and beyond; Figure 4, bottom left). However, the largest differences display with local very moderate winds from the East (Figure 4, bottom right), i.e., the forested areas with a dominance of biogenic VOCs and longest local residence time.

3.5. Effect of Reducing Input Information

So far, we have considered the influence of different parameters such as different BVOC groups, nitrogen oxides, deposition velocities and temperature on the simulation results and reproduction of observations. Finally, we check the effects of information reduction—i.e., reduction of complexity in simulation—as most models tend to be simple or belong to a larger set of processes to be calculated—competing for computational speed and space. Therefore, we omit stepwise biogenic VOC details and related processes described within the detailed box. We start by reducing two considered monoterpenes to a single one (test A), then cut down the monoterpene emission calculation from the sum of de novo and storage pool emissions to storage pool emissions according to Guenther et al. [45,46] only (test B). In the following steps, we omit the consideration of SOA (step C) and skip including monoterpenes at all (step D). Finally, all biogenic VOCs—i.e., isoprene too—are neglected (step E).

The resulting features are diverse: There is no significant change in correlation coefficients of observed and simulated ground level ozone at one hour time resolution (daily pattern), while shorter periods do vary gently and the simulated absolute amount of ozone changes notably (Table 5). At warm summer maximum temperatures of 35 °C two groups can be identified, one representing ground level ozone reducing effects such as monoterpenes (the more the less) and one referring to tropospheric ozone enhancement like SOA and isoprene (the more the more). It is quite interesting that more details on monoterpene emissions and structure do not influence ozone calculations to a significant extent (Table 5). If we assume all monoterpenes to be represented by α-pinene, there is no remarkable modification of a monoterpene mixture or β-pinene only. Furthermore, emission from storage pools or via de novo synthesis does not change the calculations with respect to ozone. But effects show up for SOA and the related feedback, which is strongest. The strength of individual effects is linked to the tree species and the dominant emission types. For an isoprene emission dominated forest type, for example in the Alsatian Vosges the isoprene related ground level ozone effect would be much more pronounced.

Table 5. Relative ozone increases at T = 35 °C compared to reference simulation.

Parametre	$\Delta O_3/O_3$(Black Forest, ref.)
MTs = α-pinene	+0.55 ± 0.10
MTs = β-pinene	+0.55 ± 0.10
MT emission from storage pools only	+0.55 ± 0.10
no SOA	−1.89 ± 0.33
no MTs	+0.56 ± 0.10
no isoprene	−0.03 ± 0.01
no BVOCs	+0.55 ± 0.09

If we investigate the behaviour at different ambient temperatures, some aspects display a rather constant contribution such as monoterpenes, which represent the major BVOC released by Black Forest trees. Other aspects, like SOA, provide a significantly changing effect with changing temperatures. At cooler temperatures of 20 to 25 °C, neglecting ozone results in overstimulation of ground level ozone because of a missing sink for gaseous VOCs, while at higher temperatures beyond 25 °C ignoring SOA leads to an underestimation, due to the missing source of VOCs re-evaporating for the particle phase (Figure 9), which is important for typical summertime conditions of 35 °C and above. At these conditions, typical for high ozone mixing ratios, several of the named tests (A)–(E) cause notable changes. As a summary, we can conclude: In the case of Black Forest species and conditions, considering no isoprene does not have significant effects on ozone because of its minor total emission rate. Neglecting monoterpenes causes approximately 70% higher ozone values at a temperature of 35 °C. It is worth mentioning that several of these effects may counteract and finally only a small to medium change (test E, "no BVOCs", Table 5) is detectable. However, this mean result is not necessarily correct for extremes such as ozone maxima and night time episodes.

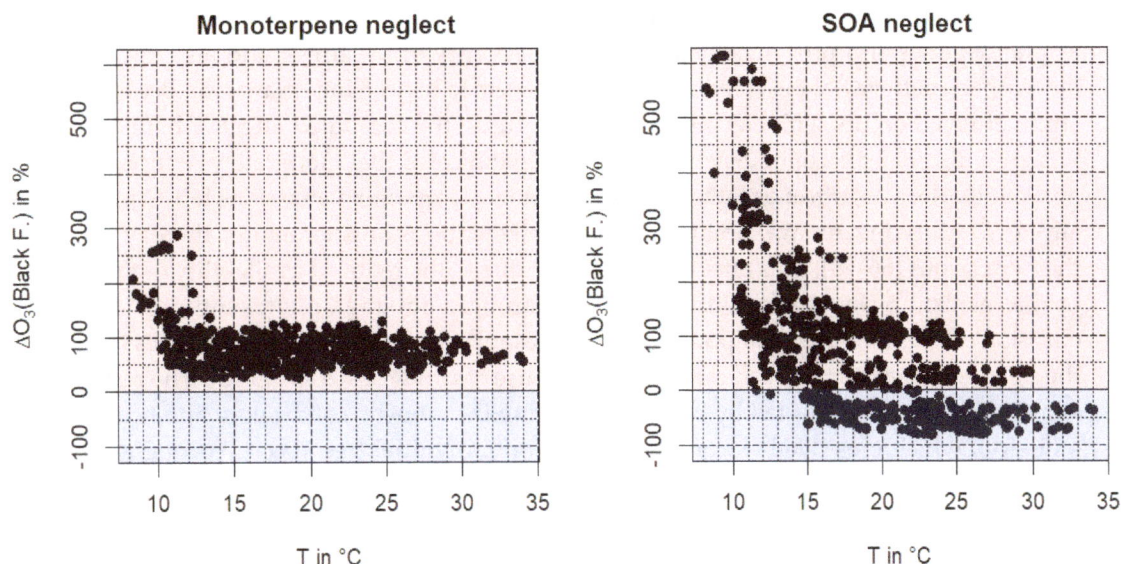

Figure 9. Effect of neglect on ozone calculated, **left**: monoterpenes; **right**: secondary organic aerosol.

4. Discussion

In this study, we have demonstrated that:

- Isoprene emission controlled forests like those in Alsace (especially the Vosges) leads to more intense tropospheric ozone formation during summer, than forests with both, i.e., monoterpene and isoprene emissions such as those in the Black Forest, if identical environmental conditions are assumed. This is caused by different oxidation product characteristics and SOA formation and is especially true for the transition from higher to lower NO_x concentrations, limiting the ozone production at the background site observation.

- The difference in formation intensifies with increasing temperatures and was found for other areas too. Churkina et al. [61] and Curci et al. [62] found a significant link between biogenic emission, summer temperatures and ground level ozone concentrations at urban [61] and remote areas [62]. Depending on the species mixture and resulting emission strengths tropospheric—i.e., ground level—ozone was found to increase notably.

- The particle formation rates from biogenic sources will remarkably intensify in cases where additional pollution does not reduce the lifetime of novel particles to a substantial extent. Additionally, a substantial local variability is to be expected, depending on tree species distribution and corresponding BVOC (mono- and sesquiterpene) emissions and varying environmental conditions [58].

- The production of SOA mass and volume differs depending on the BVOC emissions. High isoprene emissions will have different effects than high monoterpene emission rates as the oxidation degree is expected to be reduced (OH reduction) and the volatility of the products and potential SOA precursors will rise.

- Among the environmental factors influencing plant BVOC emission, water availability in the soil might be particularly important, especially when considering climate predictions [63]. The expected reduced summer precipitation, for example, in Southern and Central Europe, might considerably lower soil water availability in the future. The forecasted higher air temperatures will enhance this trend because of stimulated soil water evaporation. In a very first reaction towards drought, plants close their stomata to avoid loss of water by transpiration. Consequently, the cooling effect of transpiration is reduced leading to increased leaf temperatures, which might enhance leaf-internal VOC production [64]. Since CO_2 enters the leaves via the stomata, drought affects photosynthesis, causing lowered rates of C fixation. The effects of drought stress on tree BVOC emissions are complex and are still not fully understood, since many factors—such as plant species and provenance, duration and severity of the stress, and also the nature and biosynthetic pathway of the volatile compound—seem to play an important role. Because drought stress often co-occurs with elevated air temperatures, the combined effect of both factors also has to be taken into account. Several studies have indicated that moderately reduced soil water availability does not affect or rather slightly stimulates isoprenoid emissions [65–67]. In contrast, strongly reduced water availability affects the biosynthesis of isoprenoids, thereby decreasing their emissions [66–73]. This effect is most likely due to the reduced availability of the substrate of the relevant biosynthetic pathways due to strongly impaired photosynthesis. In contrast, emission of stress-induced compounds, such as sesquiterpenes or green leaf volatiles (products of the lipoxygenase reaction) can increase in response to drought stress [65,72]. The results of our simulations could be affected by better considering the influence of drought and heat on BVOC emission, which most likely strongly depends on individual tree species' drought sensitivity. There is certainly a lack of knowledge on the impacts of abiotic and biotic stressors, which should be resolved in order to realistically explain the capacity of ecosystems to cope with climate changes and to understand individual climate feedback process strengths.

- This box model study had the advantage of investigating processes more deeply and with a higher time resolution than regional models. However, the attribution of future changes and the feedback's most accurate 3D-regional model simulations would be favourable, although they could cause some other notable shortcomings such as reduced chemistry schemes, input information and averages for a mixed type forest! Benefits may arise for not-very-short-lived chemical species (>30 min). But challenges may arise for highly variable processes such as OH and particle formation.

5. Conclusions

To conclude, the type of forest and the contribution of individual tree species are important factors that influence the forests' survival in future climate conditions and the corresponding feedback processes that affect not only vegetation but also human health via e.g., ozone and ultrafine particulates [60,73]. So far, the Alsace and thus the corresponding Vosges forests benefit from the missing substantial traffic NO_x resources, which are elevated in the Upper Rhine valley at the Eastern part. But Black Forest areas would experience notably higher pollution levels if higher isoprene emitter species would join or replace the current species emitting a BVOCs mix.

In order to accurately simulate these feedback processes several important features are needed:

i   a chemical scheme with sufficient details, i.e., with respect to isoprene and monoterpenes: peroxy radical and major products chemistry;

ii   an incorporation of SOA formation and evaporation allowing a continuous equilibration between gas- and particle-phases and not a one directional description;

iii Inclusion of more detailed individual forest species characteristics, i.e., emissions and stress tolerance such as temperature coefficient and drought tolerance. Calculation of organic particle formations in number and mass clearly benefits from consideration of a detailed description of *de novo* production and emission from storage pools.

The latter would provide much more detailed information and feedback on the stress tolerance of the investigated forests or ecosystems. While reduced chemical schemes are essential for atmospheric chemistry effects to be calculated for approaches that are larger either spatially or in complexity, there will be shortcomings, especially with regard to the effects of monoterpene on HO_2 and SOA. A decision about complexity of these schemes should always be made within the context of the question to be answered.

Acknowledgments: The calculations are made possible by substantial data supplied by the Landesamt für Umwelt Baden-Württemberg in Stuttgart, Germany, by Pierre Robellet from ATMO Grand Est in Schiltigheim, France (ASPA) and by the Karlsruhe Institute of Technology (KIT) for server access and space for running the box model. Thanks to the FVA in Baden-Wuerttemberg, which provided the datasets of the inventory 2012 for calculating the dry weight biomass and for expert support. All the work was made possible by the Waldklimafond project "BuTaKli" (Buchen-Tannen-Mischwälder zur Anpassung von Wirtschaftswäldern an Extremereignisse des Klimawandels) (project number 28 W-C-1-069-01), which is highly acknowledged.

Author Contributions: B.B. designed and ran the experiments; F.S., T.B. and O.A. investigated and extracted the different forest compositions and parameters. J.K. supplied missing species emission rates, checked the formulation of BVOC emissions in detail and discussed water effects; R.Y. contributed and discussed the growth effects of tropospheric ozone on forests.

Conflicts of Interest: There are no conflicts of interest.

References

1. Lelieveld, J.; Evans, J.S.; Fnais, M.; Giannadaki, D.; Pozzer, A. The contribution of outdoor air pollution sources to premature mortality on a global scale. *Nature* **2015**, *525*, 367–371. [CrossRef] [PubMed]

2. Matyssek, R.; Innes, J.L. Ozone—A risk factor for trees and forests in Europe? *Water Air Soil Pollut.* **1999**, *116*, 199–226. [CrossRef]

3. Pretzsch, H.; Dieler, J.; Matyssek, R.; Wipfler, P. Tree and stand growth of mature Norway spruce and European beech under longterm ozone fumigation. *Environ. Pollut.* **2010**, *158*, 1061–1070. [CrossRef] [PubMed]

4. Augustaitis, A.; Kliučius, A.; Marozas, V.; Pilkauskas, M.; Augustaitiene, I.; Vitas, A.; Staszewski, T.; Jansons, A.; Dreimanis, A. Sensitivity of European beech trees to unfavorable environmental factors on the edge and outside of their distribution range in northeastern Europe. *iForest* **2014**, *9*, 259–269. [CrossRef]

5. Novak, K.; Schaub, M.; Fuhrer, J.; Skelly, J.M.; Frey, B.; Kräuchi, N. Ozone effects on visible foliar injury and growth of Fagus sylvatica and Viburnum lantana seedlings grown in monoculture origin mixture. *Environ. Exp. Bot.* **2008**, *62*, 212–220. [CrossRef]

6. De Vries, W.; Dobbertin, M.H.; Solberg, S.; van Dobben, H.F.; Schaub, M. Impacts of acid deposition, ozone exposure and weather conditions on forest ecosystems in Europe: An overview. *Plant Soil* **2014**, *380*, 1–45. [CrossRef]

7. Loreto, F.; Velikova, V. Isoprene produced by leaves protects the photosynthetic apparatus against ozone damage, quenches ozone products, and reduces lipid peroxidation of cellular membranes. *Plant Physiol.* **2001**, *127*, 1781–1787. [CrossRef] [PubMed]

8. Bergmann, E.; Bender, J.; Weigel, H.J. Impact of tropospheric ozone on terrestrial biodiversity: A literature analysis to identify ozone sensitive taxa. *J. Appl. Bot. Food Qual.* **2017**, *90*, 83–105. [CrossRef]

9. Jud, W.; Fischer, L.; Canaval, E.; Wohlfahrt, G.; Tissier, A.; Hansel, A. Plant surface reactions: An opportunistic ozone defence mechanism impacting atmospheric chemistry. *Atmos. Chem. Phys.* **2016**, *16*, 277–292. [CrossRef]

10. Bourtsoukidis, E.; Bonn, B.; Dittmann, A.; Hakola, H.; Hellen, H. Ozone stress as a driving force of sesquiterpene emissions: A suggested parameterization. *Biogeosciences* **2012**, *9*, 4337–4352. [CrossRef]

11. Giron-Calva, P.S.; Li, T.; Blande, J.D. Volatile-Mediated Interactions between Cabbage Plants in the Field and the Impact of Ozone Pollution. *J. Chem. Ecol.* **2017**, *43*, 339–350. [CrossRef] [PubMed]

12. Jenkin, M.E.; Young, J.C.; Rickard, A.R. The MCM v3.3.1 degradation scheme for isoprene. *Atmos. Chem. Phys.* **2015**, *15*, 11433–11459. [CrossRef]

13. Jenkin, M.E. Modelling the formation and composition of secondary organic aerosol from α- and β-pinene ozonolysis using MCM v3. *Atmos. Chem. Phys.* **2004**, *4*, 1741–1757. [CrossRef]

14. Jacobson, M.Z. *Fundamentals of Atmospheric Modeling*, 2nd ed.; Cambridge University Press: Cambridge, UK, 2005; p. 828. ISBN 978-0521548656.

15. Seinfeld, J.H.; Pandis, S.N. *Atmospheric Chemistry and Physics: From Air Pollution to Climate Change*, 2nd ed.; J. Wiley & Sons: Hoboken, NJ, USA, 2006; p. 1232. ISBN 978-0471720188.

16. Calfapietra, C.; Fares, S.; Loreto, F. Volatile organic compounds from Italian vegetation and their interaction with ozone. *Environ. Pollut.* **2009**, *157*, 1478–1486. [CrossRef] [PubMed]

17. Fiore, A.M.; Levy, H., II; Jaffe, D.A. North American isoprene influence on intercontinental ozone pollution. *Atmos. Chem. Phys.* **2011**, *11*, 1697–1710. [CrossRef]

18. Paoletti, E.; De Marco, A.; Beddows, D.C.S.; Harrison, R.M.; Manning, W.J. Ozone levels in European and USA cities are increasing more than at rural sites, while peak values are decreasing. *Environ. Pollut.* **2014**, *192*, 295–299. [CrossRef] [PubMed]

19. Sicard, P.; Serra, R.; Rossello, P. Spatiotemporal trends in ground-level ozone concentrations and metrics in France over the time period 1999–2012. *Environ. Res.* **2015**, *149*, 122–144. [CrossRef] [PubMed]

20. Sicard, P.; Augustaitis, A.; Belyazid, S.; Calfapietra, C.; de Marco, A.; Fenn, M.; Bytnerowicz, A.; Grulke, N.; He, S.; Matyssek, R.; et al. Global topics and novel approaches in the study of air pollution, climate change and forest ecosystems. *Environ. Pollut.* **2016**, *213*, 977–987. [CrossRef] [PubMed]

21. Goldstein, A.H.; Galbally, I.E. Known and Unexplored Organic Constituents in the Earth's Atmosphere. *Environ. Sci. Technol.* **2007**, *41*, 1514–1521. [CrossRef] [PubMed]

22. Jenkin, M.E.; Saunders, S.M.; Pilling, M.J. The tropospheric degradation of volatile organic compounds: A protocol for mechanism development. *Atmos. Environ.* **1997**, *31*, 81–104. [CrossRef]

23. Bloss, C.; Wagner, V.; Jenkin, M.E.; Volkamer, R.; Bloss, W.J.; Lee, J.D.; Heard, D.E.; Wirtz, K.; Martin-Reviejo, M.; Rea, G.; et al. Development of a detailed chemical mechanism (MCMv3.1) for the atmospheric oxidation of aromatic hydrocarbons. *Atmos. Chem. Phys.* **2005**, *5*, 641–664. [CrossRef]

24. Saunders, S.M.; Jenkin, M.E.; Derwent, R.G.; Pilling, M.J. Protocol for the development of the Master Chemical Mechanism, MCM v3 (Part A): Tropospheric degradation of nonaromatic volatile organic compounds. *Atmos. Chem. Phys.* **2003**, *3*, 161–180. [CrossRef]

25. Jenkin, M.E.; Wyche, K.P.; Evans, C.J.; Carr, T.; Monks, P.S.; Alfarra, M.R.; Barley, M.H.; McFiggans, G.B.; Young, J.C.; Rickard, A.R. Development and chamber evaluation of the MCM v3.2 degradation scheme for β-caryophyllene. *Atmos. Chem. Phys.* **2012**, *12*, 5275–5308. [CrossRef]

26. Stockwell, W.R.; Middleton, P.; Chang, J.S.; Tang, X. The second generation regional Acid Deposition Model chemical mechanism for regional air quality modeling. *J. Geophys. Res.* **1990**, *95*, 16343–16367. [CrossRef]

27. Stockwell, W.R.; Kirchner, F.; Kuhn, M.; Seefeld, S. A new mechanism for regional atmospheric chemistry modeling. *J. Geophys. Res.* **1997**, *102*, 25847–25879. [CrossRef]

28. Knote, C.; Tuccella, P.; Curci, G.; Emmons, L.; Orlando, J.J.; Madronich, S.; Baró, R.; Jimenez-Guerrero, P.; Luecken, D.; Hogrefe, C.; et al. Influence of the choice of gas-phase mechanism on predictions of key gaseous pollutants during the AQMEII phase-2 intercomparison. *Atmos. Environ.* **2015**, *115*, 553–568. [CrossRef]

29. Mazzuca, G.M.; Ren, X.; Loughner, C.P.; Estes, M.; Crawford, J.H.; Pickering, K.E.; Weinheimer, A.J.; Dickerson, R.R. Ozone production and its sensitivity to NOx and VOCs: Results from the DISCOVER-AQ field experiment, Houston 2013. *Atmos. Chem. Phys.* **2016**, *16*, 14463–14474. [CrossRef]

30. Kuik, F.; Lauer, A.; Churkina, G.; Denier van der Gon, H.A.C.; Fenner, D.; Mar, K.A.; Butler, T.M. Air quality modelling in the Berlin–Brandenburg region using WRF-Chem v3.7.1: Sensitivity to resolution of model grid and input data. *Geosci. Model Dev.* **2016**, *9*, 4339–4363. [CrossRef]

31. Butler, T.; Lawrence, M.; Taraborrelli, D.; Lelieveld, J. Multi-day ozone production potential of volatile organic compounds calculated with a tagging approach. *Atmos. Environ.* **2011**, *45*, 4082–4090. [CrossRef]

32. Bonn, B.; von Schneidemesser, E.; Butler, T.; Churkina, G.; Ehlers, C.; Grote, R.; Klemp, D.; Nothard, R.; Schäfer, K.; von Stülpnagel, A.; et al. Impact of vegetative emissions on urban ozone and biogenic secondary organic aerosol: Box model study for Berlin, Germany. *J. Clean. Prod.* **2017**. submitted.

33. Atkinson, R.; Baulch, D.L.; Cox, R.A.; Crowley, J.N.; Hampson, R.F.; Hynes, R.G.; Jenkin, M.E.; Rossi, M.J.; Troe, J. IUPAC Subcommittee Evaluated kinetic and photochemical data for atmospheric chemistry: Volume II—Gas phase reactions of organic species. *Atmos. Chem. Phys.* **2006**, *6*, 3625–4055. [CrossRef]

34. Pankow, J.F. An absorption-model of the gas aerosol partitioning involved in the formation of secondary organic aerosol. *Atmos. Environ.* **1994**, *28*, 189–193. [CrossRef]

35. Birmili, W.; Weinhold, K.; Rasch, F.; Sonntag, A.; Sun, J.; Merkel, M.; Wiedensohler, A.; Bastian, S.; Schladitz, A.M.; Löschau, G.; et al. Long-term observations of tropospheric particle number size distributions and equivalent black carbon mass concentrations in the German Ultrafine Aerosol Network (GUAN). *Earth Syst. Sci. Data* **2016**, *8*, 355–382. [CrossRef]

36. Prank, M.; Sofiev, M.; Tsyro, S.; Hendriks, C.; Semeena, V.; Francis, X.V.; Butler, T.; Denier van der Gon, H.; Friedrich, R.; Hendricks, J.; et al. Evaluation of the performance of four chemical transport models in predicting the aerosol chemical composition in Europe in 2005. *Atmos. Chem. Phys.* **2016**, *16*, 6041–6070. [CrossRef]

37. Inventaire Forestier. 2016. Available online: http://inventaire-forestier.ign.fr/spip/ (accessed on 20 November 2016).

38. Kändler, G.; Cullmann, D. (Eds.) *Report: Regionale Auswertung der Bundeswaldinventur 3 Wuchsgebiet Schwarzwald*; FVA: Baden-Württemberg, Germany, 2016.

39. Nabuurs, G.J.; Ravindranath, N.H.; Paustian, K.; Freibauer, A.; Hohenstein, W.; Makundi, W. LUCF sector good practice guidance. In *IPCC: Good Practice Guidance for Land Use, Land Use Change and Forestry*; Penman, J., Gytarsky, M., Hiraishi, T., Krug, T., Kruger, D., Pipatti, R., Buendia, L., Miwa, K., Ngara, T., Tanabe, K., Wagner, F., Eds.; Institute for Global Environmental Strategies: Kanagawa, Japan, 2003; ISBN 4-88788-003-0.

40. CITEPA. 2011. Available online: http://www.citepa.org/emissions (accessed on 20 November 2016).

41. Szinyei, D. Modelling and Evaluation of Ozone Dry Deposition. PhD thesis, Free University, Berlin, Germany, 2014.

42. Tiwari, S.; Grote, R.; Churkina, G.; Butler, T. Ozone damage, detoxification and the role of isoprenoids—New impetus for integrated models. *Funct. Plant Biol.* **2016**, *43*, 324–336. [CrossRef]

43. Dumont, J.; Keski-Saari, S.; Keinänen, M.; Cohen, D.; Ningre, N.; Kontunen-Soppela, S.; Baldet, P.; Gibon, Y.; Dizengremel, P.; Vaultier, M.-N.; et al. Ozone affects ascorbate and glutathione biosynthesis as well as amino acid contents in three Euramerican poplar genotypes. *Tree Physiol.* **2014**, *34*, 253–266. [CrossRef] [PubMed]

44. Leuzinger, S.; Körner, C. Tree species diversity affects canopy leaf temperatures in a mature temperate forest. *Agric. Forest Meteorol.* **2007**, *146*, 29–37. [CrossRef]

45. Guenther, A.; Karl, T.; Harley, P.; Wiedinmyer, C.; Palmer, P.I.; Geron, C. Estimates of global terrestrial isoprene emissions using MEGAN (Model of Emissions of Gases and Aerosols from Nature). *Atmos. Chem. Phys.* **2006**, *6*, 3181–3210. [CrossRef]

46. Guenther, A.; Hewitt, C.N.; Erickson, D.; Fall, R.; Geron, C.; Graedel, T.; Harley, P.; Klinger, L.; Lerdau, M.; McKay, W.A.; et al. A global model of natural volatile organic compound emissions. *J. Geophys. Res. Atmos.* **1995**, *100*, 8873–8892. [CrossRef]

47. Joback, K.G.; Reid, R.C. Estimation of pure-component properties from group-contributions. *Chem. Eng. Commun.* **1987**, *57*, 233–243. [CrossRef]

48. Stein, S.E.; Brown, R.L. Estimation of normal boiling points from group contributions. *J. Chem. Inf. Comp. Sci.* **1994**, *34*, 581–587. [CrossRef]

49. Volz-Thomas, A.; Kolahgar, B. On the budget of hydroxyl radicals at Schauinsland during the Schauinsland Ozone Precursor Experiment (SLOPE96). *J. Geophys. Res.* **2000**, *105*, 1611–1622. [CrossRef]

50. Pätz, H.W.; Corsmeier, U.; Glaser, K.; Vogt, U.; Kalthoff, N.; Klemp, D.; Kolahgar, B.; Lerner, A.; Neininger, B.; Schmitz, T.; et al. Measurements of trace gases and photolysis frequencies during SLOPE96 and a coarse estimate of the local OH concentration from HNO_3 formation. *J. Geophys. Res.* **2000**, *105*, 1563–1583. [CrossRef]

51. Klemp, D.; Kley, D.; Kramp, F.; Buers, H.J.; Pilwat, G.; Flocke, F.; Patz, H.W.; Volz-Thomas, A. Long-term measurements of light hydrocarbons (C_2-C_5) at Schauinsland (Black Forest). *J. Atmos. Chem.* **1997**, *28*, 135–171. [CrossRef]

52. Kalthoff, N.; Corsmeier, U.; Volz-Thomas, A. Influence of valley winds on transport and dispersion of airborne pollutants in the Freiburg-Schauinsland area. *J. Geophys. Res.* **2000**, *105*, 1585–1597. [CrossRef]

53. Bonn, B.; Bourtsoukidis, E.; Sun, T.S.; Bingemer, H.; Rondo, L.; Javed, U.; Li, J.; Axinte, R.; Li, X.; Brauers, T.; et al. The link between atmospheric radicals and newly formed particles at a spruce forest site in Germany. *Atmos. Chem. Phys.* **2014**, *14*, 10823–10843. [CrossRef]

54. Duhl, T.R.; Helmig, D.; Guenther, A. Sesquiterpene emissions from vegetation: A review. *Biogeosciences* **2008**, *5*, 761–777. [CrossRef]

55. Aydin, Y.M.; Yaman, B.; Koca, H.; Dasdemir, O.; Kara, M.; Altinok, H.; Dumanoglu, Y.; Bayram, A.; Tolunay, D.; Odabasi, M.; et al. Biogenic volatile organic compound (BVOC) emissions from forested areas in Turkey: Determination of specific emission rates for thirty-one tree species. *Sci. Total Environ.* **2014**, *490*, 239–253. [CrossRef] [PubMed]

56. van Meeningen, Y.; Schurgers, G.; Rinnan, R.; Holst, T. BVOC emissions from English oak (*Quercus robur*) and European beech (*Fagus sylvatica*) along a latitudinal gradient. *Biogeosciences* **2016**, *13*, 6067–6080. [CrossRef]

57. Bonn, B.; von Schneidemesser, E.; Andrich, D.; Quedenau, J.; Gerwig, H.; Lüdecke, A.; Kura, J.; Pietsch, A.; Ehlers, C.; Klemp, D.; et al. BAERLIN2014—The influence of land surface types on and the horizontal heterogeneity of air pollutant levels in Berlin. *Atmos. Chem. Phys.* **2016**, *16*, 7785–7811. [CrossRef]

58. Jacob, D.J. *Introduction to Atmospheric Chemistry*, 1st ed.; Princeton University Press: Princeton, NJ, USA, 1999; ISBN 13 978-0691001852.

59. Finlayson-Pitts, B.; Pitts, J. *Chemistry of the Upper and Lower Atmosphere*, 2nd ed.; Academic Press: Waltham, MA, USA, 1999; ISBN 978-0122570605.

60. Stocker, T.F.; Qin, D.; Plattner, G.-K.; Tignor, M.; Allen, S.K.; Boschung, J.; Nauels, A.; Xia, Y. (Eds.) *IPCC: Climate Change 2013: The Physical Science Basis. Contribution of Working Group I to the Fifth Assessment Report of the Intergovernmental Panel on Climate Change*; Cambridge University Press: Cambridge, UK; New York, NY, USA, 2013; ISBN 978-1-107-66182-0.

61. Churkina, G.; Kuik, F.; Bonn, B.; Lauer, A.; Grote, R.; Tomiak, K.; Butler, T.M. Effect of VOC Emissions from Vegetation on Air Quality in Berlin during a Heatwave. *Environ. Sci. Technol.* **2017**, *51*, 6120–6130. [CrossRef] [PubMed]

62. Curci, G.; Beekmann, M.; Vautard, R.; Smiatek, G.; Steinbrecher, R.; Theloke, J.; Friedrich, R. Modelling study of the impact of isoprene and terpene biogenic emissions on European ozone levels. *Atmos. Environ.* **2009**, *43*, 1444–1455. [CrossRef]

63. Anav, A.; de Marco, A.; Proietti, C.; Alessandi, A.; Dell'Aquila, A.; Cionni, I.; Friedlingstein, P.; Khvorostyanov, D.; Menut, L.; Paoletti, E.; et al. Comparing concentration-based (AOT40) and stomatal uptake (PODY) metrics for ozone risk assessment to European forests. *Glob. Chang. Biol.* **2016**, *22*, 1608–1622. [CrossRef] [PubMed]

64. De Marco, A.; Sicard, P.; Fares, S.; Tuovinen, J.P.; Anav, A.; Paoletti, E. Assessing the role of soil water limitation in determining the Phytotoxic Ozone Dose (PODY) thresholds. *Atmos. Environ.* **2016**, *147*, 88–97. [CrossRef]

65. Ormeno, E.; Mevy, J.P.; Vila, B.; Bousquet-Melou, A.; Greff, S.; Bonin, G.; Fernandez, C. Water deficit stress induces different monoterpene and sesquiterpene emission changes in Mediterranean species. Relationship between terpene emissions and plant water potential. *Chemosphere* **2007**, *67*, 276–284. [CrossRef] [PubMed]

66. Staudt, M.; Ennajah, A.; Florent Mouillot, F.; Joffre, R. Do volatile organic compound emissions of Tunisian cork oak populations originating from contrasting climatic conditions differ in their responses to summer drought? *Can. J. For. Res.* **2008**, *38*, 2965–2975. [CrossRef]

67. Šimpraga, M.; Verbeeck, H.; Demarcke, M.; Joó, É.; Pokorska, O.; Amelynck, C.; Schoon, N.; Dewulf, J.; Van Langenhove, H.; Heinesch, B.; et al. Clear link between drought stress, photosynthesis and biogenic volatile organic compounds in Fagus sylvatica L. *Atmos. Environ.* **2011**, *45*, 5254–5259. [CrossRef]

68. Llusià, J.; Peñuelas, J. Changes in terpene content and emission in potted Mediterranean woody plants under severe drought. *Can. J. Bot.* **1998**, *76*, 1366–1373. [CrossRef]

69. Peñuelas, J.; Filella, I.; Seco, R.; Llusià, J. Increase in isoprene and monoterpene emissions after re-watering of droughted Quercus ilex seedlings. *Biol. Plant* **2009**, *53*, 351–354. [CrossRef]

70. Nogués, I.; Medori, M.; Calfapietra, C. Limitations of monoterpene emissions and their antioxidant role in Cistus sp. under mild and severe treatments of drought and warming. *Environ. Exp. Bot.* **2015**, *119*, 76–86. [CrossRef]

71. Nogués, I.; Muzzini, V.; Loreto, F.; Bustamante, M.A. Drought and soil amendment effects on monoterpene emission in rosemary plants. *Sci. Total Environ.* **2015**, *538*, 768–778. [CrossRef] [PubMed]

72. Jud, W.; Vanzo, E.; Li, Z.; Ghirardo, A.; Zimmer, I.; Sharkey, T.D.; Hansel, A.; Schnitzler, J. Effects of heat and drought stress on post-illumination bursts of volatile organic compounds in isoprene-emitting and non-emitting poplar. *Plant Cell Environ.* **2016**, *39*, 1204–1215. [CrossRef] [PubMed]

73. Ochoa-Hueso, R.; Munzi, S.; Alonso, R.; Arróniz-Crespo, M.; Avila, A.; Bermejo, V.; Bobbink, R.; Branquinho, C.; Concostrina-Zubiri, L.; Cruz, C.; et al. Ecological impacts of atmospheric pollution and interactions with climate change in terrestrial ecosystems of the Mediterranean Basin: Current research and future directions. *Environ. Pollut.* **2017**, *227*, 194–206. [CrossRef] [PubMed]

8

Comparative Study of Monsoon Rainfall Variability over India and the Odisha State

K C Gouda [1,*], Sanjeeb Kumar Sahoo [1,2], Payoshni Samantray [1,2] and Himesh Shivappa [1]

[1] CSIR Fourth Paradigm Institute, Wind Tunnel Road, Bengaluru, Karnataka 560037, India; sanjeeb.ranjeeb@gmail.com (S.K.S.); sinisamantaray@gmail.com (P.S.); himesh@csir4pi.in (H.S.)

[2] Visvesvaraya Technological University, Belagavi, Karnataka 590018, India

* Correspondence: kcgouda@csir4pi.in

Abstract: Indian summer monsoon (ISM) plays an important role in the weather and climate system over India. The rainfall during monsoon season controls many sectors from agriculture, food, energy, and water, to the management of disasters. Being a coastal province on the eastern side of India, Odisha is one of the most important states affected by the monsoon rainfall and associated hydro-meteorological systems. The variability of monsoon rainfall is highly unpredictable at multiple scales both in space and time. In this study, the monsoon variability over the state of Odisha is studied using the daily gridded rainfall data from India Meteorological Department (IMD). A comparative analysis of the behaviour of monsoon rainfall at a larger scale (India), regional scale (Odisha), and sub-regional scale (zones of Odisha) is carried out in terms of the seasonal cycle of monsoon rainfall and its interannual variability. It is seen that there is no synchronization in the seasonal monsoon category (normal/excess/deficit) when analysed over large (India) and regional (Odisha) scales. The impact of El Niño, La Niña, and the Indian Ocean Dipole (IOD) on the monsoon rainfall at both scales (large scale and regional scale) is analysed and compared. The results show that the impact is much more for rainfall over India, but it has no such relation with the rainfall over Odisha. It is also observed that there is a positive (negative) relation of the IOD with the seasonal monsoon rainfall variability over Odisha (India). The correlation between the IAV of monsoon rainfall between the large scale and regional scale was found to be 0.46 with a phase synchronization of 63%. IAV on a sub-regional scale is also presented.

Keywords: monsoon rainfall; interannual variability; phase synchronization; correlation; Odisha

1. Introduction

The variability in the Indian summer monsoon (ISM) at different spatio-temporal scales is large, and the monsoon patterns over a larger domain, such as the whole of India, and a smaller domain, like the Odisha state, are totally different. It is very important to quantify the variability of seasonal monsoon rainfall over different spatial scales because the geographic background and local climate of the state are different from those of India. The Odisha state ($17°31'–22°31'$ N and $81°31'–87°3°'$ E) is situated on the eastern side of India, having a total area of 155,707 km^2, consisting of four physiographical regions, viz. (i) coastal plain, (ii) northern upland, (iii) central river basin, and (iv) south west hilly regions. It is seen that some of the hill peaks in the northern upland and southwest hilly regions are as high as 1000–2000 m, and the eastern part is a coast. Though the Eastern Ghats of India extend mainly into the North Odisha and South Odisha regions, very large variations in the topography affect the seasonal advance and distribution of the monsoon flow at the sub-regional scale. A map is shown in Figure 1a.

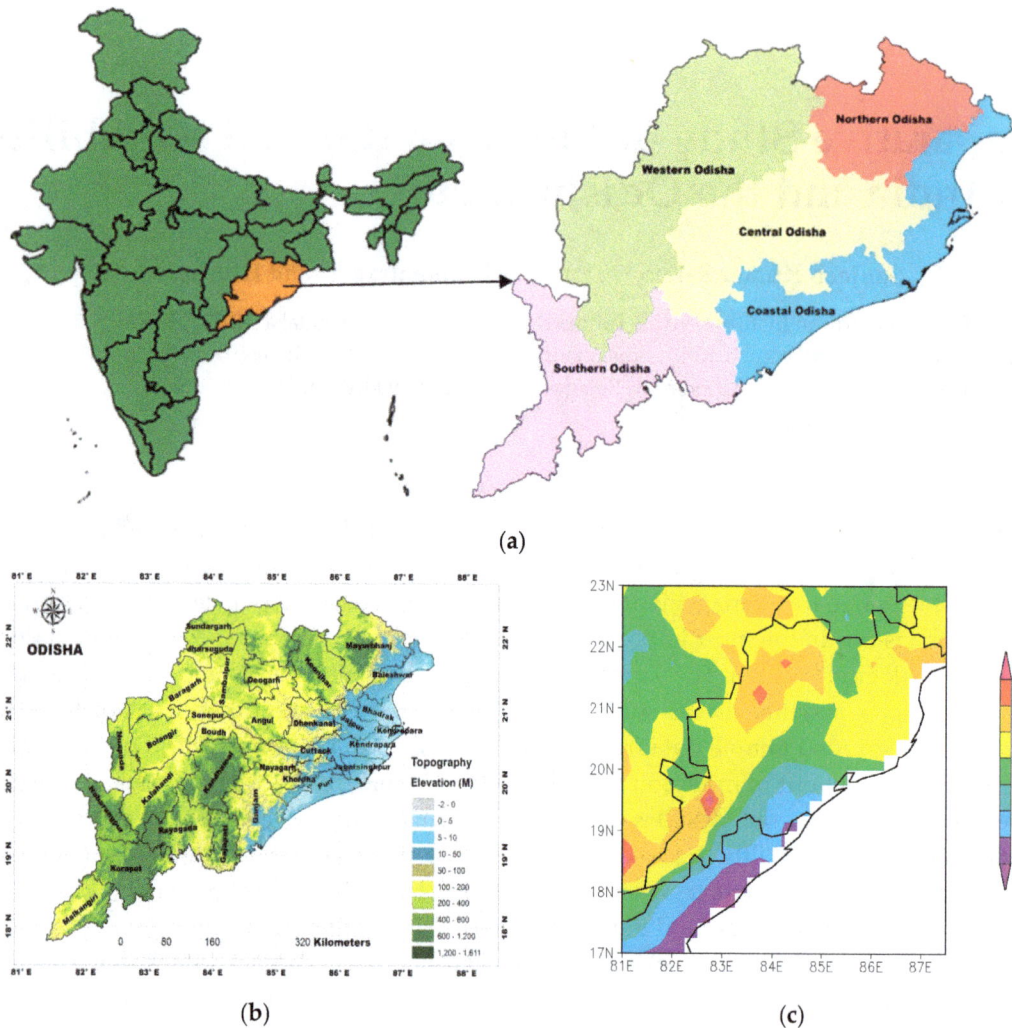

Figure 1. (a) Location of the Odisha state in India and a map showing the zones considered for study in the state of Odisha (b) topographic map of the state showing the elevation in meters; and (c) monsoon rainfall climatology over Odisha.

The complexities of the ISM have been studied by several authors emphasizing its spatial variability [1–9]. The spatial variability of ISM has also been studied by a few researchers [10–12]. However, the study of the summer monsoon over a region like Odisha is unique and important due to its location being little explored. Few studies [13,14] on the monsoon rainfall over Odisha revealed the important feature of the monsoon at the regional scale. It is also important to quantify the inter-relationship of the ISM and the monsoon rainfall over the domain of Odisha, which can be used as an indicator for understanding the monsoonal variability at the regional scale.

The Indian monsoon is a global phenomenon that is influenced by various large-scale phenomena, like El Niño, La Niña, IOD, etc. [15–19]. The study of the impact of such large-scale processes on the ISM variability is very important for better understanding of the role of large-scale processes in monsoon variability. This understanding is critical to improve the predictability of ISM using numerical models like the General Circulation Model (GCM). Several recent research efforts are focused on multi-scale modelling of monsoons and their associated processes using numerical weather models, GCM [20–24], and coupled atmosphere-land-ocean models [25–28]. It is known that monsoons exhibit variability even on inter-decadal time scales in association with many other global climate variables.

Understanding monsoon rainfall at multiple scales and advanced prediction of southwest monsoons are crucial for India. It is also important to know the inter-relationship between monsoons and large-scale ocean-atmospheric processes. Now-a-days, with the improvement in the computation power and dynamical modelling capabilities, the prediction of weather and climate at various time scales, from daily to decadal, is possible using general circulation models (GCMs), and meso-scale models.

Generally, rainfall variability is considered over the country as a whole in several studies [21,22,26], but it is very important to study and quantify the rainfall variability at regional (state) and sub-regional scales of a state, so that the application of rainfall studies will be useful. This work mainly focuses on the comparative study of monsoon variability between the regional (India) and sub-regional (Odisha). Additionally, the sub-region scales, like North, South, Central, and Coastal Odisha are analysed and compared. Section 2 gives the details of the data used and methodology adopted for the study. Results and discussion are presented in Section 3, and Section 4 describes the conclusion of the study.

2. Materials and Methods

The analysis is carried out using high-resolution gridded daily rainfall data from the India Meteorological Department (IMD). The gridded daily rainfall data are based on observations from 6955 stations [29] with a minimum 90% data availability during the analysis period (1951–2013). The station rainfall data are projected onto a rectangular grid ($0.25° \times 0.25°$) for each day for the period 1951–2013. The starting point of the grid is $6.5°$ N and $67.5°$ E.

Statistical analyses are carried out to understand the monsoon variability at the large scale (India), regional scale (Odisha), and sub-regional scale (zones of the Odisha state). The seasonal cycle, which is daily average (of N total years) rainfall over a domain, is calculated by the area average of total rainfall over the domain using the following formula:

$$R(day) = \frac{1}{im \times jm \times N} \sum_{i=1}^{im} \sum_{j=1}^{jm} \sum_{n=1}^{N} R(i,j,day,n) \tag{1}$$

where n is the year; i and j, respectively, represent the latitude and longitude of the domains; and im, and jm are the numbers of the latitude and longitude used in the domain.

Then, the inter-annual variability (IAV) is computed as the anomaly (departure from the mean) as given in the formula below:

$$R_A(i,j,n) = R(i,j,n) - \overline{R}(i,j) \tag{2}$$

where $\overline{R}(i,j) = \frac{1}{N} \sum_{n=1}^{N} R(i,j,n)$ is the (N = 63 years) mean at location (i,j) at a given time scale (daily, monthly and seasonal).

The normalized (to mean) anomaly is given by:

$$R_N(i,j,n) = \frac{R(i,j,n) - \overline{R}(i,j)}{\overline{R}(i,j)} \times 100 \tag{3}$$

Finally, the area averaged anomaly is computed for the domain, and the respective IAV values are compared in the study.

For many applications, the rainfall category, in terms of excess or deficit from the normal rainfall can be a valuable input, for seasonal rainfall at the large scale, as well as the regional and sub-regional scales. Here, three rainfall categories, based on the normalized anomaly of season (JJAS) are considered following the method adopted in a previous study [22]. The categories are as follows:

$$\text{Normal} : -10\% \leq R_N \leq 10\%$$

$$\text{Excess} : R_N > 10\% \tag{4}$$

$$\text{Deficit} : R_N < -10\%$$

The quantity used to assess the degree of variability over two domains is phase synchronization (PS), which is assigned a value 0 or 1 depending on whether the normalized anomaly over India has the same directional anomaly over Odisha.

$$I(n) = \begin{cases} 1 : \text{if} \frac{R_{NOdisha}(n)}{R_{NIndia}(n)} > 0 \\ 0 : \text{otherwise} \end{cases} \tag{5}$$

By considering only those years with $I(n) = 1$, PS is computed by using the relation $PS(\%) = \frac{p}{N} \times 100$; where p is number of years, where $I(n)$ is 1 and N is the total number of years.

Additionally, correlation analysis between the rainfall pattern over India and Odisha is carried out and discussed in this study.

3. Results and Discussion

In the present study, a spatial analysis was carried out at a larger scale (India), regional scale (the Odisha state), and sub-regional scale (different zones of Odisha). For the regional scale analysis over the state, Odisha is divided into five zones based on physiographic regions, i.e., North (two districts), South (five district), Coastal (seven districts), Western (nine districts), and Central (seven districts) Odisha. The geographical location of Odisha in India is shown in Figure 1a. Table 1 shows the details of the districts in each region, and the spatial distribution of different zones of Odisha is shown in Figure 1a.

Table 1. Detailed coverage of the different regions of Odisha.

Region	Domain
India	Continental India
Odisha	All-Odisha
	Districts considered
North Odisha	Mayurbhanj, Keonjhar
South Odisha	Malkangiri, Koraput, Rayagada, Gajpati, Nabarangpur
Coastal Odisha	Balasore, Bhadrak, Kendrapada, Jagatsinghpur, Puri, Khurdha, Ganjam
Western Odisha	Sundergarh, Jharsuguda, Sambalpur, Deogarh, Baragarh, Sonepur, Balangir, Nuapada, Kalahandi
Central Odisha	Jajpur, Dhenkanal, Anugul, Cuttack, Kandhamal, Nayagarh, Boudh

Since the topography of the region controls the rainfall tendency, it is very important to consider the topographical classification over the state. The variation of monsoon rainfall intensity and its distribution over the state of Odisha are the result of the scale interaction of the monsoon flow and the regional disturbances due to orographic variability of the Eastern Ghats in Odisha. The topographic distribution of Odisha is shown in Figure 1b. The elevation rises as low as 5 m from the coastal region to as high as 1600 m above mean sea level.

The spatial average (1951–2013) of monsoon rainfall (JJAS) over the Odisha State, as a whole, is presented in Figure 1c, which shows Central and Coastal Odisha generally receive copious rainfall compared to the rest of the state. The standard deviation (in 63 years) of the seasonal monsoon rainfall is about 2.42 mm/day over the state.

3.1. Intraseasonal Monsoon Variability

The seasonal cycle (63 year daily climatology) of daily rainfall is analysed for both India and Odisha and is shown in Figure 2a, indicating a strong correlation with a correlation co-efficient (CC) of 0.86. A trend of intense rainfall can be observed during July–August when the rainfall is more than 12 (9) mm/day averaged over Odisha (India). During September, Odisha shows maximum average rainfall as compared to India. The climatological daily rainfall in September is comparatively high over Odisha and its sub-region compared with India with the daily maximum rain received over Odisha (India) being about 11.5 mm/day (6.9 mm/day), whereas in the sub-regions of Odisha, the maximum daily rainfall is about 11–14.3 mm/day. This is apparently due to several factors, like highly variable topography and diurnal land-sea breeze circulation.

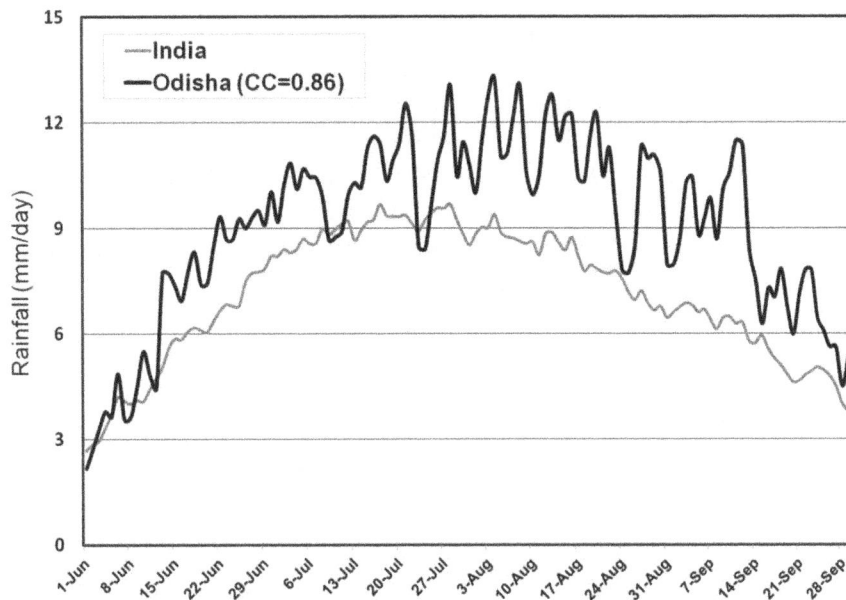

(**a**) Comparison of intraseasonal monsoon variability between a large scale (India) and regional scale (Odisha)

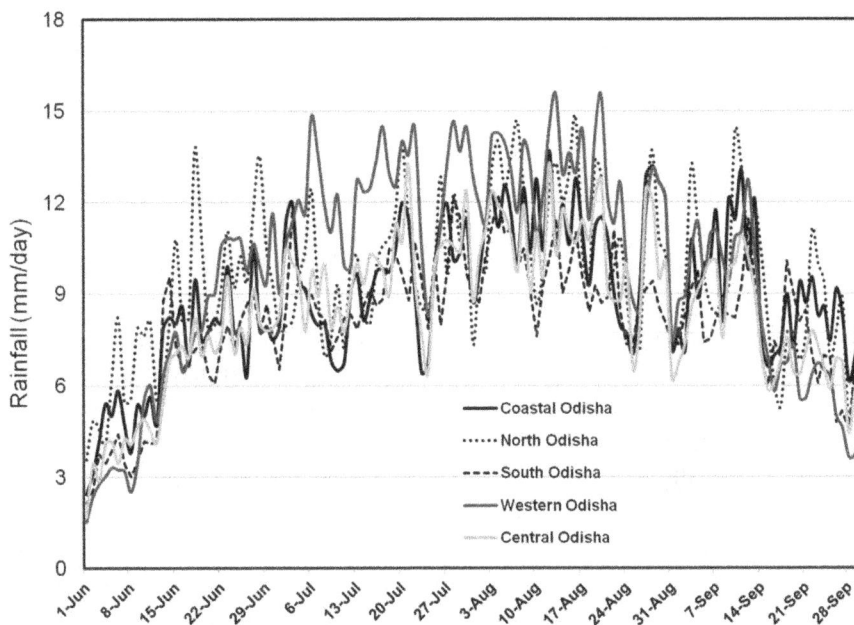

(**b**) Comparison of intraseasonal monsoon variability at a sub-regional scale (different zones of Odisha)

Figure 2. Seasonal cycle of the daily monsoon rainfall climatology averaged over (**a**) all of India and all of Odisha; and (**b**) different zones of Odisha for the period 1 June to 30 September.

The seasonal climatology of monsoon rainfall in different zones of Odisha (sub-regional scale) shows wide variation from one zone to other (Figure 2b). Western Odisha receives more rain in the first half of the monsoon compared to other sub-regions, whereas the second-half of this region receives relatively less rainfall, and Western Odisha often faces drought. The standard deviation (SD) in the daily climatology for 121 days is about 2.42 mm/day at the state scale, whereas it is higher in Western Odisha (3.07 mm/day) and lower in South Odisha (2.36 mm/day). Similarly, CC of the seasonal cycle over the zones and the state are computed, and it is observed that Western and Central Odisha have a high correlation, of about 0.95 and 0.92, whereas other zones shows 0.8 (North and Coastal Odisha) and 0.9 (South Odisha). The comparisons of annual and monsoon rainfall over India and Odisha during 1951–2013 are shown in Figure S1, which shows a CC of about 0.2 at both annual and seasonal scales.

To identify the sub-regional scale monsoon rainfall over Odisha, a climatological analysis assessing the seasonal average, daily average, and the number of rainy days in the monsoon season over different regions of Odisha is carried out, and the results are presented in Table 2. It is observed that the number of days with significantly high rainfall (>12 mm) is very high over Western Odisha, but the number of rainy days is low in this region, as compared to others. Even though Western Odisha receives the highest seasonal average rainfall (9.89 mm/day), it usually faces drought-like situations due to fewer rainy days (around 37 out of 121) during the monsoon season. These findings need to be investigated further, considering the impact of topography and wind circulation patterns on the local monsoonal flow.

Table 2. Sub-regional scale analysis of rainfall categories in daily climatology over the Odisha State.

Case	North Odisha	South Odisha	Coastal Odisha	Western Odisha	Central Odisha	All Odisha	All India
No of Days (>=12 mm)	14	4	9	37	15	8	0
No of Days (<=9 mm)	60	74	75	42	57	59	80
120 days average rainfall (in mm)	9.55	8.09	8.96	9.89	8.50	9.01	7.06

3.2. Inter-Annual Monsoon Variability

Interannual variability of monsoon rainfall is also computed for the period 1951–2013, and it is seen that the phase synchronization of IAV over Odisha and India is 63% among 63 years of variability. The CC of IAV is about 0.43, which is more significant (Figure 3a). Similarly, analyses over sub-regional zones of Odisha are presented in Figure 3b, and it is observed that, for Western Odisha, the variability is very high. Based on the IAV in the monsoon rainfall over all years, the different years are categorized into excess, deficit, and normal monsoon years for Odisha. The daily rainfall climatology for the period 1 June to 30 September composited over excess (10 years), deficit (17 years), normal (36 years), and all 63 years in the state is presented in Figure 4. It is inferred that, generally, the July rainfall is greater in excess years and August rainfall is lower in deficit years. The clear indication of the monsoon season rainfall (excess or deficit) is totally dependent on the July rainfall over the state. The role of rainfall in August can also be used as an indicator for the crop choices, in particular for the drought years.

The monsoon composites are computed, and the seasonal rainfall climatology composites over the excess monsoon (10 years) and deficit monsoon (17 years) season are shown in Figure 5. It is observed from the spatial analysis that the rainfall amount is greater (about 18 mm/day) over the western part in the composite excess year, and the rainfall amount is moderate in the central and eastern part (about 8–10 mm/day). Rainfall drops to as low as 6 mm/day throughout the state with the minimum in the central and eastern parts of Odisha in the deficit monsoon years. The contrast in the rainfall climatology in excess and deficit years is also presented in the lower panel; the contrast between excess and deficit monsoon years is very high (about 9 mm/day) in the western region and low (about 1 mm/day) in the southeastern region of Odisha.

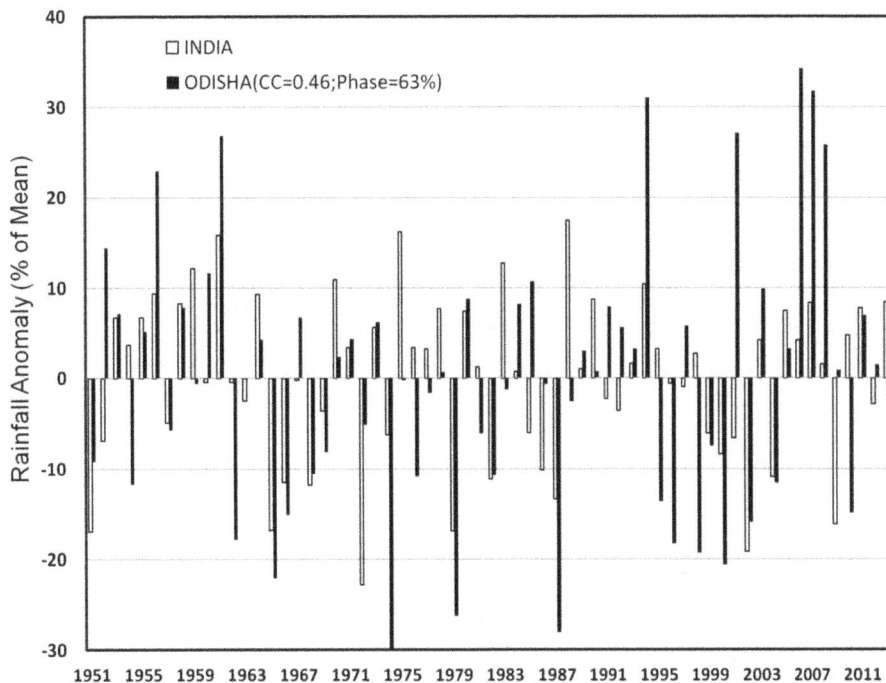

(a) Comparison of IAV of seasonal monsoon rainfall over India and Odisha

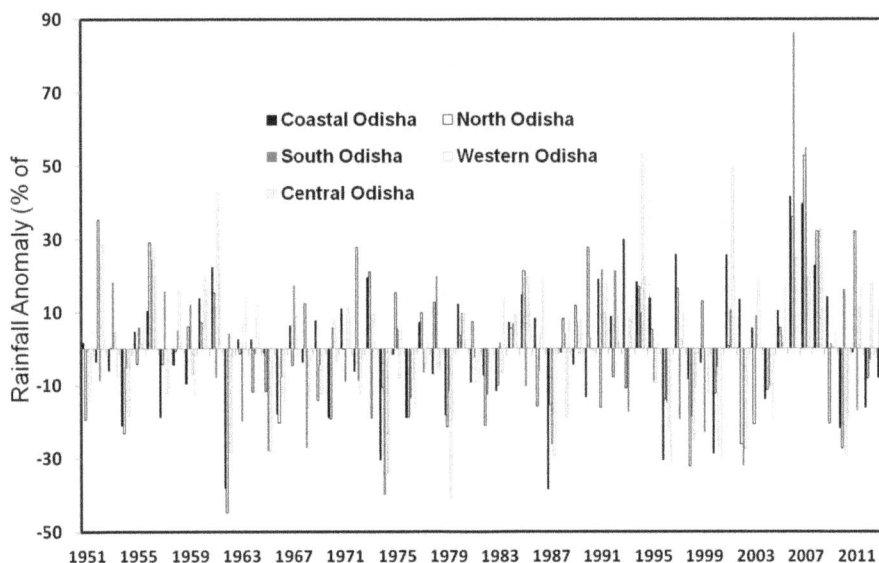

(b) Comparison of IAV of seasonal monsoon rainfall over different zones of Odisha

Figure 3. (**a**) Comparison of the interannual variability of monsoon (JJAS) rainfall anomaly over Odisha (filled bars) and India (hollow bars). The correlation co-efficient and phase of two IAVs are mentioned in parentheses; (**b**) Comparison of the interannual variability of the monsoon (JJAS) rainfall anomaly over different regions of Odisha for the period 1951–2013.

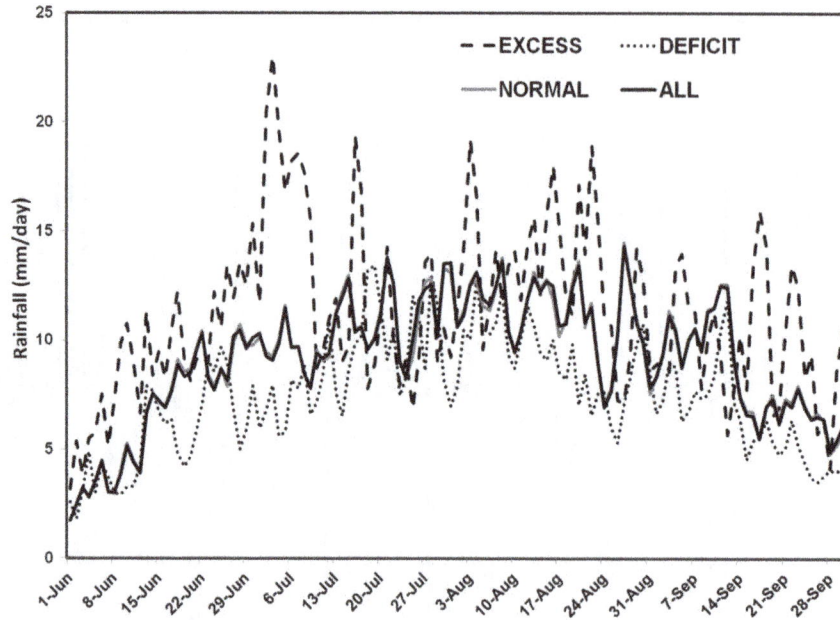

Figure 4. Seasonal cycle of the daily monsoon rainfall climatology composite over excess (dashed line), deficit (dotted line), normal (shaded line), and all 63 (thick line) years in the state of Odisha.

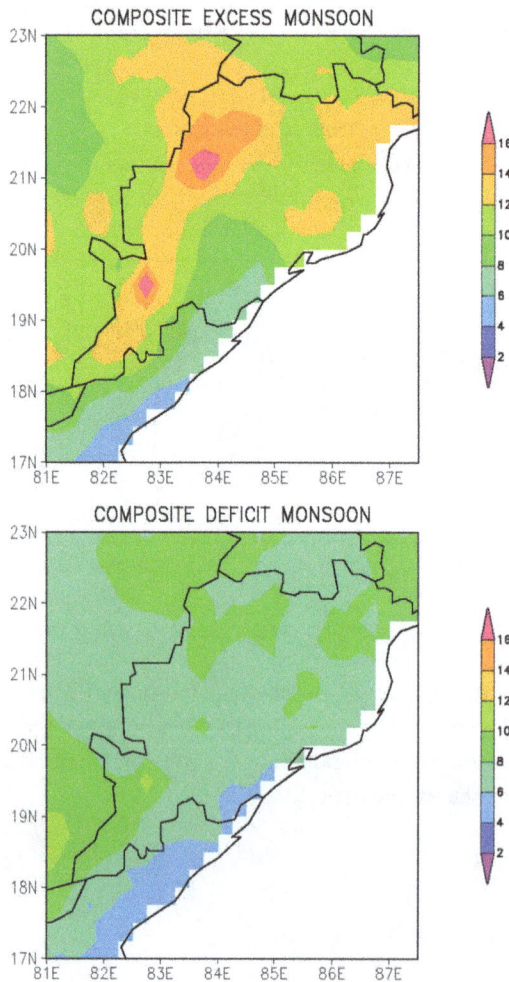

COMPOSITE EXCESS MONSOON

COMPOSITE DEFICIT MONSOON

Figure 5. *Cont.*

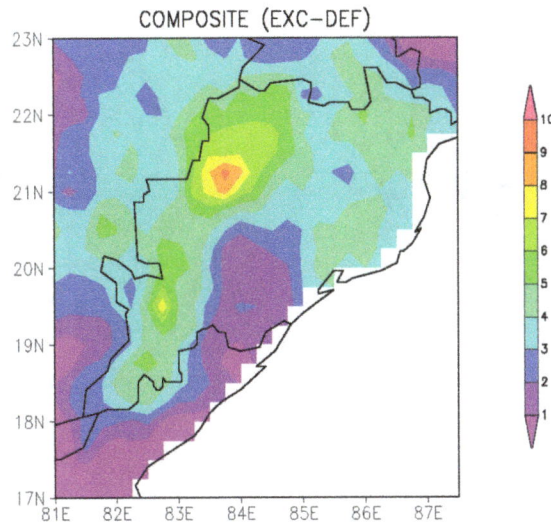

Figure 5. Comparison of seasonal (JJAS) monsoon rainfall climatology composite over excess (**top**), deficit (**middle**), and the contrast between excess and deficit years (**bottom**).

3.3. Circulation Pattern

Circulation and moisture processes associated with the monsoon are very important to understand the variability and dynamics. Analysis of wind circulation (Figure 6) and moisture or humidity (Figure 7) at two levels, i.e., 1000 hPa and 850 hPa, is carried out using NCEP Reanalysis [30] for the month of July (August), by taking the composite over excess (deficit) years. It is generally observed that the variability in the synoptic disturbances, which develop over the Bay of Bengal, results in high interannual variability of wind. The strong tropospheric Westerlies during July (excess years) are very much responsible for the excess rainfall over the state. Over the deficit years, very low intensity winds result in the low rainfalls over the state of Odisha.

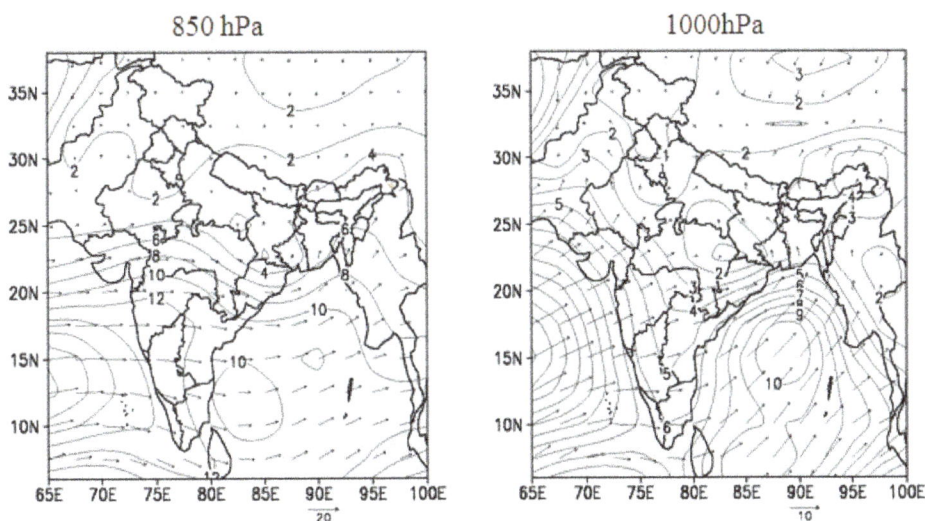

(**a**) July (Composite over 10 years of Excess Monsoon)

Figure 6. *Cont.*

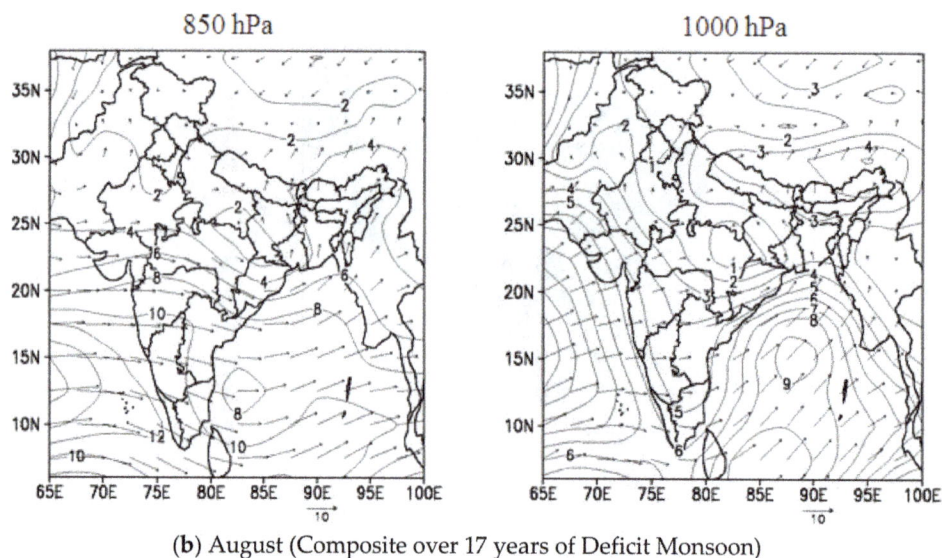

(b) August (Composite over 17 years of Deficit Monsoon)

Figure 6. Composite climatology of wind circulation using NCEP reanalysis for the month of July (**a**) and August (**b**) at 850 hPa (**left**) and 1000 hPa (**right**) levels. The July (August) analyses are composited over excess (deficit) monsoon years.

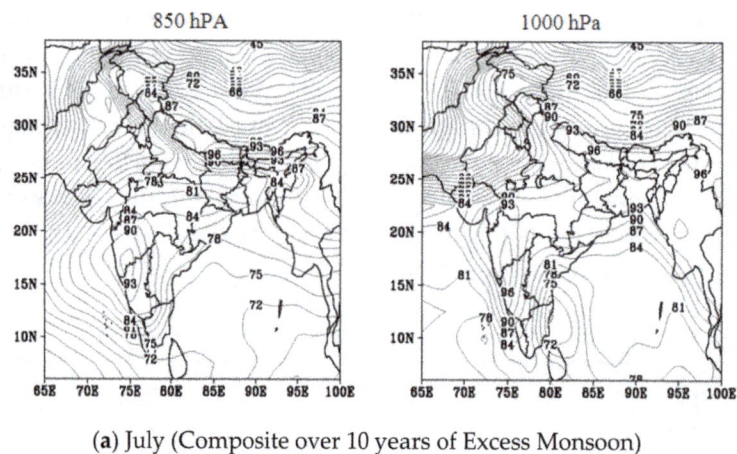

(a) July (Composite over 10 years of Excess Monsoon)

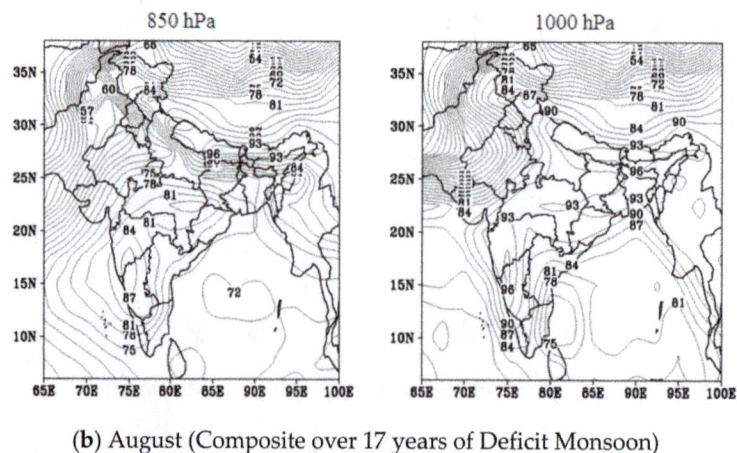

(b) August (Composite over 17 years of Deficit Monsoon)

Figure 7. Composite climatology of humidity using NCEP reanalysis for the month of July (**a**) and August (**b**) at 850 hPa (**left**) and 1000 hPa (**right**) levels. The July (August) analyses are composited over excess (deficit) monsoon years.

3.4. Impact of Global Phenomena on Monsoon Variability

The role of large-scale mechanisms, like El Niño and La Niña, can be understood as quantifying the teleconnection of these mechanisms to the ISM rainfall over the Indian subcontinent, as well as over states like Odisha. Some earlier studies showed a close association between deficit monsoon rainfall and El Niño [15–19], whereas other studies showed a link between El Niño and the weakened monsoons in the last decade and, in fact, the ISMR anomaly was positive in the recent intense warm event of 1997 [20]. In order to quantify the role of large-scale processes like El Niño and La Niña on the monsoon rainfall over India and Odisha, the seasonal rainfall climatology is analysed and shown in Figure 8. The years considered for the composite analysis of the El Niño and La Niña years presented in Table 3 are retrieved from www.ggweather.com/enso/oni.htm.

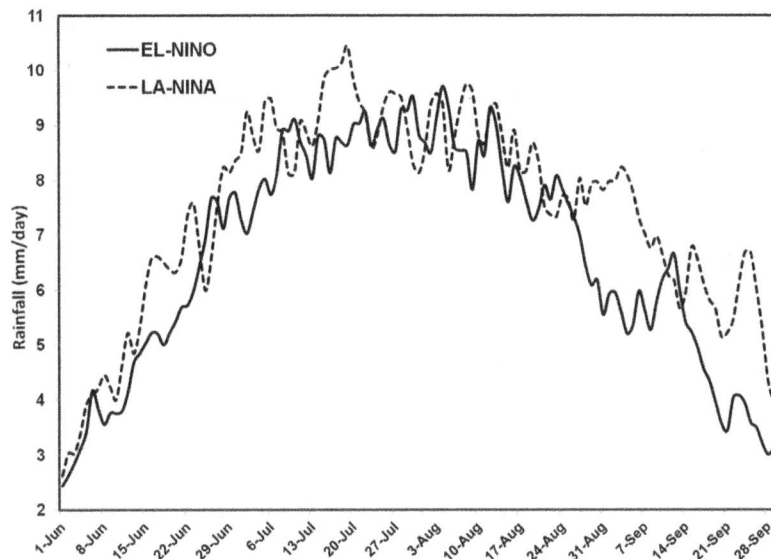

(a) Comparison of seasonal cycle of daily rainfall in El Niño and La Niña years over India

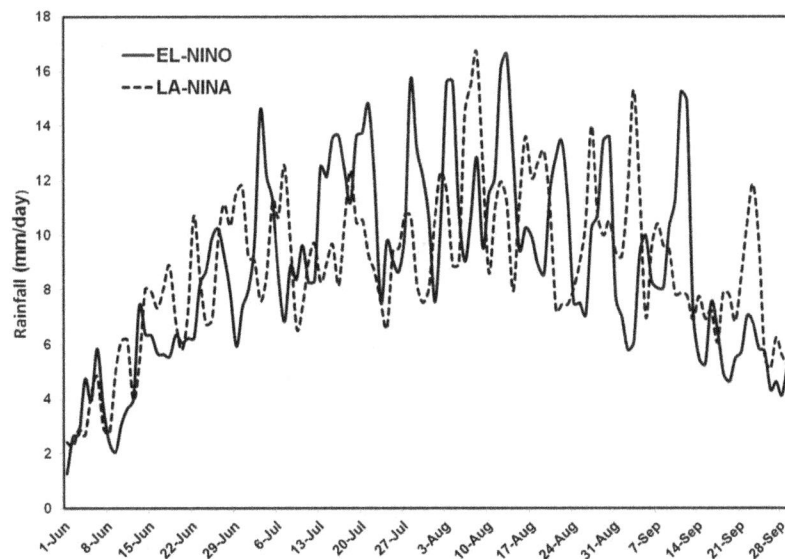

(b) Comparison of seasonal cycle of daily rainfall in El Niño and La Niña years over Odisha

Figure 8. Seasonal cycle of the daily rainfall climatology during the monsoon period (1 June to 30 September) averaged over (**a**) India and (**b**) Odisha composite over the El Niño (thick line) and La Niña (dotted line) years for the period 1 June to 32 September. The analysis is averaged over the El Niño and La Niña years separately.

Table 3. List of the years considered for large-scale phenomena (like El Niño, La Niña, positive, and negative IOD) years. The analyses are carried out for the period 1951–2013.

Events	Years	Total Number of Years
El-nino	1951, 1952, 1953, 1957, 1958, 1963, 1965, 1968, 1969, 1972, 1976, 1977, 1979, 1982, 1986, 1987, 1991, 1994, 1997, 2002, 2004, 2006, 2009	23
La-nina	1954, 1955, 1964, 1967, 1970, 1971, 1973, 1974, 1975, 1983, 1984, 1988, 1995, 1998, 1999, 2000, 2007, 2010, 2011	19
Positive IOD	1957, 1961, 1963, 1972, 1982, 1983, 1994, 1997, 2006, 2012	10
Negative IOD	1958, 1960, 1964, 1974, 1981, 1989, 1992, 1996, 1998, 2010	10

In order to find a relationship with global phenomena like ENSO, here we compared the seasonal (JJAS) rainfall, with the corresponding Niño-3.4 SST anomaly [31]. Figure 9 shows the scatterplot distribution of the seasonal (JJAS) Niño-3.4 anomaly, and the rainfall anomaly over India (top panel) and Odisha (bottom panel). The analysis shows that variations in Niño-3.4 explain about 40% and 37% of the variations of monsoon rainfall over Odisha and India, repectively.

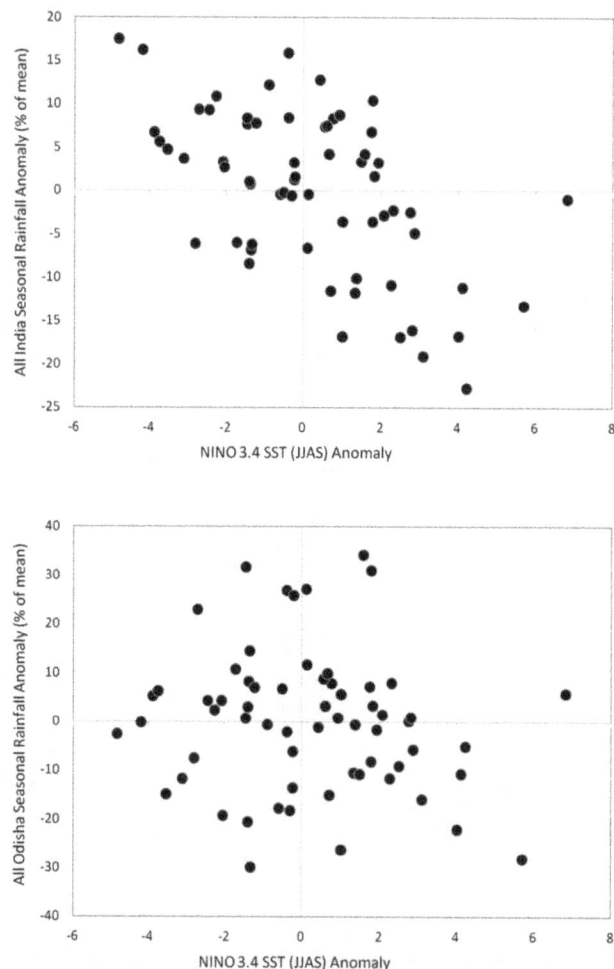

Figure 9. Scatterplot of the JJAS NINO 3.4 SST anomaly and the JJAS rainfall anomaly over all of India (**top**) and Odisha (**bottom**) for the period 1951–2013.

The difference in rainfall in El Niño and La Niña years is found to be a maximum at the larger scale (India) as compared to the regional scale (Odisha). The SD of the rainfall in 121 days is about 3 mm (6 mm) for El Niño years over India (Odisha) and about 2 mm (6mm) for La Niña years over

India (Odisha), indicating a higher variability in the rainfall during both El Niño and La Niña years at the state scale. Some studies also emphasized the impact of the Indian Ocean Dipole (IOD) mechanism on the ISM and ENSO [32,33]. In this work, the impact of IOD on the monsoon rainfall over India and Odisha is analysed and presented in Figure 10. It is clearly shown that there is no significant impact observed in the seasonal cycle, both in positive or negative IOD years, and either over India or Odisha. The phase synchronization between the monsoon rainfall at sub-regional zones (like Central Odisha, North Odisha, etc.) in the state of Odisha and the rainfall over regional (Odisha), as well as the larger (India) scale for the study period of 63 years is presented in Table 4. Here, if the sign of the interannual variability, i.e., the normalized anomaly in the smaller domains and larger domains (Odisha or India) are the same, then the phase is considered as synchronized. It is observed from the analysis (Table 4) that Western and Central Odisha domains have high percentages of phase with Odisha rainfall (89%). South Odisha, Western Odisha, and Central Odisha show high percentages of phase with all-India monsoon rainfall. Similarly, the correlation of monsoon rainfall over Western and Central Odisha is very strong with the monsoon rainfall over the larger domain. It is found that even if there is less impact of an ENSO on the interannual variability of total monsoon rainfall averaged over India, there is still a direct impact on the smaller regions like Western and Central Odisha. Similarly, there is hardly any impact of IOD on the monsoon rainfall intensity over Odisha.

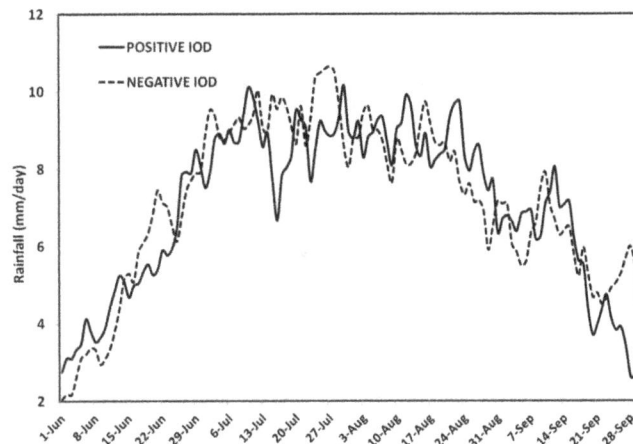

(a) Comparison of the seasonal cycle of daily rainfall in positive and negative IOD years, India

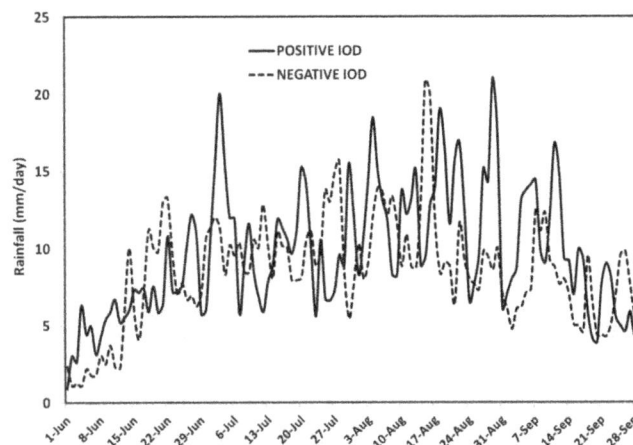

(b) Comparison of the seasonal cycle of daily rainfall in positive and negative IOD years, Odisha

Figure 10. Seasonal cycle of the daily rainfall climatology during the monsoon period (1 June to 30 September) averaged over (a) India and (b) Odisha composite over the years with positive (thick line) and negative (dotted line) IOD for the period 1 June to 30 September.

Table 4. The phase and CC analysis of rainfall in different regions of Odisha with respect to all of Odisha and all of India rainfall. The analysis was carried out for the period 1951–2013.

Domain	CC or Phase	North Odisha (2)	South Odisha (5)	Coastal Odisha (7)	Western Odisha (9)	Central Odisha (7)
All Odisha Rainfall	CC	0.82	0.76	0.88	0.95	0.92
	PS (%)	76	59	71	94	76
All India Rainfall	CC	0.59	0.58	0.76	0.89	0.79
	PS (%)	54	70	68	60	63

It is also important to find the relationship between the Dipole Mode Index (DMI) and the seasonal rainfall variability at the large scale (India) and regional scale (Odisha) to quantify the signature of the impact of IOD on the variation of seasonal rainfall at multiple spatial scales. The IOD is generally represented by DMI, which presents the anomalous SST gradient between the western equatorial Indian Ocean (50° E–70° E and 10° S–10° N) and the south-eastern equatorial Indian Ocean (90° E–110° E and 10° S–0° N) [34]. Figure 11 shows the comparison of the interannual variability of seasonal (JJAS) rainfall over India (black bars) and Odisha (shaded bars) in the principal y-axis along with the JJAS DMI calculated using the IOD SST presented on the secondary y-axis for the period 1958–2006, which shows that there is a positive (negative) relation of the DMI with the Odisha (India) seasonal rainfall.

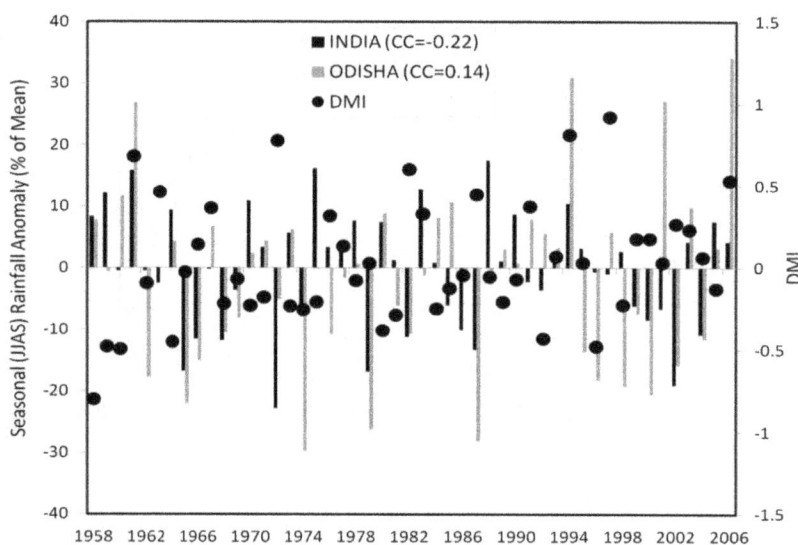

Figure 11. Inter-relation of the Dipole Mode Index (DMI) and seasonal (JJAS) rainfall anomaly over India and Odisha for the period 1958–2006. The correlation of the seasonal rainfall anomaly and the DMI anomaly are given in bracket.

4. Conclusions

The attempt at comparing the seasonality and the IAV of monsoon rainfall at a larger (India) and regional (Odisha) domain is carried out and presented here. It is observed from the present study that there is no linear or simple relation in the monsoon variability over both the scales as expected. It is also seen that there is no synchronization in the seasonal monsoon category (normal/excess/deficit) when analysed over India and Odisha. Similarly, within Odisha (different zones), there is also wide variation in the rainfall from zone to zone i.e., South Odisha shows much more variability in the rainfall distribution than the north or coastal part of Odisha. Western Odisha receives more rainfall compared to south and coastal Odisha during the monsoon, but this part of Odisha usually faces drought because of the high temperature and low humidity. The correlation of daily rainfall climatology of 63 year period during monsoon over Odisha and India is very strong (0.8), whereas the correlation between

IAV and the seasonal rainfall, as a whole, is 0.45 (99% significant) and the phase synchronization of India and Odisha monsoon is about 70% (out of 63 years). The impact of large-scale atmospheric processes like El Niño, La Niña, and IOD, etc., compared with the monsoon rainfall at both India and Odisha scales, and it is quantified that the impact is much higher for India, but it has no such relation with the Odisha rainfall. The comparative analysis of the variations in Niño-3.4 explains about 40% and 37% of the variations of monsoon rainfall over Odisha and India. These results also infer the same conclusion as drawn by previous studies [19,35,36]. The phase synchronization and correlation analysis of the monsoon rainfall averaged over sub-regions in the state of Odisha with the monsoon rainfall over regional and larger scales, i.e., Odisha and India, are variable. Western, central, and southern regions of Odisha show strong correlation and are in phase with the interannual variability of monsoon rainfall. It is also observed that there is a positive (negative) relation of the DMI (which represents the IOD) with the seasonal monsoon rainfall variability over Odisha (India). This information is critical to seasonal monsoon prediction models for reliable prediction of the seasonal monsoon cycle and category forecast, both at India and Odisha scales.

Acknowledgments: This work is supported by the projects funded by Department of Science and Technology (Grant No. SB/S4/AS-120/2013 and SERB/F/0008/2014-15), Government of India.

Author Contributions: Krushna Chandra Gouda and Himesh Shivappa conceived and designed the experiments; Sanjeeb Kumar Sahoo and Payoshni Samantray performed the experiments; Krushna Chandra Gouda, Sanjeeb Kumar Sahoo and Payoshni Samantray analyzed the data; and Krushna Chandra Gouda and Himesh Shivappa wrote the paper.

Conflicts of Interest: The authors declare no conflict of interest.

References

1. Rakhecha, P.R.; Mandal, B.N. *The Use of Empirical Orthogonal Functions for Rainfall Estimates (Rajasthan);* Cambridge University Press: Cambridge, UK, 1981.

2. Hastenrath, S.; Rosen, A. Patterns of India monsoon rainfall anomalies. *Tellus A Dyn. Meteorol. Oceanogr.* **1983**, *35*, 324–331. [CrossRef]

3. Rasmusson, E.M.; Carpenter, T.H. The relationship between eastern equatorial Pacific sea surface temperatures and rainfall over India and Sri Lanka. *Mon. Weather Rev.* **1983**, *111*, 517–528. [CrossRef]

4. Prasad, K.D.; Singh, S.V. Large scale features of Indian summer monsoon rainfall and their association with some oceanic and atmospheric variables. *Adv. Atmos. Sci.* **1988**, *5*, 499–513. [CrossRef]

5. Kripalani, R.H.; Singh, S.V.; Arkin, P.A. Large scale features of rainfall and outgoing long wave radiation over India and adjoining regions. *Contrib. Atmos. Phys.* **1991**, *64*, 159–168.

6. Parthasarathy, B.; Munot, A.A.; Kothawale, D.R. All-India monthly and seasonal rainfall series: 1871–1993. *Theor. Appl. Climatol.* **1994**, *49*, 217–224. [CrossRef]

7. Majumdar, A.B. Southwest monsoon rainfall in India: Part-1: Spatial variability. *Mausam* **1998**, *49*, 71–78.

8. Mohanty, U.C.; Mohapatra, M. Prediction of occurrence and quantity of daily summer monsoon precipitation over Orissa (India). *Meteorol. Appl.* **2007**, *14*, 95–103. [CrossRef]

9. Wang, B.; Lee, J.Y.; Xiang, B. Asian summer monsoon rainfall predictability: A predictable mode analysis. *Clim. Dyn.* **2015**, *44*, 61–74. [CrossRef]

10. Goswami, B.N.; Krishnamurthy, V.; Annmalai, H. A broad-scale circulation index for the interannual variability of the Indian summer monsoon. *Q. J. R. Meteorol. Soc.* **1999**, *125*, 611–633. [CrossRef]

11. Gadgil, S. The Indian monsoon and its variability. *Annu. Rev. Earth Planet. Sci.* **2003**, *31*, 429–467. [CrossRef]

12. Rajeevan, M.; Pai, D.S.; Kumar, R.A.; Lal, B. New statistical models for long-range forecasting of southwest monsoon rainfall over India. *Clim. Dyn.* **2007**, *28*, 813–828. [CrossRef]

13. Patra, J.P.; Mishra, A.; Singh, R.; Raghuwanshi, N.S. Detecting rainfall trends in twentieth century (1871–2006) over Orissa State, India. *Clim. Chang.* **2012**, *111*, 801–817. [CrossRef]

14. Mohapatra, M.; Mohanty, U.C. Some characteristics of low pressure systems and summer monsoon rainfall over Orissa. *Curr. Sci.* **2004**, *87*, 1245–1255.

15. Sikka, D.R. Some aspects of the large scale fluctuations of summer monsoon rainfall over India in relation to fluctuations in the planetary and regional scale circulation parameters. *Proc. Indian Acad. Sci. Earth Planet. Sci.* **1980**, *89*, 179–195. [CrossRef]

16. Pant, G.B.; Parthasarathy, S.B. Some aspects of an association between the southern oscillation and Indian summer monsoon. *Arch. Meteorol. Geophys. Bioclimatol. Ser. B* **1981**, *29*, 245–252. [CrossRef]

17. Webster, P.J.; Yang, S. Monsoon and ENSO: Selectively interactive systems. *Q. J. R. Meteorol. Soc.* **1992**, *118*, 877–926. [CrossRef]

18. Ju, J.; Slingo, J. The Asian summer monsoon and ENSO. *Q. J. R. Meteorol. Soc.* **1995**, *121*, 1133–1168. [CrossRef]

19. Krishnamurthy, V.; Goswami, B.N. Indian monsoon–ENSO relationship on interdecadal timescale. *J. Clim.* **2000**, *13*, 579–595. [CrossRef]

20. Sperber, K.R.; Palmer, T.N. Interannual tropical rainfall variability in general circulation model simulations associated with the Atmospheric Model Intercomparison Project. *J. Clim.* **1996**, *9*, 2727–2750. [CrossRef]

21. Goswami, B.N. Interannual variations of Indian summer monsoon in a GCM: External conditions versus internal feedbacks. *J. Clim.* **1998**, *11*, 501–522. [CrossRef]

22. Goswami, P.; Gouda, K.C. Comparative evaluation of two ensembles for long-range forecasting of monsoon rainfall. *Mon. Weather Rev.* **2009**, *137*, 2893–2907. [CrossRef]

23. Goswami, P.; Gouda, K.C. Evaluation of a dynamical basis for advance forecasting of the date of onset of monsoon rainfall over India. *Mon. Weather Rev.* **2010**, *138*, 3120–3141. [CrossRef]

24. Ghosh, S.; Mujumdar, P.P. Future rainfall scenario over Orissa with GCM projections by statistical downscaling. *Curr. Sci.* **2006**, *90*, 396–404.

25. Webster, P.J.; Moore, A.M.; Loschnigg, J.P.; Leben, R.R. Coupled ocean–atmosphere dynamics in the Indian Ocean during 1997–98. *Nature* **1999**, *401*, 356–360. [CrossRef] [PubMed]

26. Kang, I.S.; Jin, K.; Wang, B.; Lau, K.M.; Shukla, J.; Krishnamurthy, V.; Schubert, S.; Wailser, D.; Stern, W.; Kitoh, A.; et al. Intercomparison of the climatological variations of Asian summer monsoon precipitation simulated by 10 GCMs. *Clim. Dyn.* **2002**, *19*, 383–395.

27. Fu, X.; Wang, B. Differences of boreal summer intraseasonal oscillations simulated in an atmosphere–ocean coupled model and an atmosphere-only model. *J. Clim.* **2004**, *17*, 1263–1271. [CrossRef]

28. Goswami, B.N. South Asian monsoon. In *Intraseasonal Variability in the Atmosphere-Ocean Climate System*; Lau, K., Waliser, D., Eds.; Springer: New York, NY, USA, 2005; pp. 19–61.

29. Pai, D.S.; Latha, S.; Rajeevan, M.; Sreejith, O.P.; Satbhai, N.S.; Mukhopadhyay, B. Development of a new high spatial resolution (0.25° × 0.25°) long period (1901–2010) daily gridded rainfall data set over India and its comparison with existing data sets over the region. *Mausam* **2014**, *65*, 1–18.

30. Kalnay, E.; Kanamitsu, M.; Kistler, R.; Collins, W.; Deaven, D.; Gandin, L.; Iredell, M.; Saha, S.; White, G.; Woollen, J.; et al. The NCEP/NCAR 40-year reanalysis project. *Bull. Am. Meteorol. Soc.* **1996**, *77*, 437–471. [CrossRef]

31. Rayner, N.A.; Parker, D.E.; Horton, E.B.; Folland, C.K.; Alexander, L.V.; Rowell, D.P.; Kent, E.C.; Kaplan, A. Global analyses of sea surface temperature, sea ice, and night marine air temperature since the late nineteenth century. *J. Geophys. Res.* **2003**, *108*. [CrossRef]

32. Saji, N.H.; Goswami, B.N.; Vinayachandran, P.N.; Yamagata, T. A dipole mode in the tropical Indian Ocean. *Nature* **1999**, *401*, 360–363. [CrossRef] [PubMed]

33. Ashok, K.; Guan, Z.; Yamagata, T. Impact of the Indian Ocean dipole on the relationship between the Indian monsoon rainfall and ENSO. *Geophys. Res. Lett.* **2001**, *28*, 4499–4502. [CrossRef]

34. Saji, N.H.; Yamagata, T. Possible impacts of Indian Ocean Dipole mode events on global climate. *Clim. Res.* **2003**, *25*, 151–169. [CrossRef]

35. Krishna Kumar, K.; Rajagopalan, B.; Cane, M. On the weakening relationship between the Indian monsoon and ENSO. *Science* **1999**, *284*, 2156–2159. [CrossRef]

36. Krishna Kumar, K.; Rajagopalan, B.; Hoerling, M.; Bates, G.; Cane, M. Unraveling the mystery of Indian monsoon failure during El Niño. *Science* **2006**, *314*, 115–119. [CrossRef] [PubMed]

9

Assessing Climate Driven Malaria Variability in Ghana Using a Regional Scale Dynamical Model

Ernest O. Asare * and Leonard K. Amekudzi

Department of Physics, Kwame Nkrumah University of Science and Technology, Kumasi 00233, Ghana; lkamekudzi.cos@knust.edu.gh or leonard.amekudzi@gmail.com
* Correspondence: eoheneasare@gmail.com or eoasare.cos@knust.edu.gh

Academic Editor: Yang Zhang

Abstract: Malaria is a major public health challenge in Ghana and adversely affects the productivity and economy of the country. Although malaria is climate driven, there are limited studies linking climate variability and disease transmission across the various agro-ecological zones in Ghana. We used the VECTRI (vector-borne disease community model of the International Centre for Theoretical Physics, Trieste) model with a new surface hydrology scheme to investigate the spatio-temporal variability in malaria transmission patterns over the four agro-ecological zones in Ghana. The model is driven using temperature and rainfall datasets obtained from the GMet (Ghana Meteorological Agency) synoptic stations between 1981 and 2010. In addition, the potential of the VECTRI model to simulate seasonal pattern of local scale malaria incidence is assessed. The model results reveal that the simulated malaria transmission follows rainfall peaks with a two-month time lag. Furthermore, malaria transmission ranges from eight to twelve months, with minimum transmission occurring between February and April. The results further reveal that the intra- and inter-agro-ecological variability in terms of intensity and duration of malaria transmission are predominantly controlled by rainfall. The VECTRI simulated EIR (Entomological Inoculation Rate) tends to agree with values obtained from field surveys across the country. Furthermore, despite being a regional model, VECTRI demonstrates useful skill in reproducing monthly variations in reported malaria cases from Emena hospital (a peri urban town located within Kumasi metropolis). Although further refinements in this surface hydrology scheme may improve VECTRI performance, VECTRI still possesses the potential to provide useful information for malaria control in the tropics.

Keywords: VECTRI; malaria; EIR; surface hydrology

1. Introduction

Malaria is hyperendemic and poses a significant public health challenge in Ghana. Despite recent scaled up malaria treatment and control intervention strategies, malaria still remains the leading cause of morbidity and mortality among the entire population. For example, between 2000 and 2011, malaria alone accounted for an average of about 40% of all out-patient attendance (OPD) in public health facilities [1–3]. Similarly, in 2011, the Ghana Health Service (GHS) [3] report indicated that suspected malaria cases accounted for about 40.2% outpatient morbidity, 35.2% hospital admissions and 18.1% of all recorded death at public hospitals. Most importantly, actual malaria cases are likely to be higher than the reported cases since private health facilities and home treatment (self medication) of the disease using both orthodox and traditional medicine are not taken into account.

In addition to health implications, malaria also presents substantial economic and developmental challenges in Ghana. Asante and Asenso-Okyere [4] found a negative association between malaria cases and gross domestic product (GDP). In a related model study, Sicuri et al. [5] estimated annual total cost of malaria treatment and prevention for children under five years to be US$37.8 million in 2009. In addition, they estimated the expenditure for treating a single malaria episode to range between US$2.89 and US$123 depending on disease severity. Furthermore, a large fraction of Ghana's health budget goes to treatment and prevention of malaria. For instance, the estimated budget for National Malaria Control Programme (NMCP) strategic plan for effective malaria prevention and treatment between 2008 and 2015 was US$880 million [6]. In addition, the disease is adversely affecting sustainability of the National Health Insurance Scheme (NHIS) due to high reported cases at the various hospitals across the country [7].

On the household level, Akazili et al. [8] found the cost of treatment of malaria to be about 34% and 1% of the household's income for the poor and the wealthy, respectively, in the Kassena-Nankana district of northern Ghana. More recently, Sicuri et al. [5] estimated that about 55% of the total cost of malaria treatment in 2009, which ranged between US$ 7.99 and US$ 229.24 per malaria episode, were borne by the patient. These clearly show that successful implementation of an effective malaria control program will have a huge socio-economic and public health impact on the country.

Similar to sub-Saharan African countries, *Anopheles gambiae sensu lato complex* and *Anopheles funestus* are the main malaria vectors in Ghana [9–14]. The distribution of these vectors is heterogeneous and somehow follows climate and ecological conditions [9]. *An. gambiae s.s.*, *An. arabiensis* and *An. melas* are the three species within the *Anopheles gambiae sensu lato complex* found in Ghana [10,11]. The *An. gambiae s.s.* vector predominates the complex and distributed throughout the country [12]. However, the other two vectors have limited distribution within the country. *An. arabiensis* predominates in the savanna region while *An. melas* are confined along the coast [11,12]. Regarding *An. funestus*, Dadzie et al. [14] found *An. funestus sensu stricto* as the only malaria transmission vector in the sub group found in the country. Although *An. funestus sensu stricto* are found all over the country, they are the predominant and important vectors in the savanna ecological zone [14].

Three out of four main species of human malaria parasites are present in Ghana. *Plasmodium falciparum*, the most severe and life threatening, is predominant in the country, accounting for about 80% to 90% of all malaria infections. This is followed by *Plasmodium malariae*, which is responsible for between 20% and 36% of malaria cases while *Plasmodium ovale* is less prevalent, accounting for less than a percent (about 0.15%) of all malaria parasitemia [4,15]. Moreover, mixed infections of *Plasmodium falciparum* and *Plasmodium malariae* are also common. For instance, in Accra, Klinkenberg et al. [16] detected a single case of mixed infection of *Plasmodium falciparum* and *Plasmodium malariae* for a three-month study period among children between 6 and 60 months of age. However, 258 out of the 261 infections detected were due to *Plasmodium falciparum* with two cases of *Plasmodium malariae*. Similarly, in the Kassena-Nankana District located within the savanna zone, Koram et al. [17] identified 963, 63 and 36 cases of *Plasmodium falciparum*, *Plasmodium malariae* and mixed infections of the two, respectively. In addition, Dinko et al. [18] found all the three species in the Ahafo Ano South District of the Ashanti region, which is within the forest ecological zone.

Heterogeneities in malaria transmission dynamics across the four agro-ecological zones in Ghana have been reported. These differences in malaria incidence are due to a combination of factors such as vector and parasite distribution [12,19], climate drivers [20,21], environment and land use change [22,23], socioeconomic factors [24] and human host behavior. For instance, within the coastal, forest and transition zones with bimodal rainfall regime, malaria transmission tends to be perennial and intense but with slightly higher cases during the wet season [20,21,25]. In the savanna zone with unimodal rainfall and a long dry season, malaria transmission, although intense, shows more pronounced seasonality relative to the other zones. For example, Appawu et al. [10] observed transmission peaks between June and October in the Kassena-Nankana District in northern Ghana. Similarly, in the same district, Baird et al. [26] found malaria incidence density of five, which increased to seven infections/person/year in the dry and wet seasons, respectively, among children under two years. Despite this, non-climatic factors such as urban agriculture, open drains and irrigation, among others, introduce local hot spot transmission within the various ecological zones, which modify local disease dynamics.

Rainfall, temperature, wind speed and relative humidity are the key climate drivers that influence the spatio-temporal malaria transmission. Areas like Ghana, where mean temperatures are within the range that supports malaria transmission, variations in rainfall play a key role in understanding disease dynamics. Consequently, most studies attempt to associate malaria incidence with rainfall. However, contrasting results have been observed. For instance, the Atonsu (urban), Emena (peri-urban) and Akropong (rural) towns within the Ashanti region of Ghana, Tay et al. [27] observed weak but variable relationships between rainfall and hospital morbidity data at various time lags. Klutse et al. [28] found a poor correlation between rainfall and malaria at Winneba (coastal) and Ejura (transition) zones. Interestingly, a strong but negative correlation was observed for these two locations with a two-month lag time between malaria and rainfall. In the forest zone, Danuor et al. [29] observed a strong negative correlation between rainfall and malaria incidence. Similarly, in the forest zone, Krefis et al. [30], using a regression model, found about two-month time lag between rainfall and malaria incidence. This nonlinearity between rainfall and malaria intensity has been observed elsewhere [31,32].

Due to this strong nonlinear relationship between malaria incidence and rainfall, a model that incorporates surface hydrology (e.g., the International Centre for Theoretical Physics, Trieste (VECTRI) Tompkins and Ermert [33]) is likely to perform better in predicting malaria incidence relative to models that use rainfall as proxy for aquatic habitats. For instance, rainfall in addition to local scale hydrological conditions control mosquito developmental habitat dynamics and, to some extent, its productivity [34,35]. More importantly, in Ghana, studies linking climate fluctuations and malaria transmission across the various agro-ecological zones are limited. The few available studies are based on a single or at most two ecological zones and over a short time period [27–29]. Thus, it becomes clearly difficult to understand malaria transmission dynamics over the entire country.

The aim of this paper is to investigate the spatio-temporal variability in malaria transmission patterns over the four agro-ecological zones using the VECTRI model [33] driven by rainfall and temperature datasets obtained from the 22 synoptic stations operated by the Ghana Meteorological Agency (hereafter GMet) between 1981 and 2010. Although the potential of using VECTRI to give advance warning about malaria incidence has been explored [36], this model has never been evaluated on a local scale. Consequently, the potential of the model to predict local scale seasonal variability in malaria transmission is assessed using monthly recorded malaria cases from Emena hospital (a peri urban town located within Kumasi metropolis). Results evaluation demonstrates the ability of the VECTRI model to provide malaria early warning information over Ghana, and the model also possesses the potential to predict malaria seasonality at a local scale.

2. Method and Data

2.1. Study Area and Data

In this study, daily rainfall and maximum and minimum temperatures were obtained from GMet (Accra, Ghana). The 22 GMet synoptic stations data over the country for a 30-year period (1981–2010) were considered. The name and location of these stations across the four agro-ecological zones are shown in Figure 1. These data were used as inputs to drive the VECTRI model to simulate climate-driven malaria transmission dynamics over the country. In addition, daily observations of the same variables were obtained from GMet operated agro-meteorological station (Agromet) located at Kwame Nkrumah University of Science and Technology (KNUST) campus in Kumasi (Figure 1) to drive the model to evaluate VECTRI performance on a local scale. The Agromet station is located about 4 km from the Emena hospital.

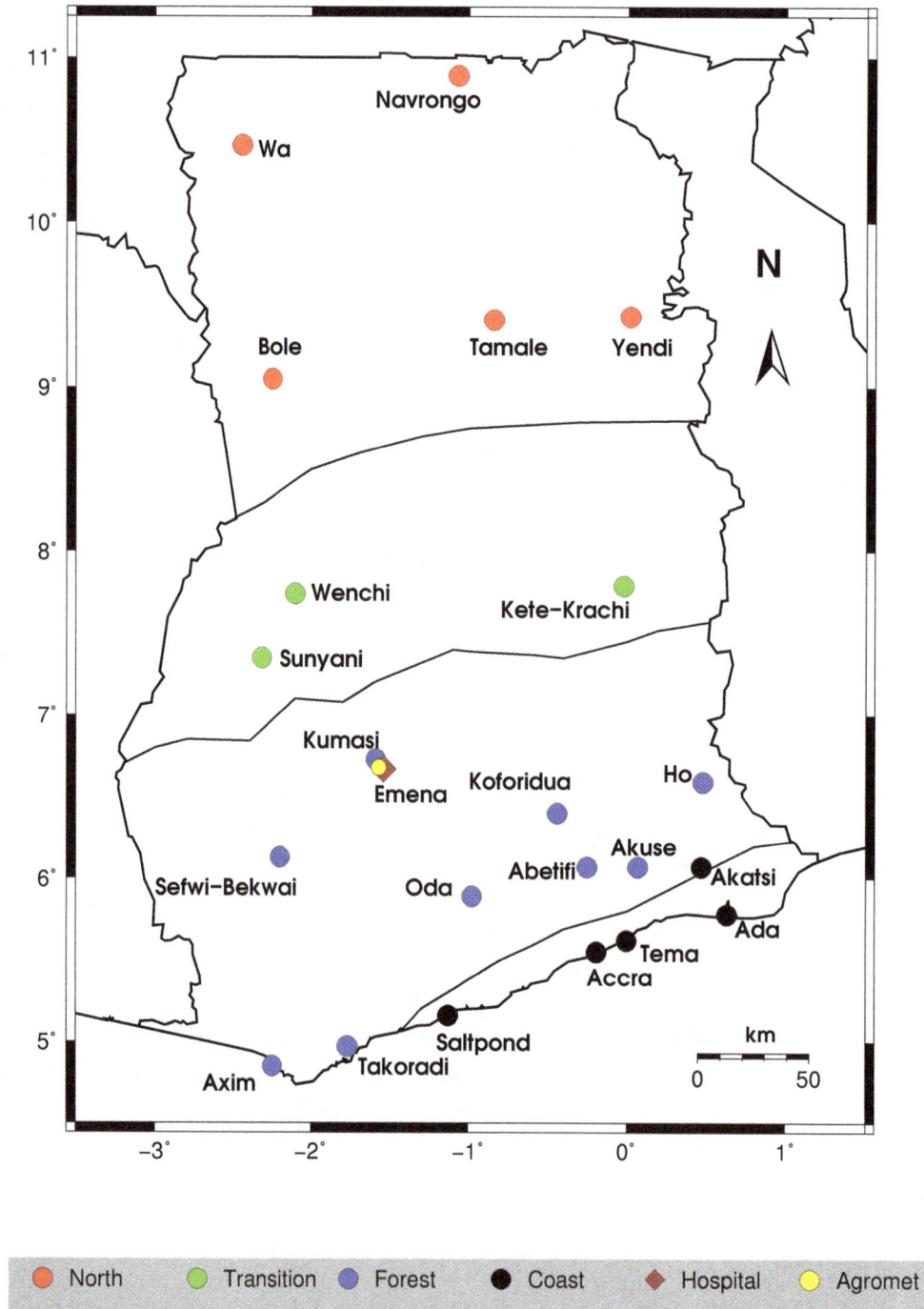

Figure 1. Map showing the 22 Ghana Meteorological Agency (GMet) synoptic stations grouped into the four agro-ecological zones. The Emena hospital and Agromet station are also shown.

Rainfall is highly variable in Ghana in terms of its onset and cessation times across different zones but exhibits less variability within the zones. These spatio-temporal variability in rainfall is mainly controlled by the north- and southward movements of the Inter-Tropical Discontinuity (ITD) [37–39], usually referred to as West African Monsoon system (WAM).

2.2. Malaria Morbidity Data

Annual malaria morbidity data between 1995 and 2010 were compiled from the Ghana Health Service annual Facts and Figures bulletin obtained from the Ghana Ministry of Health Ghana Ministry of Health [40]. These morbidity data come from only public health facilities that may not be accurate representation of the disease within the population. In addition, monthly records of confirmed malaria cases data were obtained from Emena (a peri urban town located within Kumasi metropolis) hospital from January 2010 to July 2013. The location of the Emena hospital is shown in Figure 1. These two datasets are used to evaluate the VECTRI model at national and local scales, respectively.

2.3. VECTRI Model

Tompkins et al. [33] developed VECTRI, a grid-point distributed open source dynamical model that simulates malaria transmission dynamics running with a daily integration timestep. The model uses a flexible spatial resolution that ranges from a single location to a regional scale (10–100 km) depending on the resolution of the driving climate data. The VECTRI model explicitly resolves important temperature-dependent stages such as egg-larvae-pupa, gonotrophic and sporogonic cycles. The growth stages within these cycles are presented in arrays of bins, and the process continues to advance once temperatures are within the range for growth. Complete description of the model is available in Tompkins and Ermert [33].

One novel aspect of the VECTRI model is that it incorporates the human population, which influences vector–host interaction dynamics in estimating biting rates. Consequently, the model explicitly reproduces the reduction of Entomological Inoculation Rate (EIR) with increasing population density [41]. As a result, the model is able to differentiate heterogeneities in transmission intensity between rural, peri-urban and urban areas.

VECTRI includes a simple surface hydrology scheme that estimates at each time step the fractional water coverage area in each grid cell (Equation (1)). Fractional water coverage area is a sum of both temporary and permanent developmental habitats; however, at present, spatial parametrization of permanent water bodies is not available, but is incorporated as a user defined parameter that can be tuned with knowledge of the area hydrology. Importantly, this scheme also indirectly controls habitat productivity and adult density as larvae are killed once the habitat dries out. Furthermore, although simple, the surface hydrology scheme is able to account for the negative effect of high intensity rainfall on habitat productivity through flushing away of larvae [42].

$$\frac{dw_{pond}}{dt} = K_w\Big(P(w_{max} - w_{pond}) - w_{pond}(E + I)\Big), \tag{1}$$

$$\frac{dw_{pond}}{dt} = \frac{2}{ph_{ref}}\left(\frac{w_{ref}}{w_{pond}}\right)^{p/2}\Big([Pw_{pond} + Q(w_{max} - w_{pond})]\Big(1 - \frac{w_{pond}}{w_{max}}\Big) - w_{pond}(E + fI_{max})\Big), \tag{2}$$

where w_{pond} is the daily net aggregated fractional water coverage in a grid cell, w_{max} is the maximum fractional coverage of temporary ponds, p is the pond geometry power factor, h_{ref} is the aggregated reference pond water depth, w_{ref} is the reference fractional coverage, P is the precipitation rate, E and I, which were set to a fixed constant, are evaporation rate and infiltration rate, respectively, and K_w is a linear constant, I_{max} is the maximum infiltration rate from ponds, Q is the runoff calculated from SCS formula [43] and f is the proportion of maximum pond area factor.

Recently, Asare et al. [35] developed a simplified but comprehensive prognostic surface hydrology scheme based on power-law geometrical relation that accounts for direct rainfall, pond overflow, evaporation and nonlinearities of infiltration and surface run-off terms to predict surface water area of small spatial scale mosquito developmental habitats. The scheme was further generalised to simulate, instead of individual ponds, the temporal evolution of fractional water coverage of all breeding sites within each grid-cell (Equation (2)). The scheme showed good performance in predicting both evolution dynamics of individual breeding habitats under different hydrological conditions as well as aggregated fractional coverage of all the ponds when validated against in situ pond measurements in Ghana. Asare et al. [44] implemented this scheme in the VECTRI model (which is available from VECTRI Version V1.3.0 onwards) and its performance was assessed using the 10 m resolution HYDREMATS (Hydrology, Entomology, and Malaria Transmission Simulator) model, and Bomblies et al. [45] simulations from Banizoumbou village in Niger. The newly introduced scheme (Equation (2)) demonstrated superior performance relative to the default scheme (Equation (1)) in simulating seasonal and interseasonal variability in both pond water fraction and mosquito density when compared to HYDREMATS simulations.

In this study, the VECTRI model is driven by daily rainfall and temperature measurements from various GMet stations as input data. We integrated the model using the default parameters specified (see Table 1 in [33]), except for the new parameters used for the revised surface hydrology scheme (Equation (2)). The default parameter settings for the revised surface hydrology are summarized in Table 1, and these same parameters were used for all simulations. Although all the cities where the Gmet stations are located have varied population density, VECTRI was simulated with the same population size of 500 inhabitants per km^2. For the local scale (Emena) simulation, we used a population of 150 inhabitants per km^2. The purpose of the study is to assess how climate accounts for malaria patterns across the various agro-ecological zones if all other considerations are equal.

Table 1. VECTRI (vector-borne disease community model of the International Centre for Theoretical Physics, Trieste) revised surface hydrology scheme (Equation (2)) default constants.

Symbol	Value	Units
w_{max}	0.1	
w_{ref}	0.005	
h_{ref}	250	mm
p	1.5	
I_{max}	250	mm
E	5	mm
CN	90	

CN is the curve number.

3. Results and Discussion

3.1. Rainfall and Temperature Variability

The temperature observations from various synoptic stations range from 22 °C to 34 °C, which are within the range that supports malaria transmission (Figure 2b). The high temperatures occur mostly between February and May, while low temperatures generally occur between June and October across all the various zones. The mean daily rain rates at the stations vary between 0 and 17 mm·day^{-1} (see Figure 2a). In the coastal agro-ecological zone, the major and minor rainfall peaks occurred in June and October, respectively. Similar peaks in major and minor seasons were observed over the forest agro-ecological zone with the exception of Abetifi, where the minor season peaked in September. In the transitional zone, the peaks occurred in June and September for Suyani and Kete-Krachi, respectively. However, early peaks in the major season occurred in April for Wenchi, but the minor season peak was in September. Over the savanna zone with a unimodal rainfall regime, rainfall peaked in September for Bole, Tamale and Yendi. However, rainfall onset was one month earlier at Navrongo and Wa. These variations in rainfall control spatial malaria patterns.

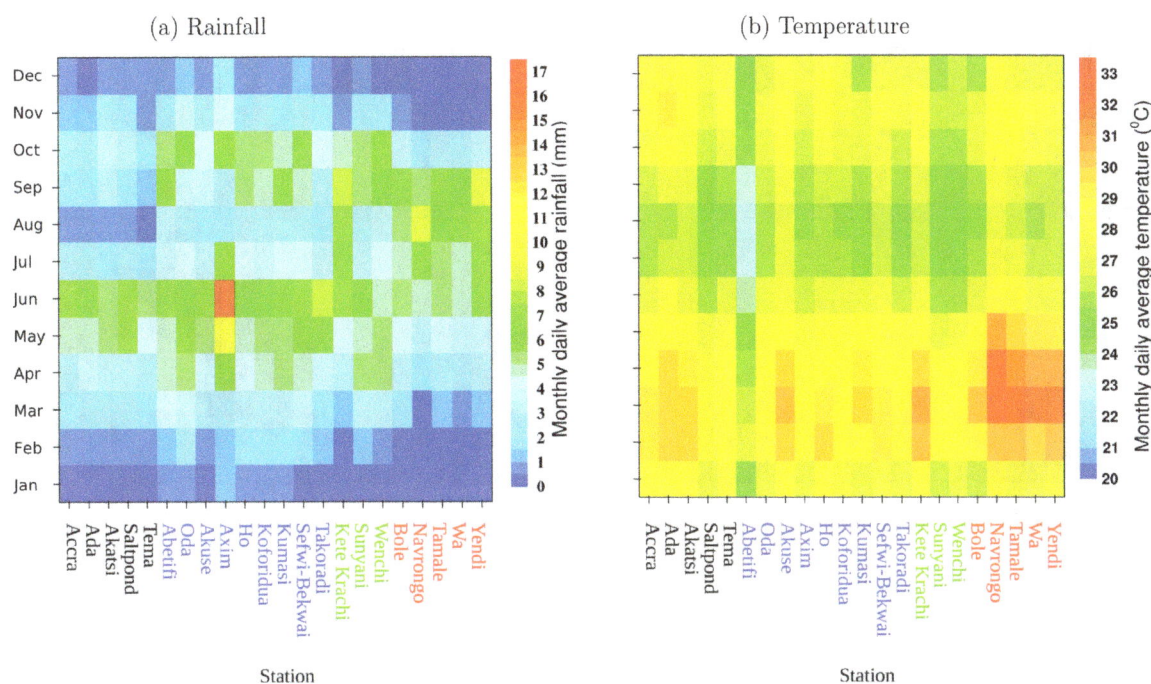

Figure 2. The monthly daily average (**a**) rainfall and (**b**) temperature between 1981 and 2010 for the various GMet synoptic weather station. Different colors are used to represent different zones (black—coastal; blue—forest; green—transition; red—north).

3.2. Model Results

Figure 3 shows VECTRI (using the V1.3.0 hydrology scheme) simulated water fraction and EIR. The results clearly show that malaria transmission generally follows rainfall patterns. The timing of peaks in the simulated EIR follows peaks in rainfall but with a lag time of approximately two months (see Figures 2b and 3b). A similar lag between rainfall and EIR have been observed in the country [21]. In the savanna zone with unimodal rainfall, the model simulated a single peak in malaria transmission, while the remaining zones with bimodal rainfall regimes simulated malaria intensity that exhibit two peaks. In the VECTRI model, malaria transmission is sustained if simulated EIR ≥ 0.01 [33]. Based on this, length of transmission is between 10 and 12 months for the coastal zone, all year transmission in the forest and transition zones and between 8 and 10 months in the savanna zone. To some extent, these results are within the range reported from field observations [46]. However, it is likely that transmission within these zones may be different from the model results due to some effects not accounted for in the VECTRI model. One such difference is permanent water bodies that can sustain transmission during the dry season.

The VECTRI simulated transmission intensity (Figure 3b) also agrees with observation studies. Appawu et al. [10] found the highest transmission between June and October for Kassena Nankana district with Navrongo as its capital. In the same district, Kasasa et al. [13] observed mosquito bites in September and Koram et al. [47] found lowest and highest transmission in May and November, respectively. The model showed a similar pattern, but the range was between July and November for Navrongo. In Accra, using hospital data, Donovan et al. [21] identified peaks in malaria either in July or August, which is consistent with EIR peak in August followed by July simulated by the model for Accra. In Kintampo in the transition zone, Dery et al. [20] found the peak month to be September followed by November. This is to some extent in agreement with VECTRI simulated EIR for the three stations located in this zone, which all peaked in November. This level of agreement between VECTRI predicted peak month and that from field observation studies point to the fact that VECTRI can provide valuable early warning information for malaria control.

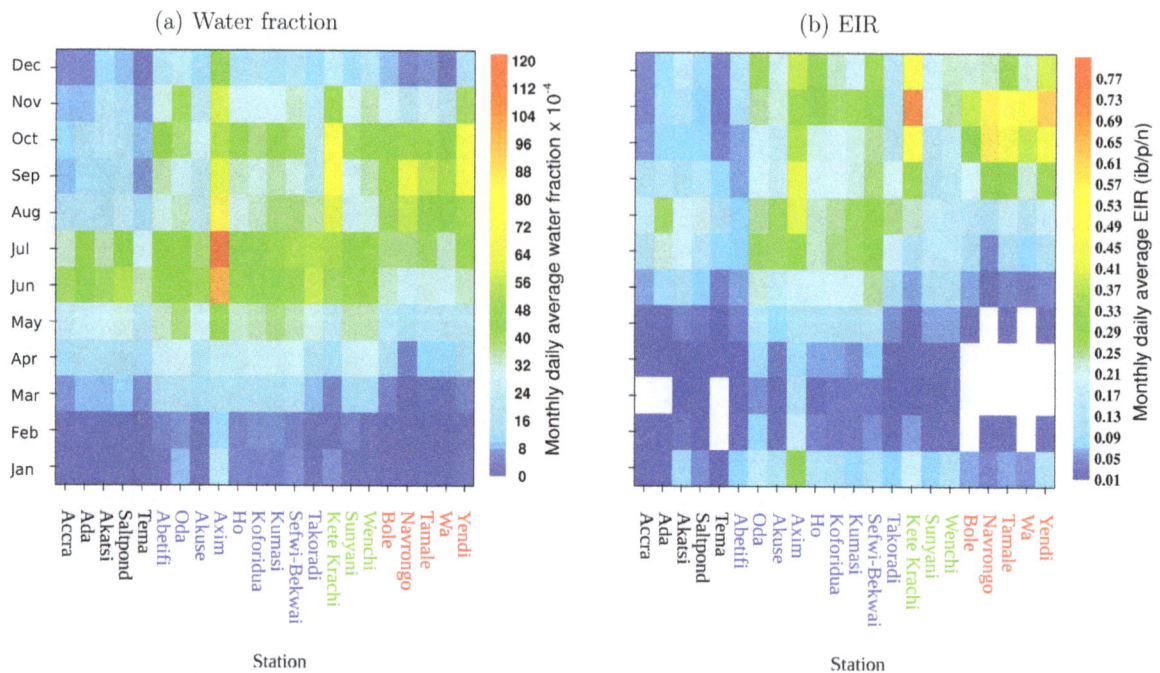

Figure 3. The monthly daily average vector-borne disease community model of the International Centre for Theoretical Physics, Trieste (VECTRI) simulated (**a**) water fraction and (**b**) Entomological Inoculation Rate (EIR) over the 30-year climatology (1981 to 2010) for various GMet synoptic weather stations. The whites spaces indicate months where EIR was less than 0.01 (no malaria transmission).

In addition to predicting the malaria peak months across the four agro-ecological zones, the VECTRI model was able to predict EIR values that tend to agree with observation based studies (Figure 3b). For instance, in Accra in the coastal zone, Klinkenberg et al. [22] found EIR values of 6.6 and 19.2 infective bites/person/year (ib/p/y) (which translates to 0.018 and 0.052 infective bites/person/night (ib/p/n), respectively, for areas located near to and far from agricultural sites. These were consistent with model values for Accra ranging between 0.014 and 0.173 with average of 0.062 ib/p/n. Similarly, the model predicted EIR values (see Figure 3b) are in good agreement with the range (0.1 and 0.7 ib/p/n) estimated by Dadzie et al. [14] from mosquitoes captured by human landing method at some locations within the country. In addition, the mean annual EIR value of 0.02 ib/p/n reported by Robert et al. [41] for urban city centers across sub-Saharan Africa is within the range of simulated EIR values. However, in Navrongo, monthly VECTRI predicted EIR values (0 to 0.587 ib/p/n) were lower but comparable to the range observed by Kasasa et al. [13] (0 to 1.06 ib/p/n). Furthermore, at Kintampo in the transition zone, Owusu-Agyei et al. [25] measured EIR that varies between 0.18 and 0.55 ib/p/n, which is within the range (0.023–0.702) of values from VECTRI for simulations for the three stations in this zone. These slight differences between the VECTRI simulated and reported EIR values from field studies may be due to the fact that the VECTRI results are based on 30-year climatology simulations, while the field studies range from a year to about five years. Despite this, VECTRI demonstrates its ability to simulate malaria patterns across the different agro-ecological zones in Ghana.

Another feature of this study is the ability of the VECTRI model to combine both rainfall and temperature effects in determining malaria transmission dynamics. While maximum rainfall and simulated water fraction were recorded at Axim (Figures 2b and 3a), the maximum simulated EIR (Figure 3b) occurred at Kete-Krachi. A combination of factors may have resulted in this observation. Firstly, the high rainfall at Axim is likely to increase overflow from the ponds, which, in effect, will reduce the productivity of the mosquito developmental habitats [42,48]. Secondly, the low temperatures recorded at Axim (mean = 26.76 °C; range between 25.02 and 28.10 °C) relative to

Kete-Krachi (mean = 28.08 °C; range between 26.17 and 30.76 °C) may have contributed as the optimum temperature for malaria transmission ranges between 28 and 32 °C [49,50]. This confirms that the VECTRI model is able to account for the non-linear relationship between rainfall and level of malaria incidence.

It should be emphasized that the VECTRI model was also able to account for the significant impact of temperature in malaria transmission dynamics. For instance, despite Abetifi (601 m elevation) and Oda (132 m elevation) stations being located about 128 km apart and having almost similar rainfall patterns (Figure 4a), the VECTRI simulated EIR shows consistently lower values at Abetifi relative to Oda (Figure 4c). This observation is significantly due to the lower temperatures recorded at Abetifi (mean = 24.68 °C; range between 22.78 and 26.50 °C) as a result of its higher altitude relative to Oda (mean = 27.05 °C; range between 25.51 and 28.51 °C) (Figure 4b). This result clearly shows the model's ability to resolve the important temperature-dependent development rate for both the vector and the parasite. In addition, the model simulated mean lower EIR value of 0.063 ib/p/n (range between 0.012 and 0.137) for Abetifi is close to the value of 0.041 ib/p/n reported by Owusu et al. [51] in Kwahu-Mpraeso about 10 km from Abetifi. This clearly shows that malaria is still prevalent in the high altitude areas in Ghana although transmission levels are low. Areas such as Abetifi are likely to experience malaria epidemics during years with anomalously warm temperature as rainfall is not limiting transmission and stands to gain a lot from advance prediction of malaria incidence that the VECTRI model is capable of providing.

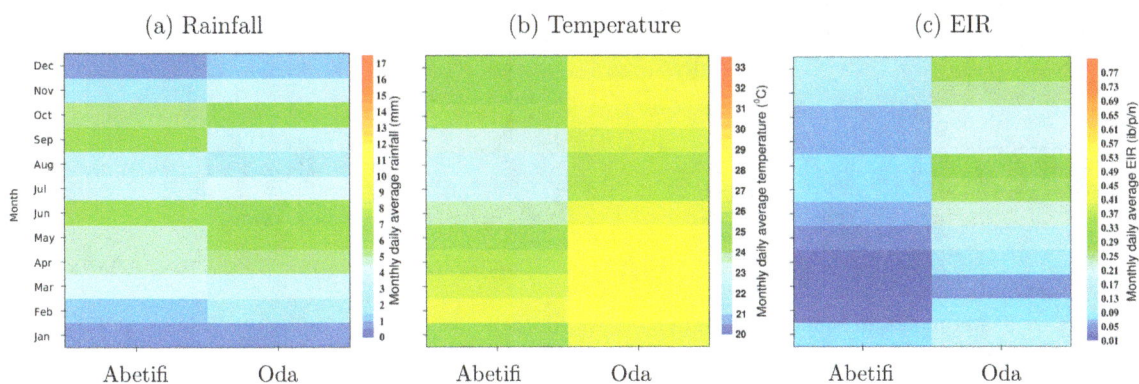

Figure 4. Comparison of observed rainfall, temperature and VECTRI simulated EIR for Abetifi and Oda stations.

3.3. VECTRI Simulated EIR and Annual Malaria Cases

Figure 5 shows anomalies in rainfall and simulated EIR for some selected stations in each of the zones. Generally, EIR closely follows the trend in rainfall across all the zones but exhibits interannual variability. This is due to the fact that the mean temperature over the whole country is above the threshold of 16 °C that supports malaria transmission. It is notable that the VECTRI simulations reveal a slight upward trend from 2003 (Figure 5), which can be attributed to rainfall increase. However, care must be taken in the interpretation of these results as critical non climatic factors were not accounted for and the model was integrated with the same population density. There is a possibility that the actual transmission patterns may differ from what is presented in Figure 5. For instance, for Yendi (Figure 5d), low EIR was predicted due to the unimodal rainfall regime with a prolonged dry season. However, actual EIR is likely to be higher if there is presence of permanent water bodies in the area, which could serve as potential breeding grounds during the dry season to sustain malaria transmission. Despite this, the simulated EIRs are within the range reported from field based studies, which point to the critical importance of rainfall in controlling spatial and temporal distribution of malaria in the country. This study serves as a baseline for future studies looking at the role of both climatic and non-climatic factors in controlling malaria transmission intensity in Ghana.

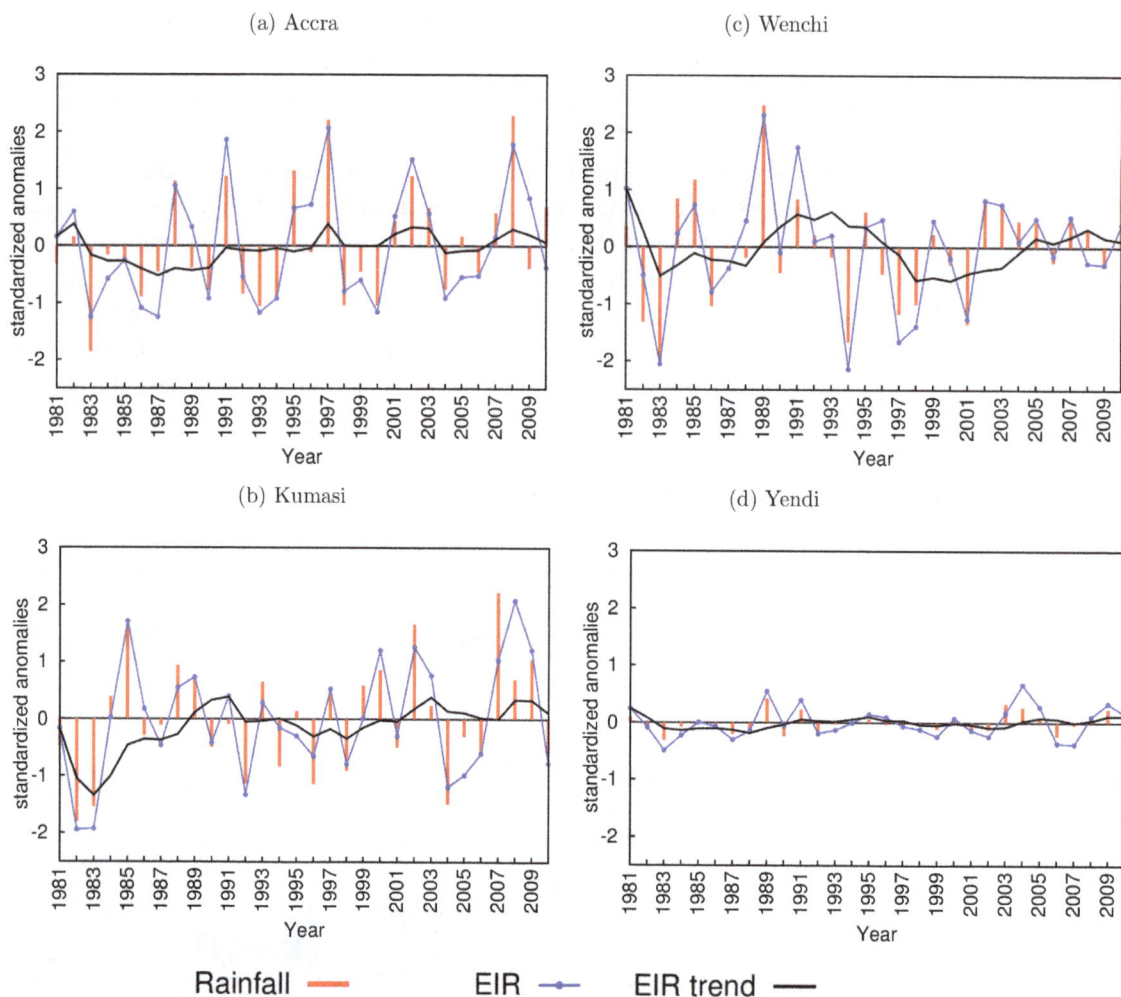

Figure 5. Anomalies of rainfall and simulated EIR for some selected synoptic stations in each of the agro-ecological zones.

The average annual malaria cases from various public health facilities and average VECTRI simulated EIR over the 22 synoptic stations between 1995 and 2010 are compared (Figure 6). Despite the disparities in trend between EIR simulations and reported malaria cases for the first seven years (up to 2001), there is a relatively good similarity between the two time series afterwards. The correlation of determination value of $R^2 = 0.53$ was observed between EIR and observed cases. It is important to note that this comparison was based on malaria case data from only public hospitals against VECTRI simulations from only 22 synoptic weather stations, and, therefore, interpretation should be done cautiously. It can be seen also from Figure 6 that there is an increase in both malaria cases and simulated EIR towards the end of the study period despite increase in intervention programs in recent times. It is worth mentioning that Ghana introduced the NHIS in 2003, which may have led to an increase in the number of people attending public health facilities for treatment, thereby increasing the number of reported malaria cases during the last periods of the study. However, this effect was not accounted for in the model, but the simulated EIR was also high during this period. This shows that there is the need to evaluate the potential of ongoing malaria control interventions in the country. The possibility of using VECTRI to address this challenge is the focus of future work.

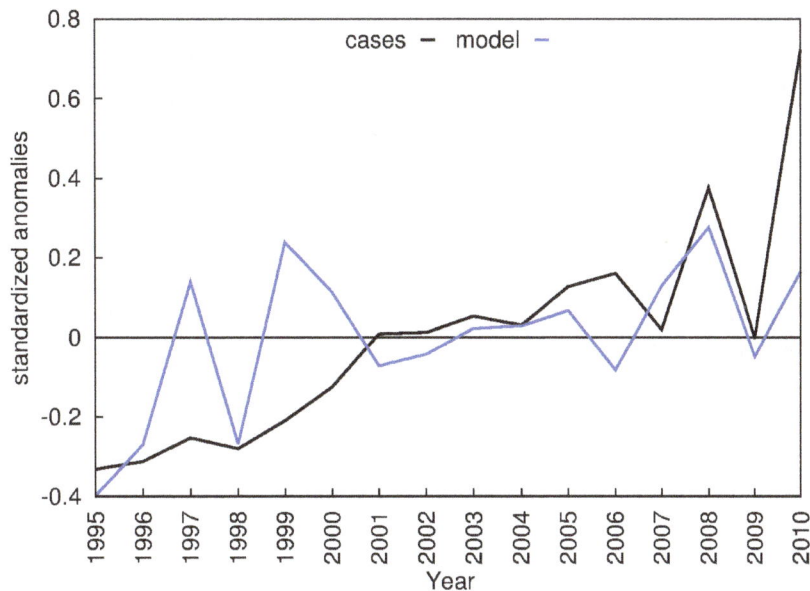

Figure 6. Comparison of average VECTRI simulated EIR and annual malaria morbidity from public health facilities.

3.4. Local Scale Malaria Transmission

The output from a single station VECTRI run is compared to the Emena monthly recorded confirmed malaria cases (Figure 7). Reported cases from the hospital indicates transmission that is slightly stable and exhibits relatively small intraannual variability. On average, there is time lag of about two months between the peaks of rainfall (June) and malaria incidence (August).

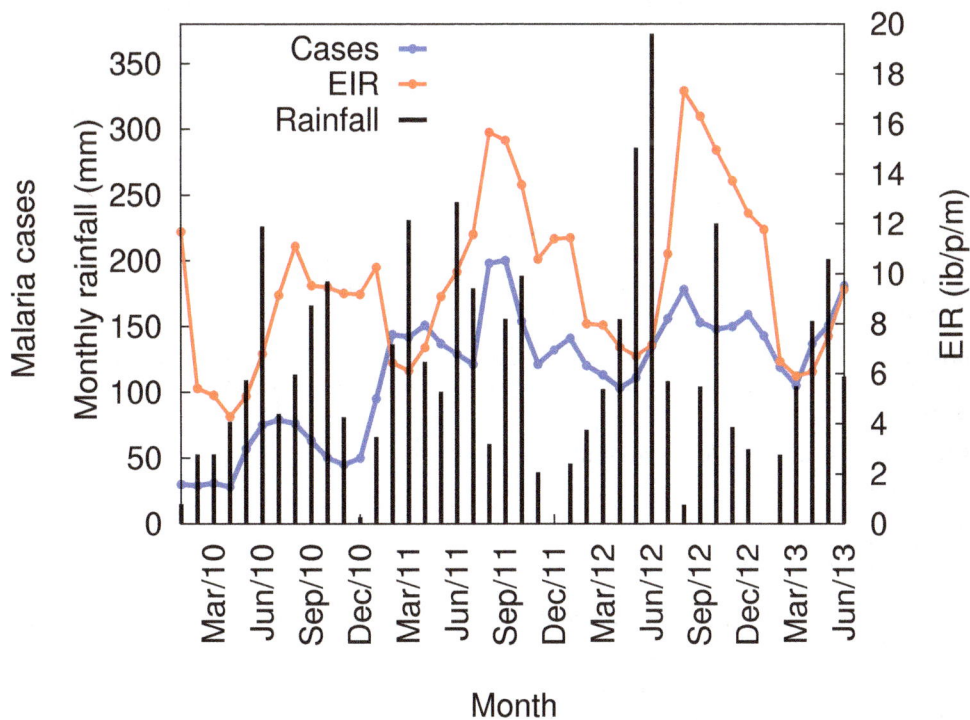

Figure 7. Comparison of monthly VECTRI simulated single location EIR and Emena hospital morbidity data.

The VECTRI using the new surface hydrology scheme reproduces a realistic trend in the reported cases with correlation of determination ($R^2 = 0.54$). Interestingly, this correlation value is higher than the best ($R^2 = 0.32$) obtained using rainfall directly. This value was obtained at a two-month time lag. A similar relatively small correlation ($R^2 = 0.29$ at a two-month lag time) between rainfall and monthly malaria cases was observed by Tay et al. [27] at this same study location. In addition, the simulated EIR (see Figure 7) tends to agree with the range 7.8 and 15 infective bites/person/month ($ib/p/m$) for the same study location (Emena) published by Wihibeturo [52] from field based studies. This further confirms the potential of the VECTRI model to simulate local scale malaria transmission dynamics despite being a regional model [44]. Nevertheless, VECTRI performance could be improved by including non-climatic factors.

4. Conclusions

In this study, we explored the potential of a regional scale dynamical model VECTRI to simulate climate driven spatio-temporal malaria transmission dynamics over the four agro-ecological zones in Ghana. The simulated results reveal intra- and inter-agro-ecological variability in terms of intensity and duration of malaria transmission that are predominantly controlled by rainfall. However, temperature was found to suppress transmission only at Abetifi, a town located on the Kwahu plateau. The correlation between annual model predicted malaria incidence (EIR) and national recorded malaria cases from public health facilities was more than 0.5. On a local scale evaluation, the correlation between monthly predicted and hospital recorded malaria cases was greater than 0.5. Interestingly, this correlation was higher than the best obtained between rainfall and malaria cases. This indicates that the VECTRI model has superior predictive ability relative to using rainfall directly.

These results demonstrate useful application of the VECTRI model to simulate malaria transmission dynamics at both national and local scales. Consequently, the VECTRI model possesses the potential to provide malaria early warning information for Ghana and should be considered by the NMCP. In addition, the model was able to discriminate between areas of low and high malaria transmission due to difference in temperature regimes and could therefore be a useful tool to study the disease patterns under future climate change. Nevertheless, improved VECTRI model performance could likely be achieved by including parametrization for permanent water bodies, topography, soil characteristics, habitat water temperature, immunity level of the population and mosquito infection status.

Acknowledgments: Ernest O. Asare was generously funded by two International Centre of Theoretical Physics (ICTP) programmes, namely the Italian government's funds-in-trust programme and the ICTP PhD Sandwich Training and Educational Programme (STEP). The study was funded by two European Union's Seventh Framework Programmes: HEALTHY FUTURES under the Grant No. 266327 and QWeCI (Quantifying Weather and Climate Impacts on health in developing countries) under Grant No. 243964.

Author Contributions: Ernest O. Asare conceived, designed and performed the experiments. Ernest O. Asare analyzed the data and wrote the manuscript. Leonard K. Amekudzi proofread the manuscript including scientific discussions.

Conflicts of Interest: The authors declare no conflict of interest.

Abbreviations

The following abbreviations are used in this manuscript:

VECTRI	vector-borne disease community model of the International Centre for Theoretical Physics, Trieste
GMet	Ghana Meteorological Agency
EIR	entomological inoculation rate
HYDREMATS	hydrology, entomology, and malaria transmission simulator model
ib/p/y	infective bites/person/year
ib/p/n	infective bites/person/night

References

1. Adams, I.; Darko, D.; Accorsi, S. Malaria: A burden explored. *Bull. Health Inf.* **2004**, *1*, 28–34.
2. National Malaria Control Programme (NMCP). *Annual Report*; National Malaria Control Programme (NMCP), Ministry of Health: Accra, Ghana, 2008.
3. Ghana Health Service (GHS). *Annual Report*; National Malaria Control Programme (NMCP), Ministry of Health: Accra, Ghana, 2011.
4. Asante, F.A.; Asenso-Okyere, K. *Economic Burden of Malaria in Ghana*; World Health Organ (WHO): Geneva, Switzerland, 2003; pp. 1–83.
5. Sicuri, E.; Vieta, A.; Lindner, L.; Constenla, D.; Sauboin, C. The economic costs of malaria in children in three sub-Saharan countries: Ghana, Tanzania and Kenya. *Malar. J.* **2013**, *12*, 307.
6. Ghana Health Service (GHS). *Strategic Plan for Malaria Control in Ghana 2008–2015*; National Malaria Control Programme (NMCP), Ministry of Health: Accra, Ghana, 2009.
7. Dontwi, I.; Dedu, V.; Aboagye, N. Ascertaining the Financial Cost of Malaria and Mitigating against It Using Actuarial Models for Financial Cost. *Int. J. Financ. Res.* **2013**, *4*, 94.
8. Akazili, J.; Aikins, M.; Binka, F.N. Malaria treatment in Northern Ghana: What is the treatment cost per case to households? *Afr. J. Health Sci.* **2008**, *14*, 70–79.
9. Appawu, M.; Baffoe-Wilmot, A.; Afari, E.; Nkrumah, F.; Petrarca, V. Species composition and inversion polymorphism of the Anopheles gambiae complex in some sites of Ghana, west Africa. *Acta Trop.* **1994**, *56*, 15–23.
10. Appawu, M.; Owusu-Agyei, S.; Dadzie, S.; Asoala, V.; Anto, F.; Koram, K.; Rogers, W.; Nkrumah, F.; Hoffman, S.L.; Fryauff, D.J. Malaria transmission dynamics at a site in northern Ghana proposed for testing malaria vaccines. *Trop. Med. Int. Health* **2004**, *9*, 164–170.
11. Yawson, A.; McCall, P.; Wilson, M.; Donnelly, M. Species abundance and insecticide resistance of Anopheles gambiae in selected areas of Ghana and Burkina Faso. *Med. Vet. Entomol.* **2004**, *18*, 372–377.
12. De Souza, D.; Kelly-Hope, L.; Lawson, B.; Wilson, M.; Boakye, D. Environmental factors associated with the distribution of Anopheles gambiae ss in Ghana; an important vector of lymphatic filariasis and malaria. *PLoS ONE* **2010**, *5*, e9927.
13. Kasasa, S.; Asoala, V.; Gosoniu, L.; Anto, F.; Adjuik, M.; Tindana, C.; Smith, T.; Owusu-Agyei, S.; Vounatsou, P. Spatio-temporal malaria transmission patterns in Navrongo demographic surveillance site, northern Ghana. *Malar. J.* **2013**, *12*, 63.
14. Dadzie, S.K.; Brenyah, R.; Appawu, M.A. Role of species composition in malaria transmission by the Anopheles funestus group (Diptera: Culicidae) in Ghana. *J. Vector Ecol.* **2013**, *38*, 105–110.
15. Afari, E.; Appawu, M.; Dunyo, S.; Baffoe-Wilmot, A.; Nkrumah, F. Malaria infection, morbidity and transmission in two ecological zones Southern Ghana. *Afr. J. Health Sci.* **1995**, *2*, 312–315.
16. Klinkenberg, E.; McCall, P.; Hastings, I.; Wilson, M.; Amerasinghe, F.; Donnelly, M. Malaria and irrigated crops, Accra, Ghana. *Emerg. Infect. Dis.* **2005**, *11*, 1290.
17. Koram, K.; Owusu-Agyei, S.; Fryauff, D.; Anto, F.; Atuguba, F.; Hodgson, A.; Hoffman, S.; Nkrumah, F. Seasonal profiles of malaria infection, anaemia, and bednet use among age groups and communities in northern Ghana. *Trop. Med. Int. Health* **2003**, *8*, 793–802.
18. Dinko, B.; Oguike, M.C.; Larbi, J.A.; Bousema, T.; Sutherland, C.J. Persistent detection of Plasmodium falciparum, P. malariae, P. ovale curtisi and P. ovale wallikeri after ACT treatment of asymptomatic Ghanaian school-children. *Int. J. Parasitol. Drugs Drug Resist.* **2013**, *3*, 45–50.
19. Afari, E.; Akanmori, B.; Nakano, T.; Ofori-Adjei, D. Plasmodium falciparum: Sensitivity to chloroquine in vivo in three ecological zones in Ghana. *Trans. R. Soc. Trop. Med. Hyg.* **1992**, *86*, 231–232.
20. Dery, D.B.; Brown, C.; Asante, K.P.; Adams, M.; Dosoo, D.; Amenga-Etego, S.; Wilson, M.; Chandramohan, D.; Greenwood, B.; Owusu-Agyei, S. Patterns and seasonality of malaria transmission in the forest-savannah transitional zones of Ghana. *Malar. J.* **2010**, *9*, 1–8.
21. Donovan, C.; Siadat, B.; Frimpong, J. Seasonal and socio-economic variations in clinical and self-reported malaria in Accra, Ghana: Evidence from facility data and a community survey. *Ghana Med. J.* **2012**, *46*, 85–94.
22. Klinkenberg, E.; McCall, P.; Wilson, M.; Amerasinghe, F.; Donnelly, M. Impact of urban agriculture on malaria vectors in Accra, Ghana. *Malar. J.* **2008**, *7*, 151.

23. Okoye, P.; Wilson, M.; Boakye, D.; Brown, C. Impact of the Okyereko irrigation project in Ghana on the risk of human malaria infection by Anopheles species (Diptera: Culicidae). *Afr. Entomol.* **2005**, *13*, 249.

24. Akpalu, W.; Codjoe, S.N.A. Economic analysis of climate variability impact on malaria prevalence: The case of Ghana. *Sustainability* **2013**, *5*, 4362–4378.

25. Owusu-Agyei, S.; Asante, K.; Adjuik, M.; Adjei, G.; Awini, E.; Adams, M.; Newton, S.; Dosoo, D.; Dery, D.; Agyeman-Budu, A.; et al. Epidemiology of malaria in the forest-savanna transitional zone of Ghana. *Malar. J.* **2009**, *8*, 220.

26. Baird, J.K.; Agyei, S.O.; Utz, G.C.; Koram, K.; Barcus, M.J.; Jones, T.R.; Fryauff, D.J.; Binka, F.N.; Hoffman, S.L.; Nkrumah, F.N. Seasonal malaria attack rates in infants and young children in northern Ghana. *Am. J. Trop. Med. Hyg.* **2002**, *66*, 280–286.

27. Tay, S.; Danuor, S.; Mensah, D.; Acheampong, G.; Abruquah, H.; Morse, A.; Caminade, C.; Badu, K.; Tompkins, A.; Hassan, H. Climate Variability and Malaria Incidence in Peri-urban, Urban and Rural Communities Around Kumasi, Ghana: A Case Study at Three Health Facilities; Emena, Atonsu and Akropong. *Int. J. Parasitol. Res.* **2012**, *4*, 83.

28. Klutse, N.A.B.; Aboagye-Antwi, F.; Owusu, K.; Ntiamoa-Baidu, Y. Assessment of Patterns of Climate Variables and Malaria Cases in Two Ecological Zones of Ghana. *Open J. Ecol.* **2014**, *4*, 764.

29. Danuor, S.; Tay, S.; Annor, T.; Forkuo, E.; Bosompem, K.; Antwi, V. The impact of climate variability on malaria incidence and prevalence in the forest zone of Ghana–A case study at two (2) hospitals located within the Kumasi Metropolitan area of the Ashanti Region of Ghana. In Proceedings of the 2nd International Conference: Climate, Sustainability and Development in Semi-Arid Regions, Fortaleza, Brazil, 16–20 August 2010; pp. 16–20.

30. Krefis, A.; Schwarz, N.; Krüger, A.; Fobil, J.; Nkrumah, B.; Acquah, S.; Loag, W.; Sarpong, N.; Adu-Sarkodie, Y.; Ranft, U.; et al. Modeling the Relationship between Precipitation and Malaria Incidence in Children from a Holoendemic Area in Ghana. *Am. J. Trop. Med. Hyg.* **2011**, *84*, 285–291.

31. Kelly-Hope, L.; Hemingway, J.; McKenzie, F. Environmental factors associated with the malaria vectors Anopheles gambiae and Anopheles funestus in Kenya. *Malar. J.* **2009**, *8*, 268.

32. Lowe, R.; Chirombo, J.; Tompkins, A.M. Relative importance of climatic, geographic and socio-economic determinants of malaria in Malawi. *Malar. J.* **2013**, *12*, 416.

33. Tompkins, A.M.; Ermert, V. A regional-scale, high resolution dynamical malaria model that accounts for population density, climate and surface hydrology. *Malar. J.* **2013**, *12*, 65.

34. Smith, M.; Macklin, M.G.; Thomas, C.J. Hydrological and geomorphological controls of malaria transmission. *Earth-Sci. Rev.* **2013**, *116*, 109–127.

35. Asare, E.O.; Tompkins, A.M.; Amekudzi, L.K.; Ermert, V. A breeding site model for regional, dynamical malaria simulations evaluated using in situ temporary ponds observations. *Geospat. Health* **2016**, *11*, 390.

36. Tompkins, A.M.; Di Giuseppe, F. Potential predictability of malaria in Africa using ECMWF monthly and seasonal climate forecasts. *J. Appl. Meteorol. Climatol.* **2015**, *54*, 521–540.

37. Owusu, K.; Waylen, P.R. The changing rainy season climatology of mid-Ghana. *Theor. Appl. Climatol.* **2013**, *112*, 419–430.

38. Manzanas, R.; Amekudzi, L.; Preko, K.; Herrera, S.; Gutiérrez, J.M. Precipitation variability and trends in Ghana: An intercomparison of observational and reanalysis products. *Clim. Chang.* **2014**, *124*, 805–819.

39. Amekudzi, L.K.; Yamba, E.I.; Preko, K.; Asare, E.O.; Aryee, J.; Baidu, M.; Codjoe, S.N. Variabilities in Rainfall Onset, Cessation and Length of Rainy Season for the Various Agro-Ecological Zones of Ghana. *Climate* **2015**, *3*, 416–434.

40. Ghana Ministry of Health. Facts and Figures. 2015. Available online: https://en.wikipedia.org/wiki/Main_Page (accessed on 3 December 2015).

41. Robert, V.; Macintyre, K.; Keating, J.; Trape, J.; Duchemin, J.; Warren, M.; Beier, J. Malaria transmission in urban sub-Saharan Africa. *Am. J. Trop. Med. Hyg.* **2003**, *68*, 169–176.

42. Paaijmans, K.; Wandago, M.; Githeko, A.; Takken, W. Unexpected high losses of Anopheles gambiae larvae due to rainfall. *PLoS ONE* **2007**, *2*, e1146.

43. United States Department of Agriculture (USDA). *National Engineering Handbook, Section 4: Hydrology*; United States Department of Agriculture: Washington, DC, USA, 1972.

44. Asare, E.O.; Tompkins, A.M.; Bomblies, A. A Regional Model for Malaria Vector Developmental Habitats Evaluated Using Explicit, Pond-Resolving Surface Hydrology Simulations. *PLoS ONE* **2016**, *11*, e0150626.

45. Bomblies, A.; Duchemin, J.; Eltahir, E. Hydrology of malaria: Model development and application to a Sahelian village. *Water Resour. Res.* **2008**, *44*, W12445.

46. Ghana Statistical Service (GSS). *Ghana Multiple Indicator Cluster Survey with an Enhanced Malaria Module and Biomarker*; Final Report; Ghana Statistical Service: Accra, Ghana, 2011.

47. Koram, K.A.; Owusu-Agyei, S.; Utz, G.; Binka, F.N.; Baird, J.K.; Hoffman, S.L.; Nkrumah, F.K. Severe anemia in young children after high and low malaria transmission seasons in the Kassena-Nankana district of northern Ghana. *Am. J. Trop. Med. Hyg.* **2000**, *62*, 670–674.

48. Dieng, H.; Rahman, G.S.; Hassan, A.A.; Salmah, M.C.; Satho, T.; Miake, F.; Boots, M.; Sazaly, A. The effects of simulated rainfall on immature population dynamics of Aedes albopictus and female oviposition. *Int. J. Biometeorol.* **2012**, *56*, 113–120.

49. Paaijmans, K.; Read, A.; Thomas, M. Understanding the link between malaria risk and climate. *Proc. Natl. Acad. Sci. USA* **2009**, *106*, 13844–13849.

50. Craig, M.; Snow, R.; Le Sueur, D. A climate-based distribution model of malaria transmission in sub-Saharan Africa. *Parasitol. Today* **1999**, *15*, 105–111.

51. Owusu, E.D.; Buabeng, V.; Dadzie, S.; Brown, C.A.; Grobusch, M.P.; Mens, P. Characteristics of asymptomatic Plasmodium spp. parasitaemia in Kwahu-Mpraeso, a malaria endemic mountainous district in Ghana, West Africa. *Malar. J.* **2016**, *15*, 1.

52. Wihibeturo, B.A. Effect of Seasonal Variability on the Incidence and Transmission Patterns of Malaria in Urban, Peri-Urban and Rural Communities around Kumasi, Ghana. Master's Thesis, Kwame Nkrumah University of Science and Technology, Kumasi, Ghana, 2014.

10

The Influence of the Antarctic Oscillation (AAO) on Cold Waves and Occurrence of Frosts in the State of Santa Catarina, Brazil

Maikon Passos A. Alves [1,*], Rafael Brito Silveira [1], Rosandro Boligon Minuzzi [2] and Alberto Elvino Franke [1]

[1]   Applied Climatology Laboratory (LabClima), Federal University of Santa Catarina (UFSC), Trindade, Florianópolis 88040-900, Brazil; rafaelbsilveirageo@gmail.com (R.B.S.); alberto.franke@ufsc.br (A.E.F.)

[2]   Agricultural Climatology Laboratory (Labclimagri), Federal University of Santa Catarina (UFSC), Admar Gonzaga Str., Itacorubi, Florianópolis 88034-000, Brazil; rbminuzzi@hotmail.com

*   Correspondence: maiconpassos@gmail.com

Academic Editors: Valdir Adilson Steinke, Charlei Aparecido da Silva and Yang Zhang

Abstract: This paper examines the relationship between the Antarctic Oscillation (AAO), cold waves and occurrence of frosts in the state of Santa Catarina, Brazil, during the winter quarter. Research on this topic can assist different spheres of society, such as public health and agriculture, since cold waves can influence and/or aggravate health problems and frosts can inflict economic losses especially in the agricultural sector. For the purpose of this paper, cold wave is considered as the event in which the daily average surface air temperature was at least two standard deviations below the average value of the series on the day and for two consecutive days or more. The data on the average air temperature and frost occurrences are provided by the Company of Agricultural Research and Rural Extension of Santa Catarina/Center for Environmental Information and Hydrometeorology (EPAGRI/CIRAM). The AAO was subjected to statistical analysis using significance tests for the averages (Student's t-test) and variances (F-test) with a significance level of $\alpha = 5\%$. The results show that cold waves are unevenly distributed in the agroecological zones of Santa Catarina. It is found that the AAO is associated with the occurrence of frosts (in the agroecological zones represented by the municipalities of Itajaí and São José) in the state of Santa Catarina.

Keywords: cold waves; frost; winter; Antarctic Oscillation; Santa Catarina; Brazil

1. Introduction

The geographical situation of the southern region of Brazil in the subtropics ensures the highest thermal amplitude in the annual cycle with a greater distinction between winter and summer. The mountain ranges and the southern plateau determine contrasts in the temperature distribution, this being the only region in Brazil with snow precipitation where noticeably cold temperatures have been registered [1]. Sharp temperature drops associated with incursions of cold air masses in the southern and southeastern regions of Brazil have great social impacts as they often cause damage, especially in the agricultural sector. These sudden temperature drops are usually accompanied by cold winds, accentuating the sensation of thermal discomfort.

Temperature drops in the state of Santa Catarina (SC), Brazil, are highly related to the state's latitude, which is submitted to the varied atmospheric systems occurring in Brazil. In winter, frontal atmospheric systems are more frequent in SC. In different seasons of the year, these systems also make temperatures drop below the expected average, since the polar air masses that come subsequently have great frequency and intensity and are responsible for the occurrences of cold and even snow [2,3].

In the state of Santa Catarina (SC) (Figure 1), incursions of polar air masses commonly occur in the early fall, causing temperature drops and favoring the formation of frosts, especially in the West and Plateau regions of SC. However, it is in the winter season that the polar anticyclone moves over Argentina towards the south of Brazil and invades the territory of Santa Catarina, causing sharp temperature drops and strong winds in the southern quadrant of the country [2]. The temperatures in the state of SC tend to increase in the south–north direction, and from the higher mountainous areas and plateau to the west and east [2,4].

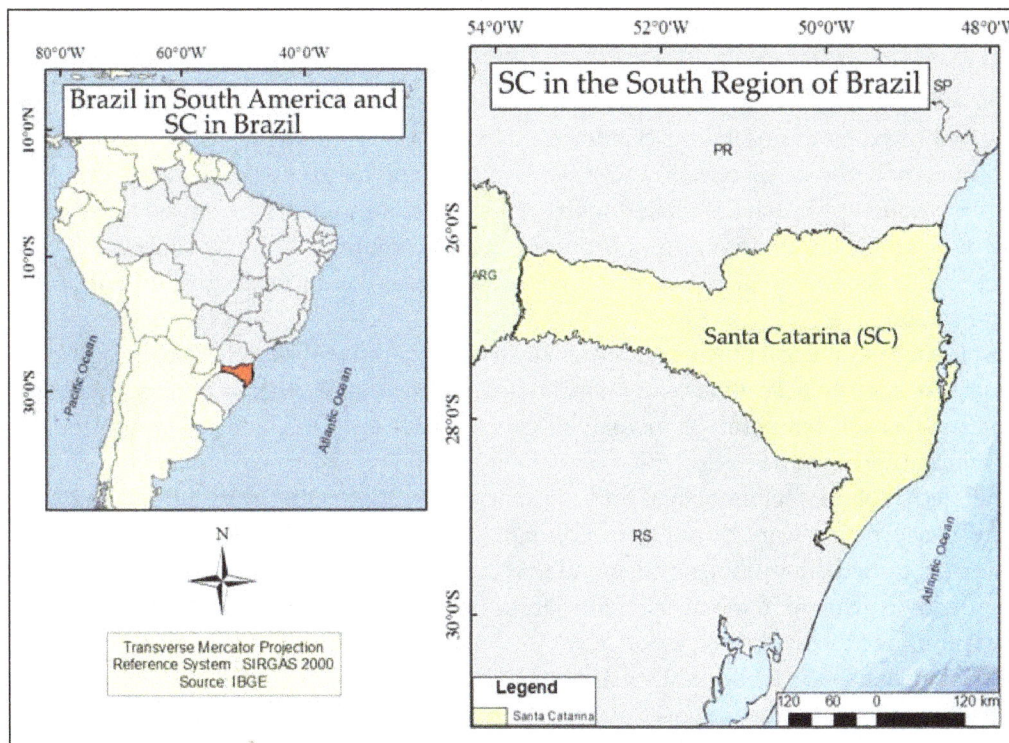

Figure 1. Location map of the state of Santa Catarina, Brazil. *Geocentric Reference System for the Americas (SIRGAS); *Brazilian Institute of Geography and Statistics (IBGE).

Santa Catarina is also known for its geographic (physical–natural) peculiarities with respect to relief. The territory has a large plateau area and approximately 20.4% of its total area consists of mountain ranges located above 900 m, so, altitude is an important factor regarding the cold in SC. These latitude peculiarities differentiate the state of SC from the state of Rio Grande do Sul (RS) and Paraná (PR), neighboring to the south and north of SC, respectively. RS is located at a higher latitude; however, it does not have areas as high as SC (although there is a mountain range). On the other hand, PR is at a lower latitude, which causes the polar masses to arrive more weakened and less frequently to the state, in general. However, these three states, RS, SC and PR, which form the Southern Region of Brazil, are directly impacted by the cold within the Brazilian context [4].

For these reasons, and the low temperatures typical of SC, the research was carried out, by choice, only in the state referring to this paper.

The social and environmental impacts of cold waves associated with oscillation rates of low-frequency variability in South America demonstrate that cold intercontinental waves sometimes cause illness—especially respiratory ones such as influenza, asthma and pneumonia—and death to people, as well as economic losses. All those factors end up impacting the development of activities in the city and countryside, damaging landscapes and cultures in different countries of the South American continent [5,6].

There is no consensus on a definition of cold wave, as there are several ways to define them [7–12]. Some studies define cold waves as a specific event, others define them as extreme or anomalous air temperature drops, without setting a threshold of temperature and frequency. In the Manual of Natural Disasters in Brazil, a cold wave is characterized by an event of rapid and big temperature drop over a wide area, lasting for several hours, days and sometimes a week or more. According to this manual, cold waves in South America occur more frequently between the months of May and September, predominantly in July and August, and typically last from four to five days [13] (p. 21).

A cold extreme is defined as the occurrence of two consecutive days or more during which the average daily surface air temperature is at least twice the standard deviation below the average local temperature in winter [11].

Cold waves differ from frosts or local coolings, as they cover large portions of the atmosphere. During a cold wave, the atmosphere is characterized by abnormally high pressures, clear sky, reduction or substitution of liquid precipitation, and periods of frost and abnormally low temperatures [5].

Frosts in southern Brazil are related with incursions of strong cold fronts, which consequently open the way to migratory anticyclones from southern Argentina [14]. The absence of cloudiness associated with a decrease in the wind speed creates conditions conducive to frost formation typical of the winter season.

Frost is a meteorological phenomenon visually recognizable by observers, usually at conventional weather stations, or its occurrence can be estimated by analyzing other registered meteorological variables such as soil temperature, hourly air temperature and/or daily minimum temperature, relative humidity, leaf wetness, etc. [15].

Other geographical factors such as latitude, maritimity, continentality, relief, soil and vegetation largely influence the intensity, duration and distribution of frosts in SC [16]. With the climatological study on frosts carried out by [16], it was found that in SC over 24 years, from 1980 to 2003, frosts were most frequently registered in the mesoregions of the South Plateau, Midwest, North Plateau and the northeast portion of the west mesoregion.

Frosts are classified as white or black. White frosts are associated with the formation of ice crystals on the ground, plants or other exposed surfaces. The process of white frost formation occurs upon dew freezing—transition from liquid to solid—when the soil temperature falls to 0 °C. However, when the air is very dry and the temperature of its dew point is below 0 °C, a frost may occur without the formation of ice crystals on the surface [17]. It is under these conditions that the so-called black frost occurs, which is more damaging to plants than the white frost as the surface temperature remains below 0 °C, freezing and burning the plant tissues [16,18].

Besides the geographical factors affecting the temperatures in SC [16], there is the impact caused by changes in the weather patterns, such as low-frequency climate variability, the El Niño-Southern Oscillation (ENSO) on an interannual scale, the Antarctic Oscillation (AAO), and the Pacific Decadal Oscillation (PDO) on a decadal scale. The climate variability timescales are as follows: intraseasonal (monthly variations occurring along the same season), interannual (annual variations of the annual or seasonal averages) and decennial/secular (variations from decade to decade or century to century) [19].

In the Southern Hemisphere (SH), the leading pattern of climate variability is the Southern Annular Mode (SAM), also known as Antarctic Oscillation (AAO) [20]. The AAO was originally identified by [21] as pressure variations in a belt that crosses Chile and Argentina, as opposed to the belt that crosses the region of the Weddell and Bellingshausen seas [22]. Subsequent studies determined that the AAO is a seasonal variation of pressure and geopotencial height between middle and high latitudes in the SH [20,23,24]. Thus, the AAO was identified as one of the modes of variability in the middle and high latitudes of the SH [23]. These authors defined the AAO as an oscillation at sea level between the pressure belts of the middle and high latitudes of the SH and a large-scale alternation of atmospheric mass between these pressure belts.

The positive phase of the AAO is associated with negative anomalies in temperature and geopotential height over the Antarctic continent and with positive anomalies in the middle latitudes.

This phase is characterized by a temperature rise in the Antarctic Peninsula and South Shetland, intensification of the cyclones over the Southern Ocean and the east winds around 60° S. The negative phase, or low polarity indices, is marked by anomalies in the opposite direction [24–28].

Research such as this can help different areas of society, such as public health and agriculture, due to the fact that cold waves can influence and/or aggravate health problems (especially respiratory and circulatory) [29–38]; in addition, the frosts can generate problems for the cultures maintained by the farmers of the state of SC [5,10,18]. Discoveries involving cold waves and low frequency climatic variabilities can help civil society to prepare for these events, also serving as a basis for the government at different levels of management, enabling the creation of policies, plans, projects and actions to reduce the problems generated.

Other studies have investigated cold waves in Brazil (e.g., [10,12]), but none of them with statistical approaches showing the influence of AAO during the occurrence of cold waves, nor in frost cases in austral winter.

One of the challenges for this paper was the lack of published studies on cold waves and frosts linked to a climatic variability, in this case the AAO. Although some publications have been internationally published, we believe that the present paper is a groundbreaker within the proposed theme in the Brazilian context.

In this sense, this paper aimed to analyze possible influences of the Antarctic Oscillation on the occurrence of cold waves and frosts in the state of Santa Catarina, Brazil, during the winter season.

2. Materials and Methods

This paper drew upon daily air temperature data from eleven conventional weather stations (Figure 2) of the EPAGRI (Company of Agricultural Research and Rural Extension of Santa Catarina) and the CIRAM (Center for Environmental Information and Hydrometeorology) and the INMET (National Institute of Meteorology). The time series data on daily air temperature, measured in degrees Celsius (°C), encompass the winter season of the years 1983–2013, totaling 31 years of data gathering.

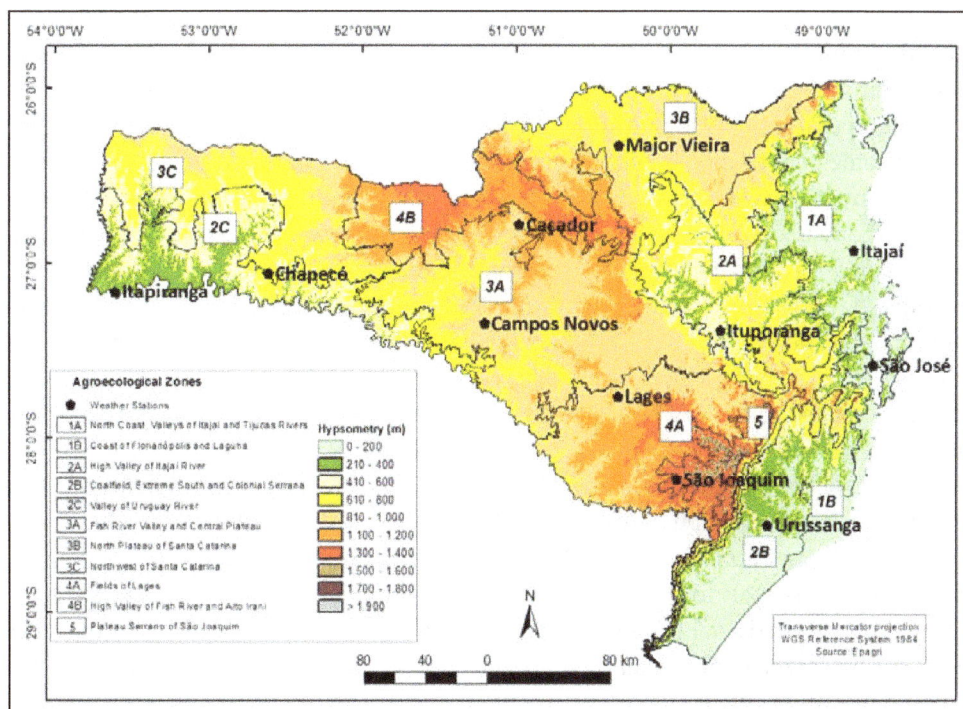

Figure 2. Hypsometric map of the state of Santa Catarina showing the locations of the weather stations and their agroecological zones. *World Geodetic System (WGS).

The choice of this time series was based on the data availability, since prior to 1983 it is difficult to find homogeneous series for all the regions analyzed, with many faults or periods without data. The 31 years analyzed represent a good data sampling; however, an ongoing updating of the data and continuation of the analysis is crucial. The analysis of the frost occurrences drew on the data of the conventional weather stations recorded by an observer at the synoptic observation times, which are 9 a.m., 3 p.m. and 9 p.m. in Brazil, Brasilia time (disregarding the Daylight Saving Time). The data recorded by a meteorological observer are advantageous for not depending upon automatic instruments with respect to their effectiveness. These data on frost occurrence recorded at the weather stations have their intensity classified from weak to exceptional, measured by the minimum temperature of the grass. The study took into account only the occurrence or non-occurrence of frosts, disregarding their intensity. The data were provided by EPAGRI/CIRAM.

For the purpose of regionalization, we used a division delimited by the agroecological zones of Santa Catarina, characterized in a relatively homogeneous way based on physical factors such as: climate, mainly regarding the thermopluviometric regime; soil; land forms; biology (flora, fauna) and socioeconomic (human activities); besides the evaluation of these same areas with respect to their sustainability potential for specific uses [39]. The state of SC has other regional classifications, among them, the political-administrative division into mesoregions by the Brazilian Institute of Geography and Statistics (IBGE), and the Köppen and Thornthwaite climate classifications. However, as presented previously, we decided to use the agroecological zones of SC for providing greater detail of the physical-natural aspects of the state [40–42].

The data on the compensated average air temperature were organized on Excel 2007 spreadsheets, and then the cold wave classification method proposed by [11] was applied. This method considers cold waves to be the event in which the average daily surface air temperature is at least twice the standard deviation below the average value for two consecutive days or more in the winter season. The standard deviation (S) is the average value of the 365 daily air temperature standard deviations. However, this paper takes into account only the 92 days of winter from June 1 to August 31. There are several methods to classify a cold wave, but there is no universal conceptualization. As an example, there are studies that use percentile techniques allied to minimum and maximum temperatures; with longer periods of time (amount of days) based on minimum temperature anomalies; percentiles with average temperature; with the wave intensity; among others (e.g., [9,43,44]). We decided not to use the minimum temperature for identification of cold waves, as in some studies, owing to the thermal amplitude that some regions may present. The method that this study used to classify the cold waves identifies only extreme waves, yet it can provide a fairly large amount of extreme events.

Thus, a cold day is classified as the day whose average air temperature is below or equal to the threshold identified for its date, and, in order to classify it as a cold wave, this needs to occur for two consecutive days or more. It can be said that this analysis can be identified as a climatological study for the classification of cold days in SC and also for identification of cold waves, given the series under analysis.

The analysis related to the AAO drew upon data on monthly averages of the Antarctic Oscillation Index (Table 1) provided by the Climate Prediction Center/National Centers for Environmental Prediction (CPC/NCEP) for the period 1983–2013. The CPC computes the Antarctic Oscillation Index daily through the projection of geopotential height anomalies at 700 hPa on the main mode of the Empirical Orthogonal Function (EOF-1) derived from monthly averages of geopotencial height anomalies at 700 hPa from 20° to 90° S. EOF-1 captures the maximum explained variance.

The AAO data allowed establishing connections of these variabilities with the occurrence of cold waves and frosts by means of graphs and comparative tables, according to the cold wave occurrence dates from each weather station.

Table 1. Monthly Antarctic Oscillation (AAO) Index[1] from 1983 to 2013—gray highlights represent winter months in the Southern Hemisphere.

YEAR	January	February	March	April	May	June	July	August	September	October	November	December
1983	−1.340	−1.081	0.166	0.149	−0.437	−0.263	1.114	0.792	−0.696	1.193	0.727	0.475
1984	−1.098	−0.544	0.251	−0.204	−1.237	0.426	0.890	−0.548	0.327	−0.009	−0.024	−1.476
1985	−0.795	0.215	−0.134	0.031	−0.066	−0.331	1.914	0.595	1.507	0.471	1.085	1.240
1986	0.158	−1.588	−0.770	−0.087	−1.847	−0.619	0.089	−0.157	0.849	0.306	−0.222	0.886
1987	−0.950	−0.708	−0.133	−0.286	0.039	−0.702	−1.531	1.485	−0.799	0.455	1.060	0.272
1988	−0.612	0.551	−0.219	−0.077	−0.749	−1.055	0.576	−0.745	−0.689	−2.314	0.401	1.074
1989	0.618	0.849	0.632	−0.573	2.691	1.995	1.458	−0.132	−0.121	0.136	0.572	−0.445
1990	−0.352	1.151	0.414	−1.879	−1.803	0.093	−1.215	0.466	1.482	0.139	−0.359	−0.312
1991	0.869	−0.852	0.522	−0.639	−0.539	−1.155	−1.220	0.036	−0.513	−0.623	−0.804	−2.067
1992	0.073	−1.627	−1.010	−0.439	−2.032	−2.193	−0.566	−0.350	0.435	−0.319	0.122	0.244
1993	−2.021	0.437	−0.378	0.087	1.260	1.218	1.957	1.083	1.061	0.748	0.324	1.028
1994	0.723	1.157	0.693	−0.052	−0.153	−1.682	−0.492	1.910	−0.947	−0.578	−0.793	0.933
1995	1.448	0.533	−0.154	0.649	1.397	−0.802	−3.010	−0.696	1.173	−0.057	0.143	1.470
1996	0.332	−0.525	0.543	0.115	0.983	−0.252	0.021	−1.502	−1.314	0.966	−1.667	−0.023
1997	0.369	−0.244	0.701	−0.458	1.028	−0.458	0.780	0.768	0.122	−0.595	−1.905	−0.835
1998	0.413	0.390	0.736	1.927	−0.038	1.031	1.450	0.904	−0.122	0.400	0.817	1.435
1999	0.999	0.456	0.180	0.949	1.639	−1.325	0.316	0.042	−0.012	1.653	0.901	1.784
2000	1.273	0.620	0.133	0.233	1.127	0.117	0.059	−0.674	−1.853	0.347	−1.537	−1.290
2001	−0.471	−0.265	−0.555	0.515	−0.262	0.386	−0.928	0.910	1.161	1.277	0.996	1.474
2002	0.747	1.334	−1.823	0.165	−2.798	−1.112	−0.591	−0.099	−0.864	−2.564	−0.924	1.308
2003	−0.988	−0.357	−0.188	0.224	0.385	−0.775	0.727	0.678	−0.323	−0.025	−0.712	−1.323
2004	0.807	−1.182	0.432	0.151	0.460	1.195	1.474	−0.071	0.254	−0.042	−0.242	−0.973
2005	−0.129	1.243	0.158	0.355	−0.297	−1.428	−0.252	0.228	0.241	0.031	−0.551	−1.968
2006	0.339	−0.211	0.501	−0.169	1.695	0.438	0.926	−1.727	−0.324	0.879	0.101	0.638
2007	−0.083	0.075	−0.570	−1.035	−0.612	−1.198	−2.631	−0.108	0.031	−0.434	−0.984	1.929
2008	1.208	1.147	0.587	−0.873	−0.490	1.348	0.320	0.087	1.386	1.215	0.920	1.194
2009	0.963	0.456	0.605	0.029	−0.733	−0.470	−1.234	−0.686	−0.017	0.085	−1.915	0.607
2010	−0.757	−0.775	0.108	0.377	1.021	2.071	2.424	1.510	0.402	1.335	1.516	0.205
2011	0.052	1.074	−0.296	−0.870	1.266	−0.099	−1.384	−1.202	−1.250	0.388	−0.908	2.573
2012	1.583	−0.283	0.275	0.666	0.153	−0.197	1.259	0.489	0.562	−0.444	−1.701	−0.764
2013	0.071	0.716	1.375	0.611	0.360	−0.271	0.945	−1.561	−1.658	−0.458	0.189	0.061

[1] Source (data provided by): [45].

Different methods are available to test the statistical significance of the difference between population means or variances, for instance the z-test, the Student's t-test, the $\chi 2$-test, and the Snedecor's F-test. For the choice of the most appropriate method, some criteria must be considered as the sample size whether the standard deviation and/or variance is known, whether the samples follow a normal distribution and whether or not the amount of samples of both populations compared is the same. So, for being more adequate for the data analysis in this study, the Student's t-test and the Snedecor's F-test were used to test the difference between means and variances, respectively, at 5% of significance level. In this study, the Student's t-test was used to compare the means of occurrence of cold waves in the positive and negative phase of AAO. In order to test the statistically significant differences between the standard deviations of the samples, the analysis of variance was applied through the F-test, at a significance level of 5%. The analyses were carried out using the statistical program PAST© [46].

3. Results

The weather stations in the municipalities of Chapecó and Caçador registered the largest number of occurrences of cold waves, 93 and 77 respectively, followed by Campos Novos (71), São Joaquim (69), Lages (64), Itajaí (58), Ituporanga (56); Urussanga (54), São José (52), Major Vieira (50) and Itapiranga (48). The average of occurrence was 62.8 cold waves among the eleven stations in SC, from 1983 to 2013 (Figure 3). It is worth mentioning that the data periods were different for some stations: Itapiranga (1987–2013), Major Vieira (1988–2013) and the stations of São Joaquim and Ituporanga (1984–2013).

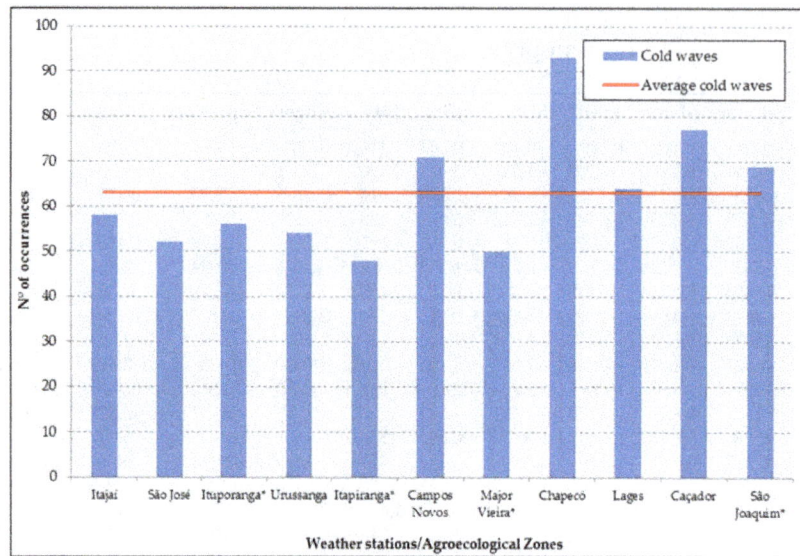

Figure 3. Number of occurrences of cold waves in winter in SC between 1983 and 2013. Note: (*) Stations with smaller data sets.

In general, the high-altitude regions presented more occurrences of cold waves, highlighting the influence of relief and altitude on the temperature behavior in SC. Another important fact that can justify this spatial distribution of cold waves are the seasonal trajectories of the polar systems, which in their mostly continental movement eventually penetrate the west/southwest regions of SC. In SC, continentality, relief, altitude and maritimity are the factors that have greater interaction with the atmospheric systems and therefore are the most influential ones [4] (p. 5).

By relating cold wave occurrences in SC with the AAO in the austral winter season (Figure 4), it can be stated that, except for Ituporanga, Itapiranga and Campos Novos, the stations presented more occurrences of cold waves in the positive phase of the AAO. Regarding the AAO, its positive phase is the one that contributes to negative temperature anomalies in southern Brazil [5,24,27,28,47,48].

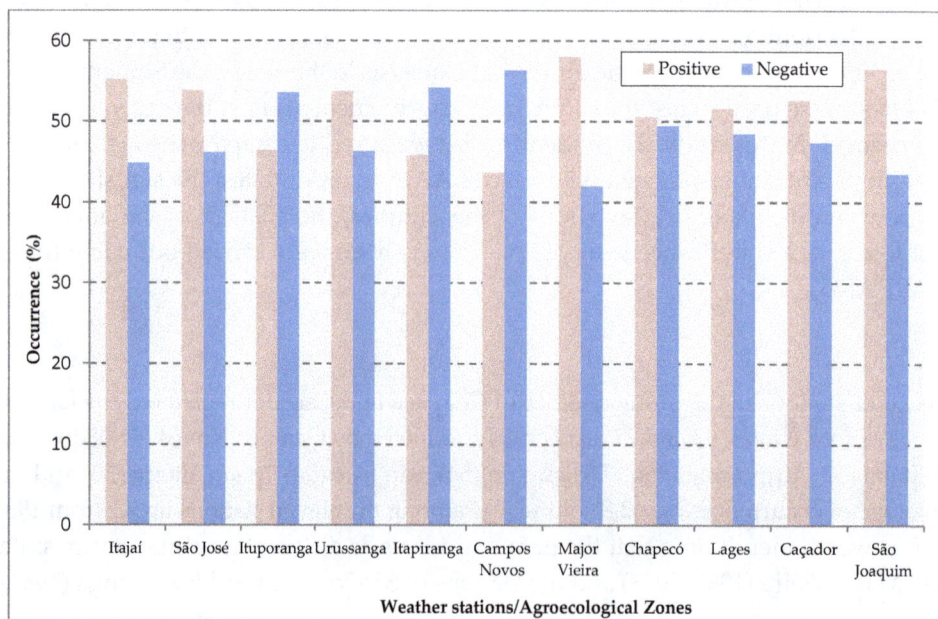

Figure 4. Percentage of occurrence of cold waves per weather station in the AAO phases in SC in the winter period from 1983 to 2013.

When considering the occurrence of frosts during the winter cold waves in relation to the AAO phases (Figure 5), it appears that in ten seasons the occurrence of frosts is frequent in the positive phase of the AAO, especially in Itajaí (70%), a result similar to those for cold waves. The Ituporanga station was the only station that, during the negative period of the analyzed variability, obtained a higher occurrence (60.9%).

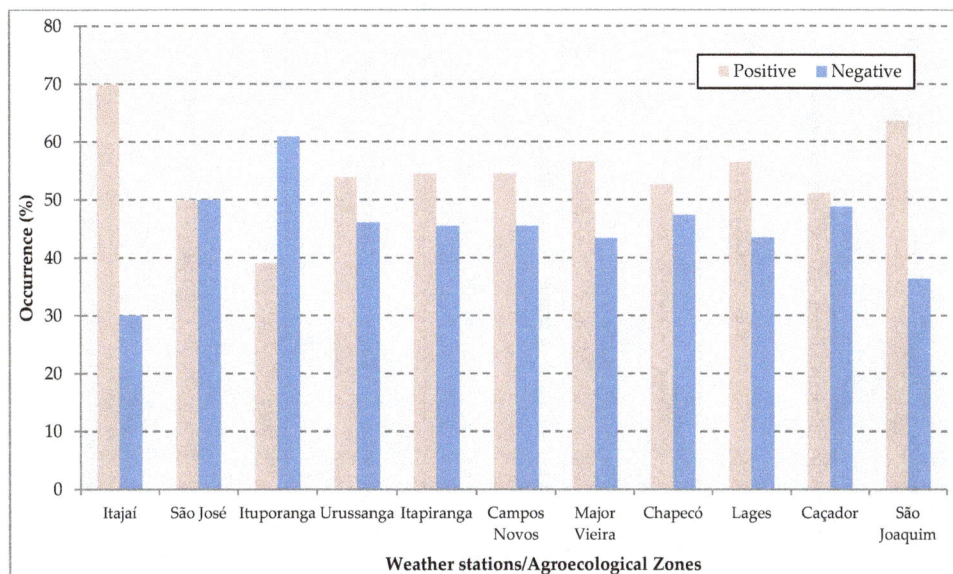

Figure 5. Percentage of occurrence of frosts during the cold waves per winter season in the AAO phases in SC in the winter period from 1983 to 2013.

Positive anomalies in the Antarctic Oscillation Index cause abnormally dry conditions in southern South America, Australia and South Africa, while strong zonal winds increase the isolation of Antarctica, reducing heat exchange with the tropics and causing a cooling of the continent and the seas surrounding it, generating the air masses that are responsible for the cold waves, low temperatures, and subsequent occurrence of frosts [5]. According to the literature review, the AAO was in line with the expectations for its phases in relation to the occurrence of cold weather (cold waves and occurrence of frosts).

3.1. Statistical Analysis between Cold Waves and the AAO

During cold waves, in most of the stations the highest average values prevailed in the negative phase of the AAO. However, these averages do not show statistical significance to any of the stations analyzed (Table 2).

Table 3 shows that, just as with the averages, the variances did not differ either among themselves in any station, with well distributed dispersions among the stations in the positive and negative phases of the AAO.

3.2. Statistical Analysis between Frosts and the AAO

The highest averages related to frost occurrences during the AAO are well distributed in the positive and negative phases among the weather stations (Table 4). Thus, the average occurrence of frosts in SC in cold wave events is the same among the phases of the AAO.

When testing the variances, the weather stations of Itajaí and São José showed a 5% statistical significance (Table 5), demonstrating that after years of positive AAO, a greater dispersion of frost occurrences than that of the negative phase is expected.

Table 2. Average values[1] of the total cold waves in the AAO phases in SC in the winter period from 1983 to 2013.

Phase of the AAO	Weather Stations/Agroecological Zones										
	Itajaí	São José	Chapecó	Caçador	Campos Novos	São Joaquim	Ituporanga	Lages	Urussanga	Major Vieira	Itapiranga
Positive	1.88a	1.64a	2.76a	2.35a	2.06a	2.43a	1.73a	1.94a	1.70a	2.07a	1.57a
Negative	1.85a	1.71a	3.28a	2.57a	2.85a	2.14a	2.14a	2.21a	1.78a	1.75a	2.00a

[1] Averages followed by the same letter in the columns do not differ among themselves by the Student's t-test at the 5% significance level.

Table 3. Results of the variances[1] for the total cold waves in the AAO phases in SC in the winter period from 1983 to 2013.

Phase of the AAO	Weather Stations/Agroecological Zones										
	Itajaí	São José	Chapecó	Caçador	Campos Novos	São Joaquim	Ituporanga	Lages	Urussanga	Major Vieira	Itapiranga
Positive	2.11a	2.86a	1.44a	1.99a	2.20a	2.12a	1.92a	1.05a	2.09a	3.30a	1.80a
Negative	2.13a	1.14a	2.52a	2.10a	1.51a	2.13a	1.82a	2.02a	1.10a	2.20a	1.16a

[1] Variances followed by the same letters in the columns do not differ among themselves by the F-test at the 5% significance level.

Table 4. Average values[1] of frost occurrences during cold waves in the AAO phases in SC in the winter period from 1983 to 2013.

Phase of the AAO	Weather Stations/Agroecological Zones										
	Itajaí	São José	Chapecó	Caçador	Campos Novos	São Joaquim	Ituporanga	Lages	Urussanga	Major Vieira	Itapiranga
Positive	0.82a	0.52a	2.94a	3.70a	3.66a	4.93a	1.80a	3.58a	2.41a	4.28a	1.28a
Negative	0.42a	0.64a	3.21a	4.28a	3.28a	3.21a	3.00a	3.35a	2.50a	3.83a	1.15a

[1] Averages followed by the same letter in the columns do not differ among themselves by the Student's t-test at the 5% significance level.

Table 5. Results of the variances[1] for frost occurrences during cold waves in the AAO phases in SC in the winter period from 1983 to 2013.

Phase of the AAO	Weather Stations/Agroecological Zones										
	Itajaí	São José	Chapecó	Caçador	Campos Novos	São Joaquim	Ituporanga	Lages	Urussanga	Major Vieira	Itapiranga
Positive	3.90a	2.88a	3.05a	5.72a	9.23a	13.26a	2.88a	5.38a	9.00a	15.29a	2.98a
Negative	1.18b	0.55b	5.71a	9.14a	5.91a	7.25a	5.69a	5.78a	4.57a	11.24a	1.97a

[1] Variances followed by the same letters in the columns do not differ among themselves by the F-test at the 5% significance level.

4. Discussion and Conclusions

During the positive phase of the AAO, low pressure anomalies occur in Antarctica and high pressure anomalies occur in the middle latitudes. Thus, during the positive phase there is a greater chance of cold events in the middle latitudes, where SC is situated [24,27,28]. A positive anomaly in the Antarctic Oscillation Index causes abnormally dry conditions in southern South America, Australia and southern Africa, while strong zonal winds increase Antarctic insulation by reducing heat exchange with the tropics and cooling the continent and surrounding seas, giving rise to air masses which are responsible for the cold waves, low temperatures and consequent occurrence of frosts [5].

In this sense, the results found in this study are in agreement with the literature on the subject, as it was verified that the positive phase of the AAO presented more occurrences of cold waves and frosts in SC during winter. Based on the Student's t-test and the F-test, with a significance level of 5%, it can be stated that the AAO does not have a statistically significant relationship with the frosts episodes in SC, the only exceptions are the agroecological zones represented by the weather stations of the municipalities of Itajaí and São José (greater variability in frost occurrences in the positive phase of the AAO). As for cold wave occurrences, the AAO has no statistically significant influence.

It was found that cold waves are unevenly distributed in the agroecological zones of SC due to the local geographic factors, the variability of the weather systems, and the trajectories of the polar systems in the continental winter. The weather stations in the municipalities of Chapecó, Caçador, Campos Novos and São Joaquim had the highest number of cold wave events in SC during winter. It is worth emphasizing the influence of relief and altitude on the air temperatures in the state, which may justify the uneven distribution of cold waves and frosts in the agroecological zones, as well as the influence of the AAO, as they are interconnected. The studies conducted by [49] on the relationship between altitude and temperature in the climate zones of SC demonstrated that altitude, latitude and longitude, in this order, influence the average air temperature. The average thermal gradient obtained for the set of all weather stations analyzed in SC was $-1\,°C/213$ m, which is equivalent to a reduction of approximately $0.48\,°C$ per 100 m of altitude.

The relationship between altitude and temperature is especially important for tropical and subtropical regions where an altitude difference of a few hundred meters and slope exposure tend to cause significant changes in the environment.

It is recommended that further research be carried out with an extended data series, from 2013 on, and including all seasons. Considering that other low-frequency climate variabilities exert influence on air temperatures in various parts of the Earth, it is suggested that cold waves and frosts be investigated in this regard. Low-frequency climate variabilities can also exert influence on temperature in a coupled way, possibly generating anomalies even in their classical signatures along their phases. In addition, the method employed here can be applied to other locations in the Southern Cone of South America.

Acknowledgments: Special thanks to the LabClima (Applied Climatology Laboratory) of the Department of Geosciences at UFSC represented by Professor Magaly Mendonça in charge of the project "Climas regionais e locais em Santa Catarina e América do Sul: mudanças, variabilidade, eventos extremos e impactos socioambientais" (Regional and local climates in Santa Catarina and South America: changes, variability, extreme events and socio-environmental impacts) in which the present study is inserted. We also thank the Geography Graduate Program at UFSC and the EPAGRI/CIRAM for providing the meteorological data.

Author Contributions: This paper originated from the master's research carried out by Maikon P. A. Alves; however, all authors contributed in the various stages of preparation of this paper. Maikon and Rafael B. Silveira contributed more actively to structuring and writing; and Rosandro B. Minuzzi and Alberto E. Franke contributed mainly to the critical review of the research.

Conflicts of Interest: The authors declare no conflict of interest.

References

1.   Grimm, A.M. Variabilidade Interanual do Clima no Brasil. In *Tempo e Clima no Brasil*, 1st ed.; Cavalcanti, I.F.A., Ed.; Oficina de texto: São Paulo, Brasil, 2009; Volume 1, pp. 353–373.

2.   Monteiro, M. Caracterização climática do estado de Santa Catarina: Uma abordagem dos principais sistemas atmosféricos que atuam durante o ano. *Rev. Geosul.* **2001**, *16*, 69–78.

3.   Murara, P.G.; Fuentes, M.V. Neve. In *Atlas de Desastres Naturais do Estado de Santa Catarina: Período de 1980 a 2010*, 2nd ed.; Cavalcanti, I.F.A., Ed.; IHGSC-GCN/UFSC: Florianópolis, Brasil, 2014; Volume 1, pp. 163–166.

4.   Monteiro, M.; Mendonça, M. Dinâmica Atmosférica no Estado de Santa Catarina. In *Atlas de Desastres Naturais do Estado de Santa Catarina: Período de 1980 a 2010*, 2nd ed.; Cavalcanti, I.F.A., Ed.; IHGSC-GCN/UFSC: Florianópolis, Brasil, 2014; Volume 1, pp. 5–12.

5.   Mendonça, M.; Romero, H. Ondas de frio, índices de oscilação e impactos socioambientais das variabilidades climáticas de baixa frequência na América do Sul. *Rev. Acta Geogr.* **2012**, *2*, 185–203. [CrossRef]

6.   BRASIL. Ministério da Saúde. Exposição ao frio Pode Agravar Doenças Respiratórias. Portal Brasil/Portal da Saúde, 2014. Available online: http://www.brasil.gov.br/saude/2014/04/exposicao-ao-frio-pode-agravar-doencas-respiratorias (accessed on 24 November 2016).

7.   Boyle, J.S. Comparison of the synoptic conditions in midlatitudes accompanying cold surges over eastern Asia for the months of December 1974 and 1978. Part I: Monthly mean fields and individual events. *Mon. Weather Rev.* **1986**, *114*, 903–918. [CrossRef]

8.   Konrad, C.E. Relationships between the intensity of cold-air outbreaks and the evolution of synoptic and planetary-scale features over North America. *Mon. Weather Rev.* **1996**, *124*, 1067–1083. [CrossRef]

9.   Walsh, J.E.; Phillips, A.S.; Portis, D.H.; Chapman, W.L. Extreme cold outbreaks in the United States and Europe, 1948–99. *J. Clim.* **2001**, *14*, 2642–2658. [CrossRef]

10.   Escobar, G.C.J. Padrões sinóticos associados a ondas de frio na cidade de São Paulo. *Rev. Bras. Meteor.* **2007**, *22*, 241–254. [CrossRef]

11.   Vavrus, S.; Walsh, J.E.; Chapman, W.L.; Portis, D. The behavior of extreme cold air outbreaks under greenhouse warming. *Int. J. Climatol.* **2006**, *26*, 1133–1147. [CrossRef]

12.   Firpo, M.A.F. Climatologia das ondas de frio e calor para o Rio Grande do Sul e sua Relação com o El Niño e La Niña. Master's Thesis, Federal University of Pelotas, Pelotas, Brazil, 2008; p. 120.

13.   Castro, A.L.C.; Calheiros, L.B.; Cunha, M.I.R.; Bringel, M.L.N.C. *Manual de Desastres Naturais*, 1st ed.; Ministério do Planejamento e Orçamento, Secretaria Especial—Defesa Civil: Brasília, Brasil, 1995; p. 182.

14.   Seluchi, M.E.; Nery, J.T. Condiciones meteorológicas associadas a la ocurrencia de heladas en la region de Maringá. *Rev. Bras. Meteor.* **1992**, *7*, 523–534.

15.   Minuzzi, B.M. Ocorrências de Geada no Estado de Santa Catarina. In *Anais de Resumos Expandidos [Recurso Eletrônico]*; II Simpósio Brasileiro de Agropecuária Sustentável, Viçosa, Brasil, 23–26 September 2010; Lana, R.P., Guimarães, G., Eds.; Viçosa, Brasil, 2010; Available online: https://www.alice.cnptia.embrapa.br/alice/bitstream/doc/874384/1/anaissimbras2010.pdf (accessed on 1 October 2016).

16.   Aguiar, D.; Mendonça, M. Climatologia das Geadas em Santa Catarina. In *Anais do SIBRADEN*; I Simpósio Brasileiro de Desastres Naturais, Florianópolis, Brasil, 26–30 September 2004; Herrmann, M.L.P., Ed.; GEDN/UFSC (CD-ROM): Florianópolis, Brasil, 2004; pp. 762–773.

17.   Tubelis, A.; Nascimento, F.J.L. *Meteorologia Descritiva: Fundamentos e Aplicações Brasileiras*, 3rd ed.; Nobel: São Paulo, Brasil, 1992; p. 374.

18.   Spinelli, K.; Alves, D.B. Geada. In *Atlas de Desastres Naturais do Estado de Santa Catarina: Período de 1980 a 2010*, 2nd ed.; Hermann, M.L.P., Ed.; IHGSC-GCN/UFSC: Florianópolis, Brasil, 2014; Volume 1, pp. 155–162.

19.   Wallace, J.M.; Hobbs, P. *Atmospheric Science: An Introductory Survey*, 2nd ed.; Academic Press: New York, NY, USA, 2006.

20.   Carvalho, L.M.V.; Jones, C.; Ambrizzi, T. Opposite phases of the Antarctic Oscillation and relationships with intraseasonal to interannual activity in the tropics during the austral summer. *J. Clim.* **2005**, *18*, 702–718. [CrossRef]

21.   Walker, G.T. World weather. *Q.J.R. Meteorol. Soc.* **1928**, *54*, 79–87. [CrossRef]

22.   Boiaski, N.T. Extremos Intra-Sazonais de Temperatura na Península Antártica e Mecanismos Atmosféricos Associados. Master's Thesis, University of São Paulo, São Paulo, Brazil, 2007; p. 119.

23.   Gong, D.; Wang, S. Definition of Antarctic Oscillation. *Geophys. Res. Lett.* **1999**, *26*, 459–462. [CrossRef]

24.   Thompson, D.W.J.; Wallace, J.M. Annular modes in the extratropical circulation. Part I: Month-to-month variability. *J. Clim.* **2000**, *13*, 1000–1016. [CrossRef]

25.   Thompson, D.W.J.; Solomon, S. Interpretation of recent southern hemisphere climate change. *Science* **2002**, *296*, 895–899. [CrossRef] [PubMed]

26. Marshall, G.J. Half-century seasonal relationships between the Southern Annular Mode and Antarctic temperatures. *Int. J. Climatol.* **2007**, *27*, 373–383. [CrossRef]

27. Justino, F.B.; Peltier, W.R. Climate anomalies induced by the Artic and Antarctics Oscillations: Glacial maximum and presente-day perpectives. *J. Clim.* **2008**, *21*, 459–475. [CrossRef]

28. Lindemann, D.S. Variações de Temperatura no Continente Antártico: Observações e Reanálises. Master's Thesis, Federal University of Viçosa, Viçosa, Brazil, 2012; p. 105.

29. Feigin, V.L.; Nikitin, Y.U.; Bots, M.L.; Vinogradova, T.E.; Grobbee, D.E. A population-based study of the associations of stroke occurrence with weather parameters in Siberia, Russia (1982–92). *Eur. J. Neurol.* **2000**, *7*, 171–178. [CrossRef] [PubMed]

30. Wilkinson, P.; Landon, M.; Armstrong, B.; Stevenson, S.; McKee, M. *Cold Comfort: The Social and Environmental Determinants of Excess Winter Death in England, 1986–1996*, 1st ed.; Joseph Rowntree Foundation, Ed.; The Policy Press: York, UK, 2001.

31. Healy, J.D. Excess winter mortality in Europe: A cross country analysis identifying key risk factors. *J. Epidemiol. Community Health* **2003**, *57*, 784–789. [CrossRef] [PubMed]

32. WHO. *Extreme Weather and Climate Events and Public Health Responses: Report on a WHO Meeting*; European Environment Agency: Bratislava, Slovakia, 2004.

33. Lomborg, B. *The Skeptical Environmentalist: Measuring the Real State of the World*, 1st ed.; Cambridge University Press: Cambridge, UK, 2003.

34. Morabito, M.; Crisci, A.; Grifoni, D.; Orlandini, S.; Cecchi, L.; Bacci, L.; Modesti, P.; Gensini, G.; Maracchi, G. Winter air-mass-based synoptic climatological approach and hospital admissions for myocardial infarction in Florence, Italy. *Environ. Res.* **2006**, *102*, 52–60. [CrossRef] [PubMed]

35. Carson, C.; Hajat, S.; Armstrong, B.; Wilkinson, P. Declining vulnerability to temperature-related mortality in London over the 20th century. *Am. J. Epidemiol.* **2006**, *164*, 77–84. [CrossRef] [PubMed]

36. Guerreiro, V.I.V. Mortalidade e Conforto Bioclimático em Coimbra—Estudo da Vulnerabilidade das Populações ao frio. Master's Thesis, University of Coimbra, Coimbra, Portugal, 2011.

37. Barnett, A.G.; Hajat, S.; Gasparrini, A.; Rocklöv, J. Cold and heat waves in the United States. *Environ. Res.* **2012**, *112*, 218–224. [CrossRef] [PubMed]

38. Gasparrini, A.; Guo, Y.; Hashizume, M.; Lavigne, E.; Zanobetti, A.; Schwartz, J.; Tobias, A.; Tong, S.; Rocklöv, J.; Forsberg, B.; et al. Mortality risk attributable to high and low ambient temperature: A multicountry observational study. *Lancet* **2015**, *386*, 369–375. [CrossRef]

39. Thomé, V.M.R.; Zampieri, S.; Braga, H.J.; Pandolfo, C.; Silva Júnior, V.P.; Bacic, I.; Laus Neto, J.; Soldateli, D.; Gebler, E.; Ore, J.D.; et al. *Zoneamento Agroecológico e Socioeconômico de Santa Catarina*, 1st ed.; (CD-ROM) Epagri: Florianópolis, Brasil, 1999.

40. Instituto Brasileiro de Geografia e Estatística—IBGE. Centro de Referência em Nomes Geográficos—Referência e Identidade. Available online: http://www.ngb.ibge.gov.br/?pagina=meso (accessed on 25 November 2016).

41. Thornthwaite, C.W. An approach toward a rational classification of climate. *Geogr. Rev.* **1948**, *38*, 55–94. [CrossRef]

42. Alvares, C.A.; Stape, J.L.; Sentelhas, P.C.; de Moraes, G.; Leonardo, J.; Sparovek, G. Köppen's climate classification map for Brazil. *Meteorol. Z.* **2013**, *22*, 711–728.

43. Landa, A.L.G. Olas de Frío en la Zona Central del Estado de Veracruz. Doctor's Thesis, University of Veracruz, Revolucion, Mexico, 2012.

44. Reboita, M.S.; Escobar, G.; Lopes, V.S. Climatologia sinótica de eventos de ondas de frio sobre a região sul de Minas Gerais. *Rev. Bras. Climatol.* **2015**, *16*, 72–92. [CrossRef]

45. NOAA/CPC. Monitoring Weather & Climate. Monthly mean AAO index since January 1979. Available online: http://www.cpc.ncep.noaa.gov/products/precip/CWlink/daily_ao_index/aao/aao.shtml (accessed on 4 August 2016).

46. Hammer, Ø.; Harper, D.A.T.; Ryan, P.D. Past: PAleontological STatistics software package for education and data analysis. *Palaeontol. Electron.* **2001**, *4*, 1–9.

47. Grimm, A.M.; Togatlian, I.M. Relação Entre Eventos El Nino/La Nina e Frequência de Extremos Frios e Quentes de Temperatura no Cone Sul da América do Sul. In *Anais do Congresso Brasileiro de Meteorologia*; XII Congresso Brasileiro de Meteorologia, Foz do Iguaçu, Brasil, 2002; SBMET: Foz do Iguaçu, Brasil, 2002; pp. 1192–1197.

48. Firpo, M.A.F.; Sansigolo, C.A.E.; Assis, S.V. Climatologia e variabilidade sazonal do número de ondas de calor e de frio no Rio Grande do Sul associadas ao ENOS. *Rev. Bras. Meteor.* **2012**, *27*, 95–106. [CrossRef]

49. Fritzsons, E.; Mantovani, L.E.; Wrege, M.S. Relationship between altitude and temperature: A contribution to climatic zonning for the state of Santa Catarina, Brazil. *Rev. Bras. Climatol.* **2016**, *18*, 80–92.

11

Urban Land Use Land Cover Changes and Their Effect on Land Surface Temperature: Case Study Using Dohuk City in the Kurdistan Region of Iraq

Gaylan Rasul Faqe Ibrahim [1,2]

[1] Geography Department, Faculty of Arts, Soran University, Soran 44008, Iraq; gailan.faqe@soran.edu.iq

[2] Tourism Department, Rawandz Private Technical Institute, Soran 44008, Iraq

Academic Editor: Yang Zhang

Abstract: The growth of urban areas has a significant impact on land use by replacing areas of vegetation with residential and commercial areas and their related infrastructure; this escalates the land surface temperature (LST). Rapid urban growth has occurred in Duhok City due to enhanced political and economic growth during the period of this study. The objective is to investigate the effect of land use changes on LST; this study depends on data from three Landsat images (two Landsat 5-TM and Landsat OLI_TIRS-8) from 1990, 2000 and 2016. Supervised classification was used to compute land use/cover categories, and to generate the land surface temperature (LST) maps the Mono-window algorithm was used. Images were also used to create the normalized difference vegetation index (NDVI), normalized difference built-up index (NDBI), normalized difference bareness index (NDBAI) and normalized difference water index (NDWI) maps. Linear regression analysis was used to generate relationships between LST with NDVI, NDBI, NDBAI and NDWI. The study outcome proves that the changes in land use/cover have a significant role in the escalation of land surface temperatures. The highest temperatures are associated with barren land and built-up areas, ranging from 47°C, 50°C, 56°C while lower temperatures are related to water bodies and forests, ranging from 25°C, 26°C, 29°C respectively, in 1990, 2000 and 2016. This study also proves that NDVI and NDWI correlate negatively with low temperatures while NDBI and NDBAI correlate positively with high temperatures.

Keywords: land use cover/change; LST; NDVI; NDBI; NDWI regression analysis

1. Introduction

In the last decade, climate researchers' attention was increasingly drawn to local and regional climate under anthropogenic influences to better understand the increasing change in the climate's driving factors [1]. One of the main causes of global climate change is increasing industrialization and urbanization. Currently, the most crucial problem that urban areas suffer from is rising surface temperatures caused by the loss of areas of vegetation and the increase of impermeable non-transpiring, non-evaporating, hard land surfaces [2–6]. One of the most noticeable effects of the modifications of terrestrial ecosystems by human activity is the change in land use/land cover (LULC) as it has greatly impacted the environment locally, regionally and globally [7–9]. The amount of humidity in the air is greatly affected by the change of natural land surfaces to built-up areas as vegetation is a major source of humidity [10]. For all surface materials, certain internal properties such as inertia, conductivity and heat capacity have an immense impact on balancing the body temperature with its surroundings [11]. Higher thermal capability for releasing daytime heat at night and greater solar radiation absorption are usually caused in urban areas by replacing vegetative areas with paved surfaces such as buildings, parking lots, roads, etc., thus causing 'heat islands' (UHI) which is the contrast of temperature between the warmer urban areas and the colder surrounding rural areas often resulting from this process [12,13].

Environmental and urban climate studies use land surface temperature (LST) and emissivity data for numerous purposes but mainly to analyze LST patterns and how they are connected to surface characteristics, urban heat island forecasts and for the relationship of LSTs with surface energy fluxes so that landscape procedures, properties, and patterns can be characterized [12,14]. LST can be utilized to represent and control the biological, physical and chemical processes of earth systems; it is also a good indicator of the earth's surface energy [15,16]. Awareness of LST supplies knowledge of spatial and temporal variations on the state of surface stability and therefore is essential in many applications [17]. A wide variety of studies employ LST as it is useful in many fields including hydrological cycles, urban climate, climate change, evapotranspiration, vegetation observations, as well as environmental observations [18–22]. It has been recognized by, among others, the International Geosphere and Biosphere Program (IGBP) as a high-priority parameter [20]. Land use classification, thermal environment, urban heat island research and hydrological investigation in urban growth, or even on a larger scale, utilize the LST satellite-derived images [23]. Land surface temperature (LST) assisted by the thermal infrared bands of remote sensing data of space-borne sensors, which analyze the relationship between urban thermal patterns, spatial structure and urban surface characteristics, is a major application of remote sensing in urban climate studies, as it helps land use and occupation planning [24]. LST information on regional and global scales is obtained by thermal infrared (TIR) remote sensing; it is a unique approach as sensors in this spectral region detect the energy that is emitted directly from the land surface [25].

Researchers A and Devadas, 2009 [26]; Abdullah, 2012 [10]; Fu and Weng, 2016 [27]; Lv and Zhou, 2011 [28]; Xiao et al., 2007 [29] utilized remote sensing images using Landsat images to generate land use and surface temperature maps and to monitor land use changes [30–33] for commercial and business centers, government offices, residential areas and public amenities which are replacing green spaces, forest and unused lands. The Klang Valley Region in Malaysia contained the most noticeable LULC change. For sustainable development to be implemented, monitoring the changes in land use can be considered as alternative good governance for administration [34]. Studies noticed an increase in urban growth with a related decrease in vegetation, which resulted in an alteration of urban microclimates [6]. Another study determined the land surface temperature and vegetation abundance relationship. Different indices of vegetation indicate an abundance of vegetation, such as fractional vegetation cover, and the normalized vegetation index (NDVI). A negative connection between the NDVI and land surface temperature was revealed, as well as the green area's cooling effect [35,36] due to soil moisture variations, land surface emissivity, albedo, and profusion of vegetation, resulting in the fall of the variable temperatures of dense vegetation [37]. The authors of [7,38] proved that political and socio-economic developments are essential factors impacting urban growth. Their results show that the urban area of their case study corresponded to sites of key economic progress. Therefore, the example of Duhok City in Iraqi Kurdistan, a fast-growing urban area, was selected to employ updated methodology to address the following:

(1)   To evaluate urban land use/cover changes in Duhok City and to analyze the impact of land use/cover on LST.
(2)   To examine the relationship between LST with NDVI, NDWI NDBAI and NDBI values.

2. Materials and Methods

2.1. Study Area

The study site covers the capital of Dohuk Province, Dohuk City, in the north of Iraqi Kurdistan, located between latitudes 37°00′00″ N and 37°07′30″ N and longitudes 42°27′30″ E and 42°47′30″ E [39], and 585 m above mean sea-level [40] Figure 1. The study area was chosen due to its strategic site on the international transport links connecting the Kurdistan Region of Iraq to Turkey as well as Syria.

Duhok city is located between two opposing mountains ranges, the Bekher Mountains in the north and Zawa Mountains in the south. As the surrounding mountains are of relatively high altitudes, the climate is similar to that of the Mediterranean region [41] in that the Mediterranean climate is characterized by dry summers and winters with reasonable precipitation. The summers are hot with low moisture and bright sunshine. In contrast, winters have a noticeably higher humidity and lower temperatures. In the winter season the climate is characterized by its low temperatures and snowfall on the high mountains [42]. Occasional drought seasons that are repeated over periods of time lead to an underground water recharge deficiency. A significant amount of rainfall as well as cold temperatures characterizes the spring seasons.

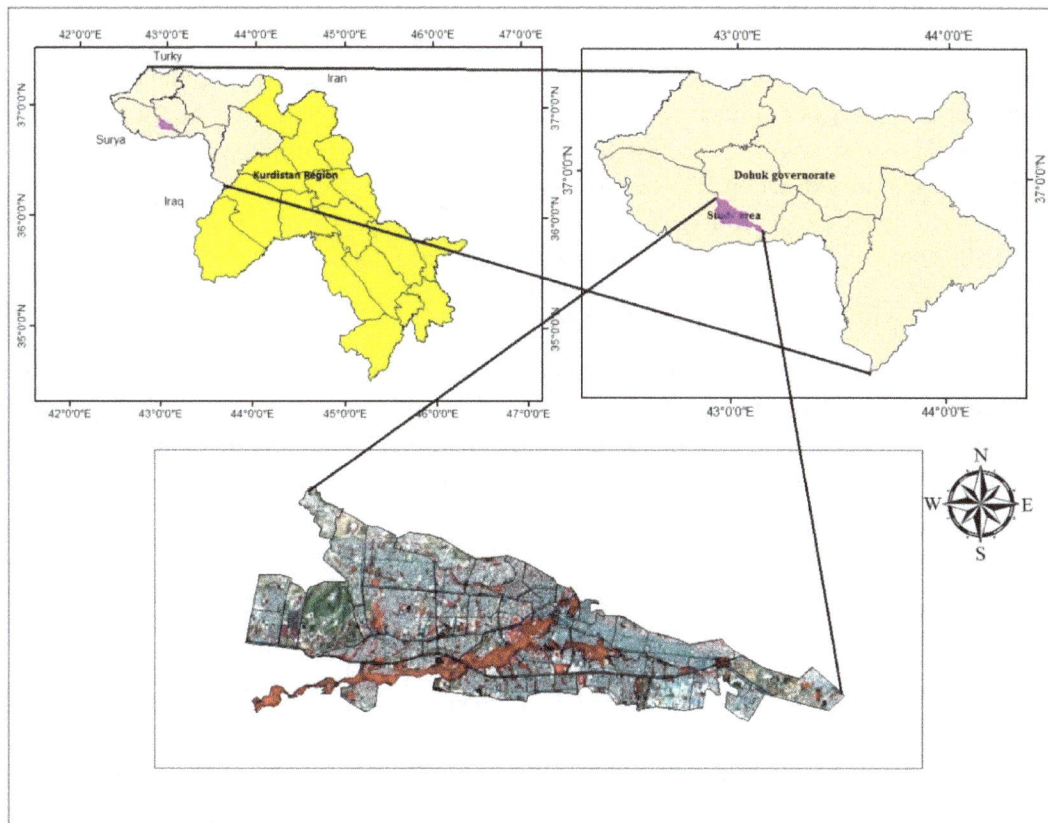

Figure 1. Illustration of the location of the study area, Duhok City.

2.2. Data Used

Primary and secondary data are both adapted in the study in order to efficiently detect how land surface temperature (LST) is affected by the alteration in land use/cover. United States Geological Survey (USGS) Gloves provided the primary data of three Landsat images with the spatial resolution of 30 m, 100 m and 120 m. The first Landsat TM-5 is dated 11 October 1990, second Landsat TM-5 is dated 21 August 2000, and the third image of Landsat OLI_TIRS-8 is dated 1 August 2016. All bands were used in this study, in particular thermal bands which are popular for identifying LST (Table 1). Secondary data such as municipal boundaries, geographical wards and the master plan map were sourced from the governorate of Duhok.

2.3. Methodology

Different processes for analyzing the Landsat images were used in this study: (1) Classification of the images; (2) derivation of NDVI, NDWI, NDBI and NDBAI; (3) LST for each image was retrieved; (4) All files were entered into GIS, after being converted to vector files to calculate and manipulate through attribute tables in ArcGIS, as shown in Figure 2.

2.3.1. Image Classification and Accuracy Assessment

In order to detect the changes in land use during the period of the study, LULC classification is essential to study the effects of human actions on a regional scale. Landsat images mapped LULC changes for 1990, 2000 and 2016. Built up areas, water, barren land and vegetation lands are the four selected LULC types. The images were analyzed according to their spectral and spatial profiles so that training sites could be developed, based on ancillary information and reference data from various sources. This study designated 40 training samples of 40 pixels for each land cover class. However, Lillesand et al, 2008 [43] noted the need for 20 training samples of 40 pixels for each land cover category. The statistical characteristics of the land cover categories were developed once the training sites were digitized. Landsat images were then classified by utilizing the maximum likelihood algorithm with a supervised signature extraction. The three classified maps were assessed on accuracy by stratified random sampling methods. From each LULC class, fifty samples were chosen. Apart from field checked LULC maps, a field survey was also used as reference data.

2.3.2. Computation of NDVI, NDWI, NDBI and NDBAI

LST studies widely use the NDVI parameter because NDVI is less sensitive to the changes in atmospheric conditions than other indices; it has, therefore, become very popular to monitor vegetation statuses [44]. NDVI was used to present the relationship between LST and vegetation area in this study by linear regression correlation. In order to compute an NDVI image this formula was used:

$$\text{NDVI} = \frac{NIR_{um} - Red_{um}}{NIR_{um} + Red_{um}} \tag{1}$$

NDBI is a widely-used index for evaluation built up statuses [45,46]. NDBI values can, depending on the spectral signature, range from medium infra-red to near infra-red band. As well as being useful for mapping human settlements [47], it is also useful for some elements of surrounding constructions. NDBAI is therefore reformulated for mapping Normalized Difference Bareness Index. The water state of vegetation and the water content within vegetation is implied by the Normalized Difference Water Index (NDWI) [48]. The values of NDBI, NDBAI and NDWI can vary from −1 to +1. Positive indicates water bodies and highly built up areas, whilst other land cover types are represented by negative values. The formula for calculating this index is:

$$\text{NDBI} = \frac{\text{MIR}_{um} - \text{NIR}_{um}}{\text{MIR}_{um} + \text{NIR}_{um}} \tag{2}$$

$$\text{NDWI} = \frac{\text{NIR}_{um} - \text{MIR}_{um}}{\text{NIR}_{um} + \text{MIR}_{um}} \tag{3}$$

$$\text{NDBAI} = \frac{\text{MIR}_{um} - \text{TIR}_{um}}{\text{MIR}_{um} + \text{TIR}_{um}} \tag{4}$$

Table 1. Details of Landsat satellite images.

Details of Landsat 5-TM Satellite Images			
Band Number	**Spectral Range μm**	**Spatial Resolution (m)**	**Band Name**
1	0.450–0.515	30	Blue
2	0.525–0.605	30	Green
3	0.630–0.690	30	Red
4	0.760–0.900	30	Near IR
5	1.550–1.750	30	Mid IR
6	10.40–12.5	120	Thermal
7	2.080–2.35	30	Mid IR
Details of Landsat-8 OLI Satellite Images			
Band Number	**Spectral Range μm**	**Spatial Resolution (m)**	**Band Name**
1	0.435–0.451	30	Coastal/Aerosol
2	0.452–0.512	30	Blue
3	0.533–0.590	30	Green
4	0.636–0.673	30	Red
5	0.851–0.879	30	NIR
6	1.566–1.651	30	SWIR-1
7	2.107–2.294	30	SWIR-2
8	0.503–0.676	15	Pan
9	1.363–1.384	30	Cirrus
10	10.60–11.19	100	TIR-1
11	11.50–12.51	100	TIR-2

Source: http://landsat.gsfc.nasa.gov/landsat-data-continuity-mission/.

2.3.3. Computation of Land Surface Temperature LST

The study employed the Mono-window algorithm developed by Qin et al., 2001 [49], to generate the Land Surface Temperature (LST) maps from Landsat satellites thermal infrared with 100 m and 120 m Spatial resolution. Radiation from the surface of the earth was recorded by the thermal infrared band, with a spectral range between 10.4 and 12.5 μm [50,51]. Derived LST requires three steps: first, spectral radiance was gained from DN of Landsat images with this formula:

$$L(\lambda) = \text{gain} * \text{DN} + \text{offset} \tag{5}$$

This can also be stated as

$$L(\lambda) = (\text{LMAX} - \text{LMIN})/255 \times \text{DN} + \text{LMI} \tag{6}$$

where

$L(\lambda)$ = Spectral radiance $\text{w} \cdot \text{sr}^{-1} \cdot \text{m}^{-3}$
$LMIN$ = 1.238 (Spectral radiance of DN value 1)
$LMAX$ = 15.600 (Spectral radiance of DN value 255)
DN = Digital Number

The next step is to transform Spectral Radiance to Temperature in Kelvin with the following formula:

$$TB = \frac{K_2}{In\frac{K_1}{R} + 1} \tag{7}$$

where

K_1 = Calibration Constant 1 (607.76)
K_2 = Calibration Constant 2 (1260.56)

R = Radiance values W/m² SRμm

TB = Surface Temperature °C

In the final step, Kelvin is converted to Celsius with the following formula:

$$TB = TB - 273$$

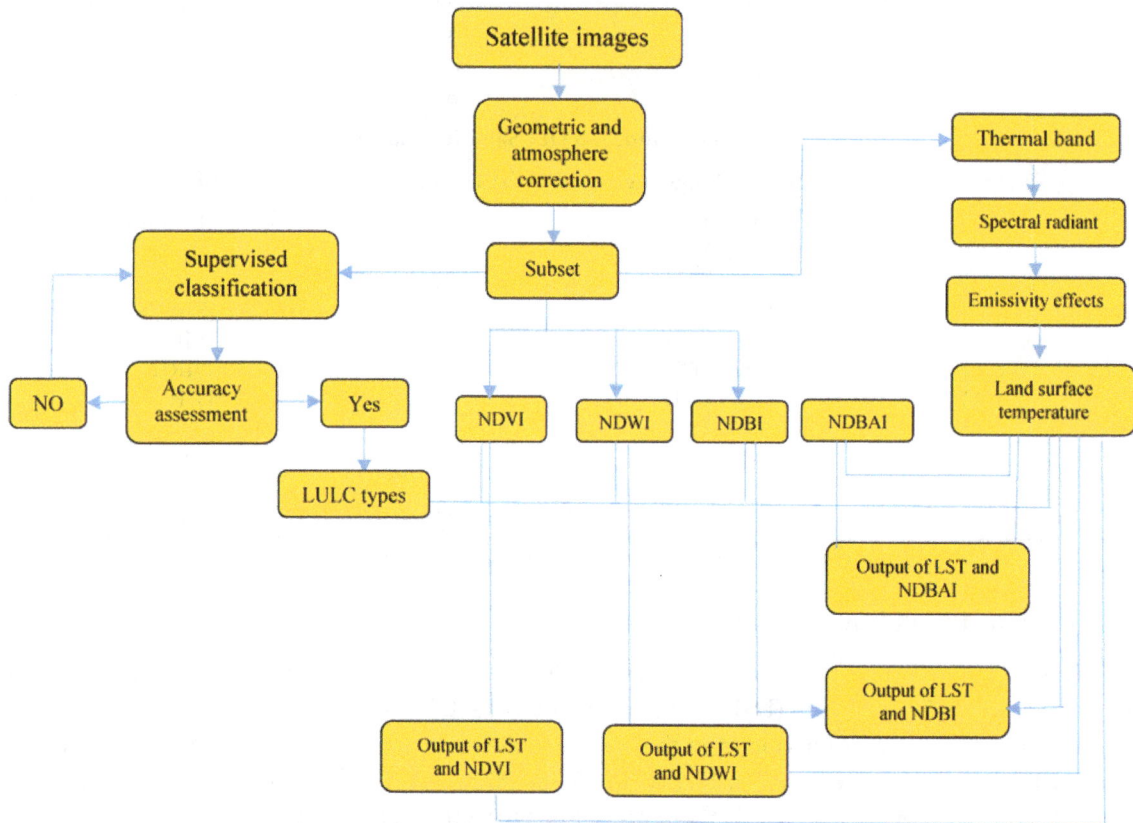

Figure 2. Flowchart showing the methodology.

3. Results and Discussion

Land use/land cover maps, land surface temperature distribution and the NDVI, NDWI, NDBAI and NDBI of the study area are the three main subsections in which the results of this study are presented:

3.1. Land Use/Land Cover Maps

The supervised classification maximum likelihood was applied to generate the LULC map in 1990, 2000 and 2016 with high accuracy as seen in Table 2. The total area of interest is approximately 17,007.25 hectares; in Table 3 and Figure 3, the exact area of the LULC of this study is listed. LST changes were caused by alternations of LULC, specifically in urbanized areas which have increased noticeably.

Table 2 shows that the built-up categories (residential, commercial and administrative buildings) increased slightly by 0.86% from 1095.77 ha to 1241.55 ha between the years 1990 and 2000, while a significant increase was recorded between the years 2000 to 2016, growing by 11.2% from 1241.55 ha to 3140.01 ha, respectively. The total area of built-up land increased from 1095.77 ha to 3140.01 ha between the years 1990 and 2016.

Table 2. Accuracy assessment of land use/cover between 1990, 2000 and 2016.

Years	1990	2000	2016
Overall accuracy %	88	91	87
Kappa Index %	85	90	84

There are many factors contributing to the increase in urbanized areas; in old parts of the city major alternations have occurred. The study area has seen remarkable changes since 2000 (Table 3) due to political and socio-economic factors. When Saddam Hussein was forced out of power, the political and socio-economic situation improved. The outcome of the study endorses the findings of [7,38] who found that both political and economic factors contributed to urban growth. The government and/or private companies developed a great deal of these areas for various retail, industrial, and residential purposes; this has been a further cause of the reduction of barren land surrounding the city, as much land was developed into large buildings and skyscrapers. Impermeable materials such as steel frames and concrete were used in the construction of these buildings. On the other hand, barren land increased from 13,141.13 ha to 13,420.7 ha between the year 1990 to 2000, although this rate lowered by 12.2% from 13,420.7 ha to 11,342.2 ha between the years 2000 and 2016, while land coverage by vegetation and water decreased by 2.48% and 0.01% from 2629.78 ha to 2206.26 ha, and from 240.57 ha to 138.74 ha from 1990 to 2000. In addition, the outcomes of the study indicate that barren land and green areas dropped from 13,141.13 ha and 11,342.2 ha in 1990 to 2629.78 ha and 2381.13 ha in 2016, while water bodies increased from 140.57 ha to 143.91 ha.

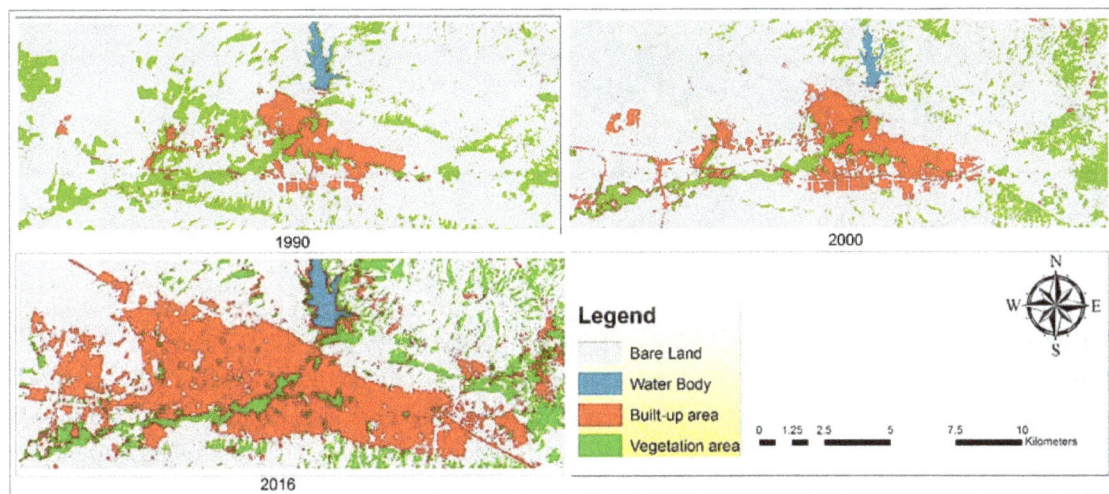

Figure 3. Supervised classification of land use/cover map.

Table 3. Shows the quantity of land use change.

Class Name	Area Hectares 1990	Area % 1990	Area Hectares 2000	Area % 2000	Area Hectares 2016	Area % 2016
Barren Land	13,141.13	77.27	13,420.7	78.9	11,342.2	66.7
Vegetation Land	2629.78	15.46	2206.26	12.98	2381.13	14
Built-up Land	1095.77	6.44	1241.55	7.3	3140.01	18.5
Water	140.57	0.83	138.74	0.82	143.91	0.84
Total	17,007.25	100	17,007.25	100	17,007.25	100

3.2. Land Surface Temperature Retrieval (LST)

The outcome of the research has been to produce a map of the study area's absolute LST. The computed LST map is illustrated in Figure 4. Respectively, in the years 1990, 2000 and 2016, LST values showed ranges between 25–47 °C, 25–50 °C and 29–56 °C. This study revealed that the maximum LST for the whole area went up by 9 °C from 1990, 2000 and 2016, which were 47 °C, 50 °C and 56 °C; during the same period of time, the minimum temperature increased by 4 °C from 25 °C, 26 °C and 29 °C, shown in Figure 4. Reasons for this increase in the range values include the different times the images were captured, meaning that different times of the year affected the results. The 1990 images were captured on 11 September 2000, the 2000 images were captured on 21 August 2000 and the 2016 images were captured on 1 August 2016. In addition, these changes could be the result of climate change. Extreme seasons have a great effect on this phenomenon. The study area experienced drought seasons particularly in 1998 and 2000; the percentage of droughts was 56% [52].

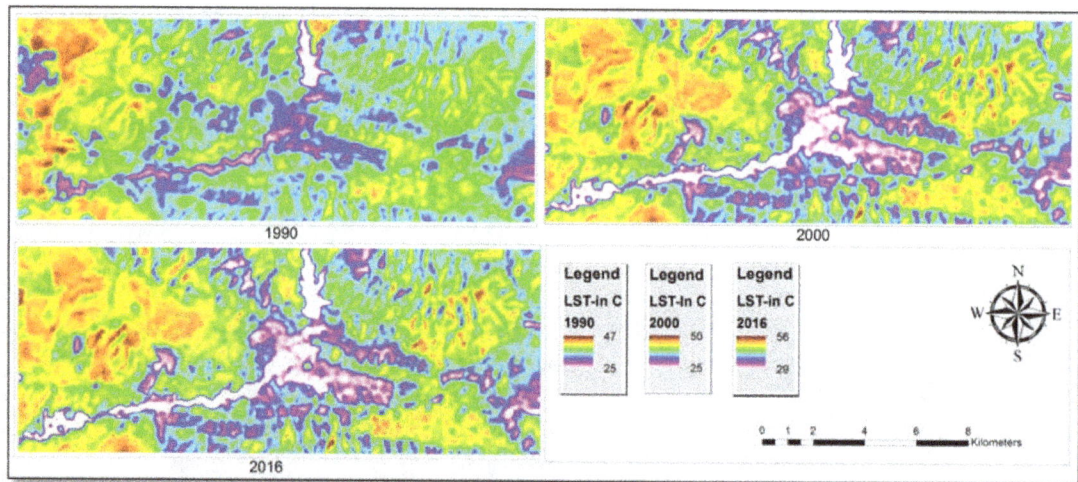

Figure 4. Land surface temperature map extract in thermal band.

Figures 5–7 display the spatial distribution of LST; higher temperatures are detected outside the city rather than at the outskirts. The LST in Duhok ranged from 25 °C to 47 °C, from 26 °C to 50 °C and from 29 °C to 56 °C in 1990, 2000 and 2016, respectively. The city has a number of LULC categories including vegetation cover, water bodies, barren land, as well as high-density, high-rise buildings in the city, interspersed with large areas covered with high-density housing. The highest temperatures around and in the city were 47 °C, 50 °C and 56 °C, and were shown in large areas of barren land and built-up areas with concrete surfaces. Most of the study site possesses densely built-up areas which cause high temperatures in contrast to the water and vegetation areas. The highest temperature of 47 °C from 1990 was recorded in Lower Malta, Meda, Shakhka, Shandokha, and Razato in the west of the study area, as well as in a part of Mazi and Pishazazi. The highest temperature of 50 °C in 2000 was noted in Zanko, Upper Malta, Lower Malta, Media, Shandokha and Raza, in the west of the city. The highest temperature of 56 °C in 2016 was recorded in Zanko, Masike and a part of Etite. In 1990 the LST of 37 °C to 43 °C was recorded in the north, south, east and west of the study area including Upper Malta, Zanko, Sarbasti, Mahabad and Mazi. In 2000 the LST of 37 °C to 44 °C was recorded in the center, north and east of the city including Shorsh, Gre Base, Shahidan, Gall, Shele, Khabat and Sarhaldan, whereas in 2016 a moderate temperature was recorded in the whole study area, apart from Zanko and Etite in the west and east of the study area, respectively, shown in Figures 5–7. The Duhok dam and the area of vegetation had mainly a lower LST between 25 °C and 29 °C and are surrounded by water bodies and greener areas. The zones previously mentioned present a moderate range of temperatures as they are along built-up areas.

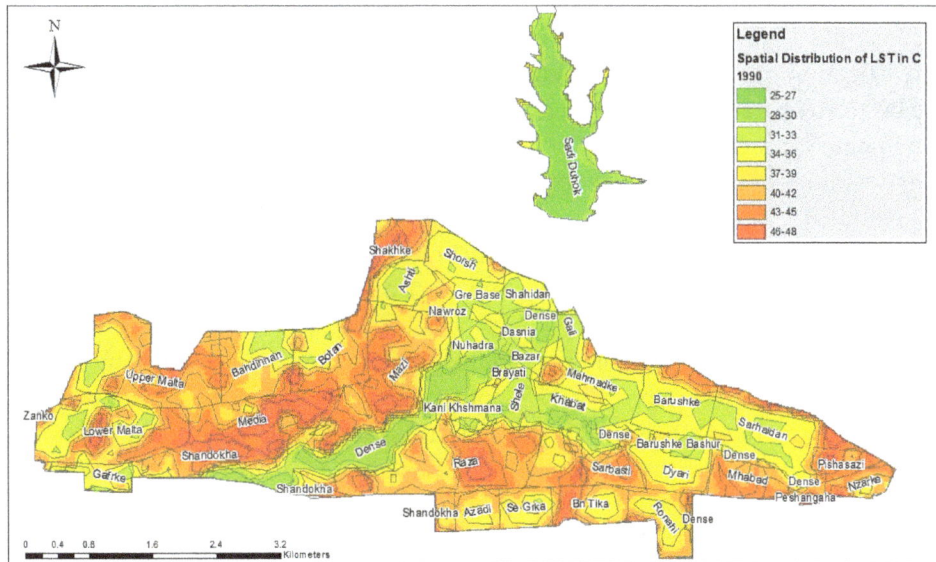

Figure 5. Spatial distribution of land surface temperature (LST) for 1990.

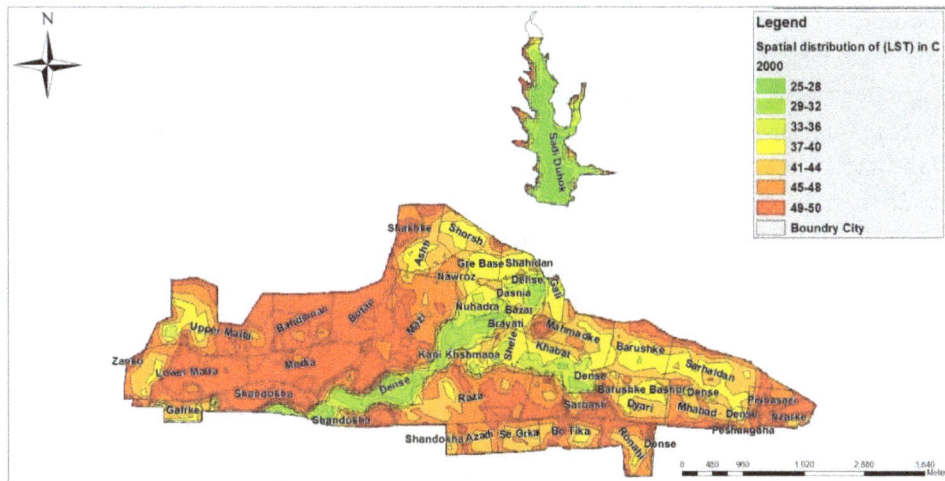

Figure 6. Spatial distribution of land surface temperature (LST) for 2000.

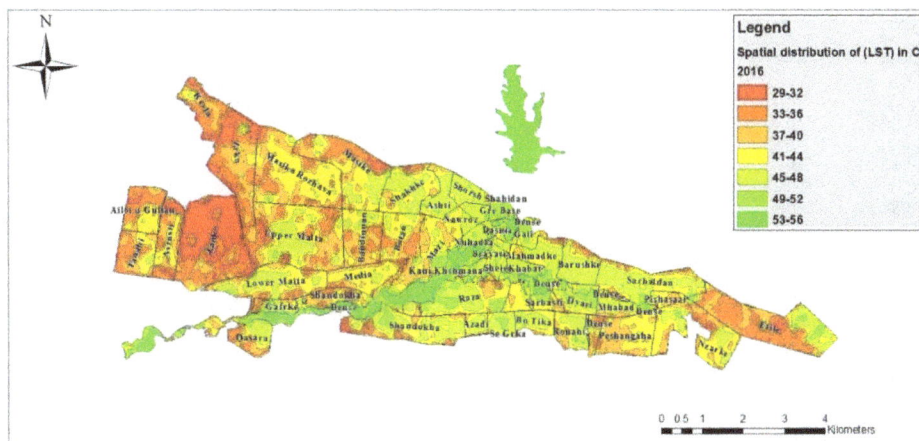

Figure 7. Spatial distribution of land surface temperature (LST) for 2016.

3.3. Relationship between Land Surface Temperature and Different Land Covers

The investigation of the thermal signature of each LULC type is essential to understand the relationship between LST and land cover [13]. Therefore, a comparison of LULC and LST was carried out; sampling points for each LULC category in the study area were selected to compare the LST values. The mean temperature of each land use/cover category was calculated by averaging all consistent pixels of a given LULC category. The results indicated the highest LST in the rock outcrops while the lowest was recorded for water bodies. Cold anchor pixels were observed in vegetated areas and water bodies, while the warmest were rock, built-up areas or bare soils. The surface temperature pixels ranged from 25 °C to 56 °C (Figure 7).

This study detected higher temperatures in the outskirts and the non-built up areas of the city rather than inside the city. Therefore, the LST outcomes of this study may disagree with previous studies [6,53,54] which show higher LST values in urban areas than in the areas surrounding and outside cities. In the period studied, Duhok City showed a lower LST in urban areas than in the suburbs (Figure 8); this is due to the sun's heat in surrounding areas being absorbed directly into the ground, causing it to heat up faster than in other land cover categories. In contrast, roads, pavements, buildings, concrete and other features that make up urban surfaces tend to release the absorbed heat slowly. In other words, built-up land has a tendency to retain the heat longer than other land cover classes such as barren land on the outskirts that does not retain heat for as long. The results of this study prove that the surrounding areas/barren lands have higher temperatures than urban areas; this outcome could be a result of the timing of the Landsat images captured. At approximately 7 a.m. the sun is just beginning to heat up the ground. Urban surfaces take in temperature more slowly, so the features in built-up areas warm up and cool down slower than other land cover categories such as barren land, which is why lower LST values were recorded in built-up areas compared to barren areas. Despite that, the changing of the LST is also caused by the land changes, since each type of land has its own qualities in terms of energy radiation and absorption. Built-up lands possesses lower albedo and higher absorption than barren lands due to the surrounding areas/ barren lands having higher temperatures than urban areas. These outcomes conform to the findings of [2], who noticed that areas with bare soil and built-up areas show a higher LST while other categories, such as water bodies, agriculture and vegetation, have lower LST values during daytime. In contrast, during the night built-up and barren lands have lower LST values, while water bodies and vegetation are found to have higher LST values.

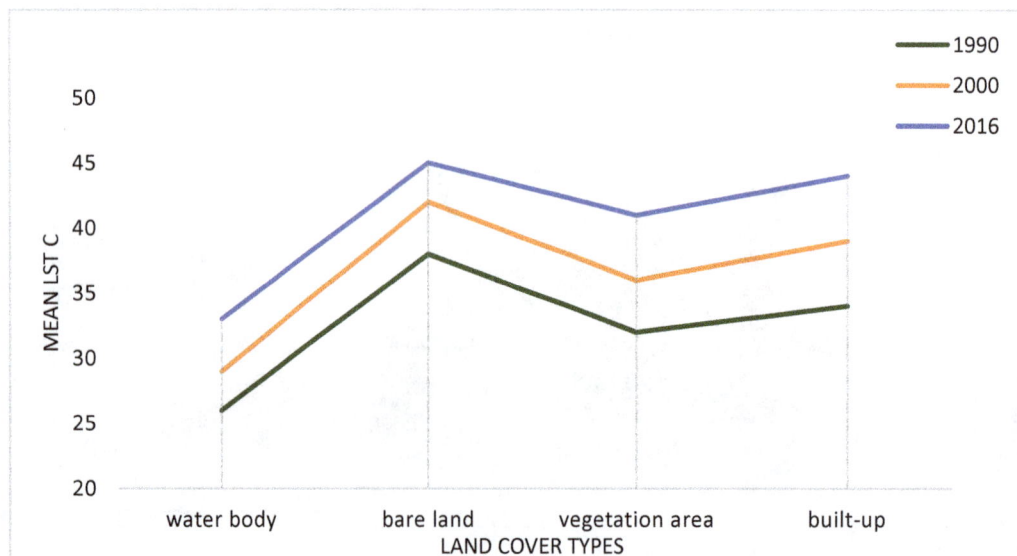

Figure 8. The differences of mean LST over variations of land cover types in 1990, 2000 and 2016.

The LST of each LULC class therefore depends on its particular characteristic. Weng (2001) [7] showed that studying the relationship between land cover types and thermal signatures is the most efficient approach in understanding the way LST is affected by LULC changes. To investigate the connection of LST to NDVI, NDWI and NDBI derived from the Landsat TM-5 1990, 2000 and Landsat OLI_TIRS-8 2016, respectively, a sample point method using 50 randomly selected points was applied. The four transect lines in Figures 9–13 clearly demonstrate the degree of correlation and the relationships of the LST, NDVI, NDWI, NDBI and NDBAI. These relationships were investigated in the performance of the Pearson's correlation coefficient analysis and correlation analysis. The result shows that lower NDVI and NDWI values were detected in areas characterized by higher temperature and higher NDBI and NDBAI. However, a positive relationship between NDBI and LST existed, with a correlation coefficient of $R^2 = 0.8714$, $R^2 = 0.848$ and $R^2 = 0.9397$ indicated in all images, between NDBI-derived built-up fractions and the surface temperature (LST), as shown in Figure 9. The results of the linear relationship detected a positive correlation between NDBAI-derived bare land fractions and LST with correlation coefficients of $R^2 = 0.8137$, $R^2 = 0.8027$ and $R^2 = 0.841$, as shown in Figure 10.

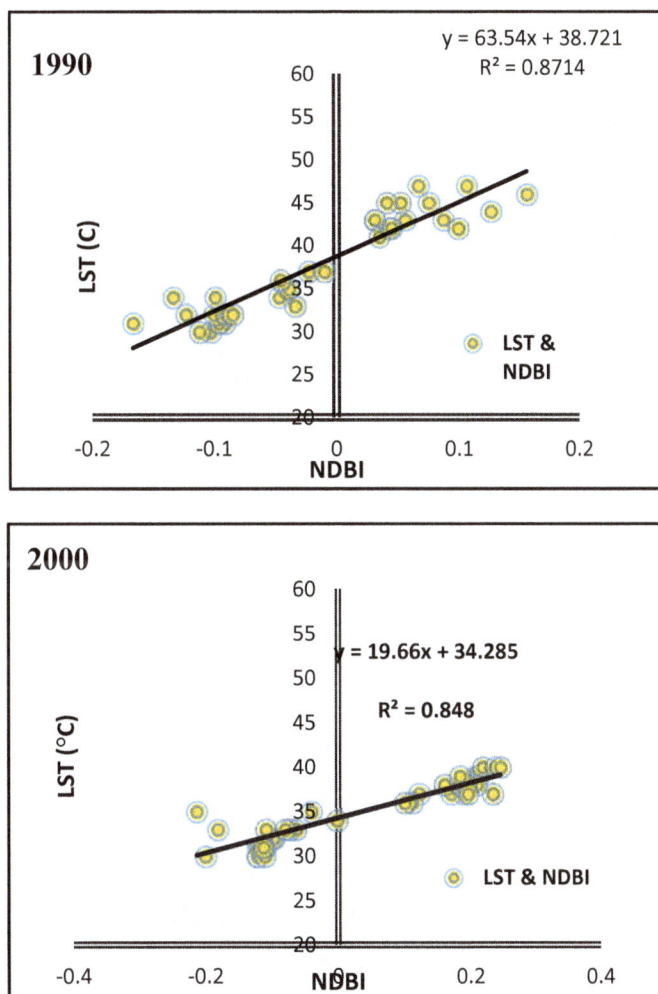

Figure 9. *Cont.*

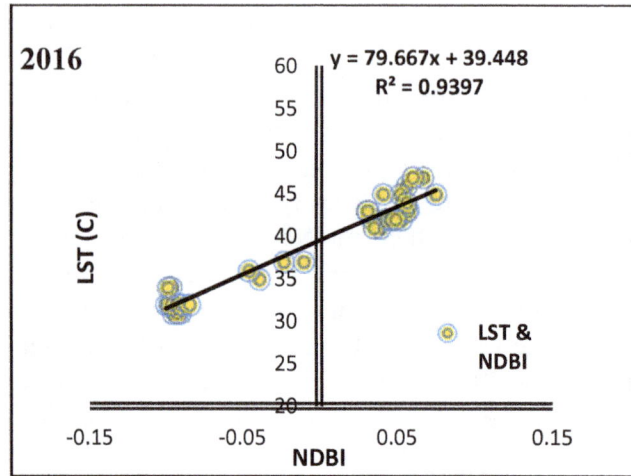

Figure 9. Correlation between NDBI and LST in the period of the study (1990, 2000 and 2016).

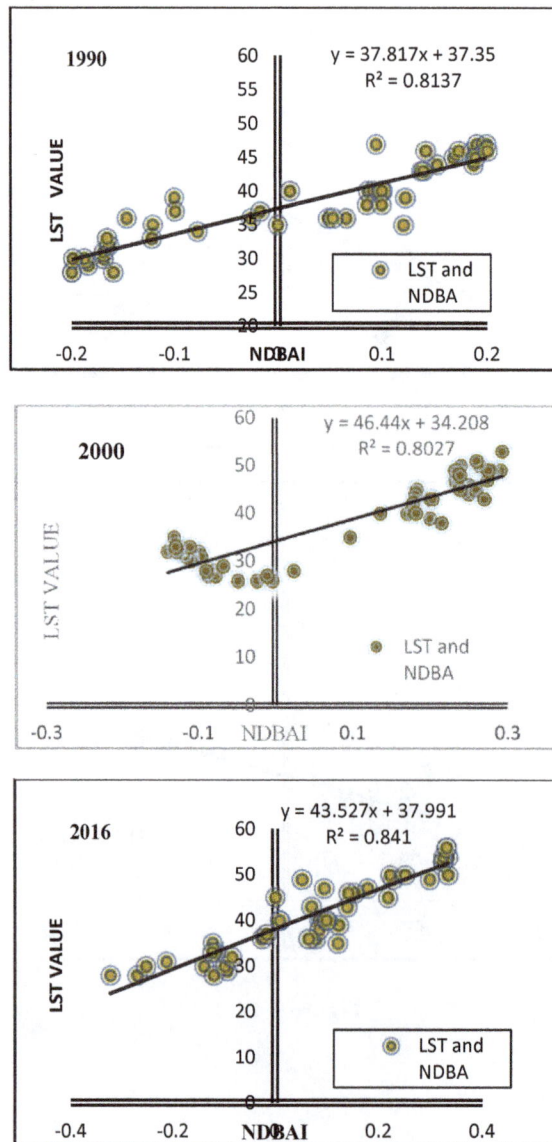

Figure 10. Correlation between NDBAI and LST in the period of the study (1990, 2000 and 2016).

In addition to a negative correlation between NDVI and LST, a negative relationship between LST and NDVI-derived vegetation fractions was shown in the results (Figure 11) of the linear relationship, with the correlation coefficients of $R^2 = 0.9038$, $R^2 = 0.8641$ and $R^2 = 0.8963$ (Figure 12). The linear relationship results detected a negative correlation between NDWI-derived water fractions and LST, with correlation coefficients of $R^2 = 0.8503$, $R^2 = 0.9026$ and $R^2 = 0.887$, as shown in Figure 13. This is a negative correlation with regard to physical changes, ground surfaces, increased soil moisture in the irrigated areas, land surface emissivity, albedo, profusion of vegetation, etc., that has a great effect on the heating of the ground surface [10]. This study's results matched the discoveries of [55], which leaned towards weak evaporation feedback of bare soils, open shrub lands and a highly possible relation to soil moisture levels. Likewise, [15,56] regarded lower temperatures in vegetation area due to processes such as transpiration and evapotranspiration.

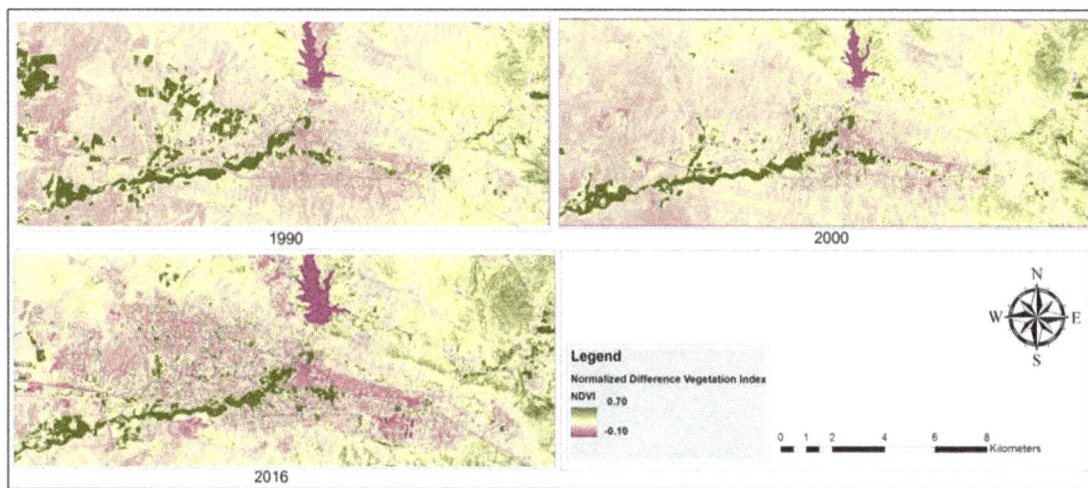

Figure 11. Normalized difference vegetation index (NDVI) in 1990, 2000 and 2016.

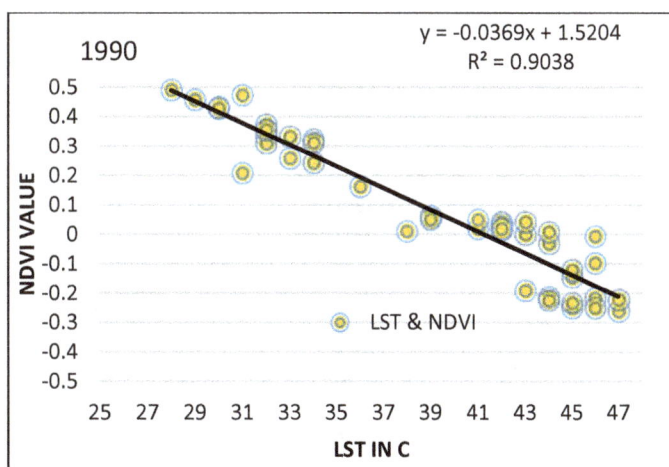

Figure 12. *Cont.*

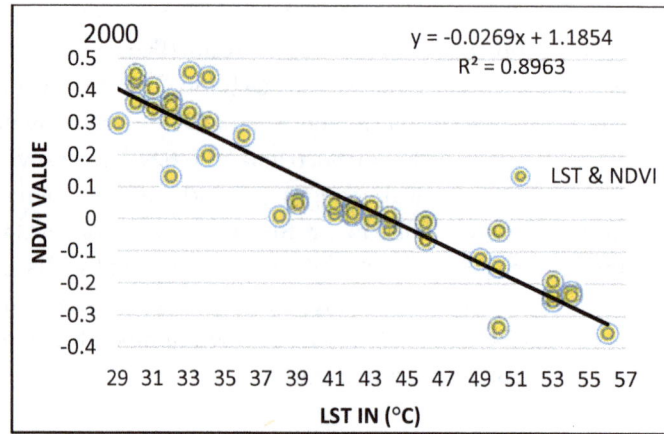

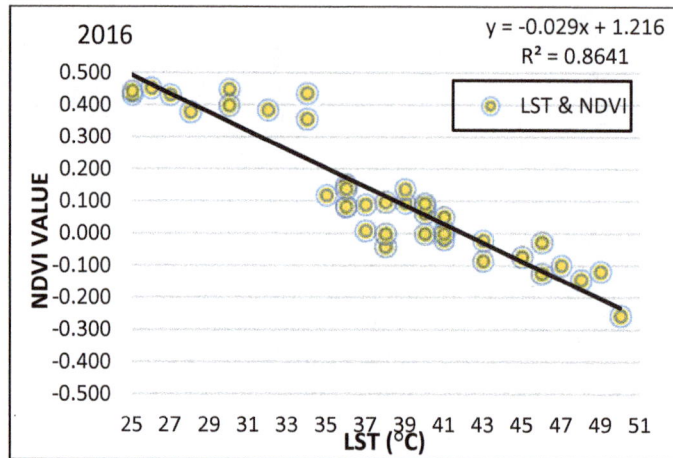

Figure 12. Correlation between NDVI and LST in the period of the study (1990, 2000 to 2016).

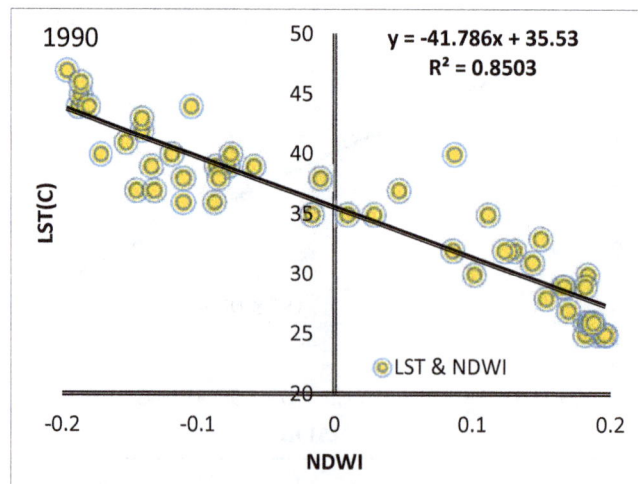

Figure 13. *Cont.*

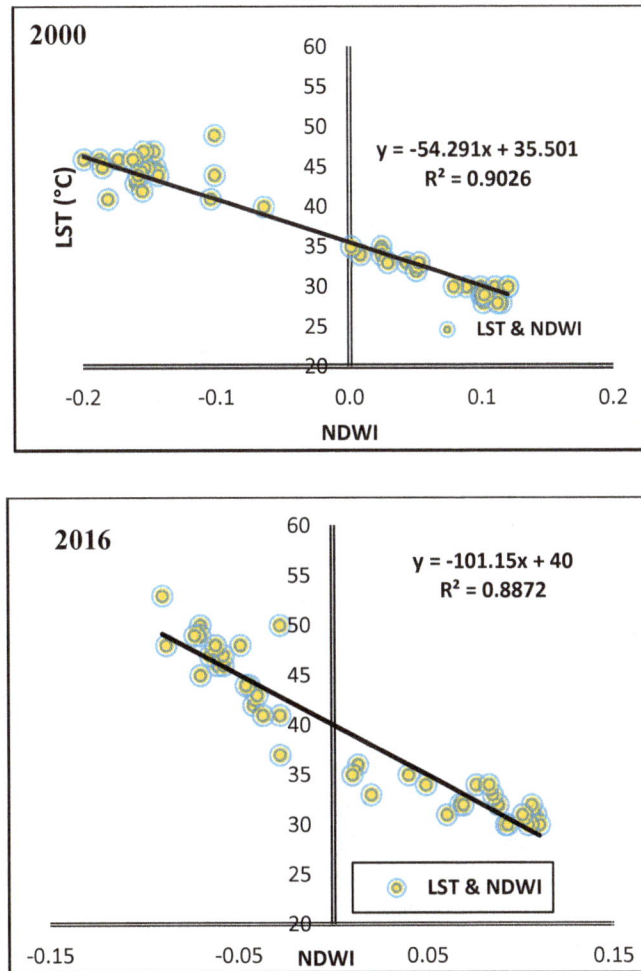

Figure 13. Correlation between NDWI and LST in the period of the study (1990, 2000 and 2016).

4. Conclusions

This paper applied, and depends on, multi-temporal remote sensing data to monitor changes in land use/cover and how it impacts the LST in Dohuk City. The applied approaches utilized in this study were very efficient in achieving the aims of this project. The study attempted to identify the changes in land use classes and their effects on LST. The study area was classified into four categories: urban areas, barren land, areas of vegetation and water bodies. The outcome of the land cover classification showed that the built-up areas and water bodies increased by 12.02% and 0.1%, respectively, while the barren land and vegetation decreased by 1.63% and 1.46%, respectively, during the study period, due to political and socio-economic factors. LST and LULC have a strongly connected relationship. The research proved that the LST value varied over the different categories, for example barren land and urban areas had increased radiant temperature. Higher temperatures on the borders and non-built-up areas of the city, rather than inside the city, may disagree with previous studies that reported higher LST values in urban areas than in the areas surrounding and outside of urban areas. This is due to the city's high temperatures, particularly in the summer. The environment of the city, being semi-arid, is the main reason that urban expansion had the opposite impact on the LST, with alternations in natural and physical characteristics of land cover, including the replacement of vegetation in built-up areas. In addition, the study found that the vegetation area (NDVI) and water bodies (NDWI) have a negative relationship with the land surface temperature. The LST was highly influenced by the LULC, and very sensitive to vegetation and soil moisture; specifically, the amount of vegetation was discovered to be the main factor on which this relationship is built. Higher LST is seen in areas with less vegetated LULC, and vice versa, although it showed a positive relationship between NDBI, NDBAI and LST.

Conflicts of Interest: The author declares no conflict of interest.

References

1. Adegoke, J.O.; Pielke, R.A., Sr.; Eastman, J.; Mahmood, R.; Hubbard, K.G. Impact of irrigation on midsummer surface fluxes and temperature under dry synoptic conditions: A regional atmospheric model study of the US High Plains. *Mon. Weather Rev.* **2003**, *131*, 556–564. [CrossRef]
2. Kant, Y.; Bharath, B.D.; Mallick, J.; Atzberger, C.; Kerle, N. Satellite-based analysis of the role of land use/land cover and vegetation density on surface temperature regime of Delhi, India. *J. Indian Soc. Remote Sens.* **2009**, *37*, 201–214. [CrossRef]
3. Hussain, A.; Bhalla, P.; Palria, S. Remote sensing based analysis of the role of land use/land cover on surface temperature and temporal changes in temperature; A case study of Ajmer District, Rajasthan. *Int. Arch. Photogramm. Remote Sens. Spat. Inf. Sci.* **2014**, *8*, 1447–1454. [CrossRef]
4. Fall, S.; Niyogi, D.; Gluhovsky, A.; Pielke, R.A.; Kalnay, E.; Rochon, G. Impacts of land use land cover on temperature trends over the continental United States: Assessment using the North American Regional Reanalysis. *Int. J. Climatol.* **2010**, *30*, 1980–1993. [CrossRef]
5. Kumar, K.S.; Bhaskar, P.U.; Padmakumari, K. Estimation of land surface temperature to study urban heat island effect using Landsat ETM+ image. *Int. J. Eng. Sci. Technol.* **2012**, *4*, 771–778.
6. Buyadi, S.N.A.; Mohd, W.M.N.W.; Misni, A. Impact of land use changes on the surface temperature distribution of area surrounding the National Botanic Garden, Shah Alam. *Procedia Soc. Behav. Sci.* **2013**, *101*, 516–525. [CrossRef]
7. Weng, Q. A remote sensing? GIS evaluation of urban expansion and its impact on surface temperature in the Zhujiang Delta, China. *Int. J. Remote Sens.* **2001**, *22*, 1999–2014. [CrossRef]
8. Xiao, H.; Weng, Q. The impact of land use and land cover changes on land surface temperature in a karst area of China. *J. Environ. Manag.* **2007**, *85*, 245–257. [CrossRef] [PubMed]
9. Guo, Z.; Wang, S.D.; Cheng, M.M.; Shu, Y. Procedia Environmental sciences assess the effect of different degrees of urbanization on land surface temperature using remote sensing images. *Procedia Environ. Sci.* **2012**, *13*, 935–942. [CrossRef]
10. Abdullah, H.J. The Use of Landsat-5 TM Imagery to Detect Urban Expansion and Its Impact on Land Surface Temperatures in The City of Erbil. Iraqi Kurdistan. Master's Thesis, Leicester University, Leicester, UK, 2012.
11. Campbell, J.B. *Introduction to Remote Sensing*, 3rd ed.; The Guilford Press: New York, NY, USA, 2002.
12. Quattrochi, D.A.; Luvall, J.C. Thermal Infrared Remote sensing for analysis of landscape ecological processes: methods and applications. *Landsc. Ecol.* **1999**, *14*, 577–598. [CrossRef]
13. Weng, Q.; Lu, D.; Schubring, J. Estimation of Land Surface Temperature—Vegetation abundance relationship for urban heat island studies. *Remote Sens. Environ.* **2004**, *89*, 467–483. [CrossRef]
14. Weng, Q. Thermal infrared remote sensing for urban climate and environmental studies: Methods, applications, and trends. *ISPRS J. Photogramm. Remote Sens.* **2009**, *64*, 335–344. [CrossRef]
15. Vorovencii, I.; Oprea, L.; Ienciu, I.; Popescu, C. Evaluation of land surface temperature for different land cover using Landsat TM Thermal Infrared band. *Ann. West Univ. Timis. Ser. Chem.* **2013**, *22*, 1–6.
16. Feizizadeh, B.; Blaschke, T.; Nazmfar, H.; Akbari, E.; Kohbanani, H.R. Monitoring land surface temperature relationship to land use/land cover from satellite imagery in Maraqeh County, Iran. *J. Environ. Plan. Manag.* **2013**, *56*, 1290–1315. [CrossRef]
17. Kerr, Y.H.; Lagouarde, J.P.; Nerry, F.; Ottlé, C. Land surface temperature retrieval techniques and applications. In *Thermal Remote Sensing in Land Surface Processes*; Quattrochi, D.A., Luvall, J.C., Eds.; CRC Press: Boca Raton, FL, USA, 2000; pp. 33–109.
18. Bendib, A.; Dridi, H.; Kalla, M.I. Contribution of Landsat 8 data for the estimation of Land Surface Temperature in Batna city, Eastern Algeria. *Geocarto Int.* **2016**, *6049*, 1–11. [CrossRef]
19. Kogan, F.N. Operational space technology for global vegetation assessment. *Bull. Am. Meteorol. Soc.* **2001**, *82*, 1949–1964. [CrossRef]

20. Li, Z.-L.; Tang, B.H.; Wu, H.; Ren, H.; Yan, G.; Wan, Z.; Trigo, I.F.; Sobrino, J.A. Satellite-derived Land Surface Temperature: Current status and perspectives. *Remote Sens. Environ.* **2013**, *131*, 14–37. [CrossRef]

21. Rozenstein, O.; Qin, Z.; Derimian, Y.; Karnieli, A. Derivation of Land Surface Temperature for Landsat-8 TIRS using a split window algorithm. *Sensors* **2014**, *14*, 5768–5780. [CrossRef] [PubMed]

22. Randrianjatovo, R.N.; Rakotondraompiana, S.; Rakotoniaina, S. Estimation of Land Surface Temperature over Reunion Island using the thermal infrared channels of Landsat-8. In Proceeding of the 2014 IEEE Canada International Humanitarian Technology Conference-(IHTC), Montréal, QC, Canada, 1–4 June 2014.

23. Shi, T.; Huang, Y.; Wang, H.; Shi, C.E.; Yang, Y.J. Influence of urbanization on the thermal environment of meteorological station: Satellite-observed evidence. *Adv. Clim. Chang. Res.* **2015**, *6*, 7–15. [CrossRef]

24. Chen, X.L.; Zhao, H.M.; Li, P.X.; Yin, Z.Y. Remote sensing image-based analysis of the relationship between urban heat island and land use/cover changes. *Remote Sens. Environ.* **2006**, *104*, 133–146. [CrossRef]

25. Yu, X.; Guo, X.; Wu, Z. Land surface temperature retrieval from LANDSAT 8 TIRS-comparison between radiative transfer equation-based method, split window algorithm and single channel method. *Remote Sens.* **2014**, *6*, 9829–9852. [CrossRef]

26. Rose, L.; Devadas, M.D. Analysis Of Land Surface Temperature And Land Use/Land Cover Types Using Remote Sensing Imagery—A Case In Chennai city, India. In Proceeding of the Seventh International Conference on Urban Climate, Yokohama, Japan, 29 June–3 July 2009; pp. 1998–2001.

27. Fu, P.; Weng, Q. A time series analysis of urbanization induced Land Use and Land Cover change and its impact on Land Surface Temperature with Landsat imagery. *Remote Sens. Environ.* **2016**, *175*, 205–214. [CrossRef]

28. Lv, Z.; Zhou, Q. Utility of Landsat image in the study of Land Cover and Land Surface Temperature change. *Procedia Environ. Sci.* **2011**, *10*, 1287–1292. [CrossRef]

29. Xiao, R.B.; Ouyang, Z.Y.; Zheng, H.; Li, W.F.; Schienke, E.W.; Wang, X.K. Spatial pattern of impervious surfaces and their impacts on Land Surface Temperature in Beijing, China. *J. Environ. Sci.* **2007**, *19*, 250–256. [CrossRef]

30. Dewan, A.M.; Yamaguchi, Y. Using remote sensing and GIS to detect and monitor Land Use and Land Cover change in Dhaka Metropolitan of Bangladesh during 1960–2005. *Environ. Monit. Assess.* **2009**, *150*, 237–249. [CrossRef] [PubMed]

31. Belal, A.A. Detecting urban growth using remote sensing and GIS techniques in Al Gharbiya governorate, Egypt. *Egypt. J. Remote Sens. Space Sci.* **2011**, *14*, 73–79. [CrossRef]

32. Rahman, A.; Kumar, S.; Fazal, S. Assessment of Land use/Land Cover Change in the North-West District of Delhi Using remote sensing and gis techniques. *J. Indian Soc. Remote Sens.* **2012**, *40*, 689–697. [CrossRef]

33. Takeuchi, W.; Hashim, N.; Thet, K.M. Application of remote sensing and GIS for monitoring urban heat island in Kuala Lumpur Metropolitan area. In Proceedings of the Map Asia 2010 and the International Symposium and Exhibition on Geo-information, Kuala Lumpur, Malaysia, 26–28 July 2010.

34. Zurina, M.; Hukil, S. Appraising Good Governance in Malaysia Based on Sustainable Development Values. *J. Asian Behav. Stud. Sustain. Sci. Manag.* **2012**, *7*, 247–253.

35. Jiang, J.; Tian, G. Analysis of the impact of Land use/Land cover change on Land Surface Temperature with remote sensing. *Proc. Environ. Sci.* **2010**, *2*, 571–575. [CrossRef]

36. Carlson, T.N.; Arthur, S.T. The impact of land Use Land Cover changes due to urbanization on surface microclimate and hydrology: A satellite perspective. *Glob. Planet. Chang.* **2000**, *25*, 49–65. [CrossRef]

37. Huang, J.; Wang, R.; Li, F.; Yang, W.; Zhou, C.; Jin, J.; Shi, Y. Simulation of thermal effects due to different amounts of urban vegetation within the built-up area of Beijing, China. *Int. J. Sustain. Dev. World Ecol.* **2009**, *16*, 67–76. [CrossRef]

38. Chen, S.; Zeng, S.; Xle, C. Remote Sensing and GIS for urban growth analysis in China. *Photogramm. Eng. Remote Sens.* **2000**, *66*, 593–598.

39. Abdulla, H.H. Rock Slop Analysis in Duhok Governorate/Bekhair anticline by using GIS technique. *J. Al-Nahrain Univ.* **2013**, *16*, 46–54.

40. Mohammed, H.D.; Ali, M.A. Monitoring and prediction of urban growth using GIS techniques: A Case study of Dohuk City Kurdistan Region of Iraq. *Int. J. Sci. Eng. Res.* **2014**, *5*, 1480–1488.

41. Koeppe, C.E.; De Long, G.C. *Weather and Climate*; McGraw-Hill: New York, NY, USA, 1958.

42. Mohammed, J. Land use and cover change assessment using Remote Sensing and GIS: Dohuk City, Kurdistan, Iraq (1998–2011). *Int. J. Geomat. Geosci.* **2013**, *3*, 552–569.

43. Lillesand, T.; Kiefer, R.W.; Chipman, J. *Remote Sensing and Image Interpretation;* John Wiley and Sons: Hoboken, NJ, USA, 2008.

44. Raynolds, M.K.; Comiso, J.C.; Walker, D.A.; Verbyla, D. Relationship between satellite-derived Land Surface Temperatures, arctic vegetation types, and NDVI. *Remote Sens. Environ.* **2008**, *112*, 1884–1894. [CrossRef]

45. Grover, A. Monitoring Spatial patterns of Land Surface Temperature and urban heat island for sustainable megacity: A case study of Mumbai, India, using Landsat TM data. *Environ. Urban Asia* **2016**, *7*, 38–54. [CrossRef]

46. Jalili, S.Y. The effect of land use on land surface temperature in The Netherlands. In Proceedings of the 2015 International Conference on Sensors & Models in Remote Sensing & Photogrammetry, Kish Island, Iran, 23–25 November 2015.

47. He, C.; Shi, P.; Xie, D.; Zhao, Y. Improving the normalized difference built-up index to map urban built-up areas using a semiautomatic segmentation approach. *Remote Sens. Lett.* **2010**, *1*, 213–221. [CrossRef]

48. Ahmed, B.; Kamruzzaman, M.; Zhu, X.; Rahman, M.S.; Choi, K. Simulating land cover changes and their impacts on Land Surface Temperature in Dhaka, Bangladesh. *Remote Sens.* **2013**, *5*, 5969–5998. [CrossRef]

49. Qin, Z.; Karnieli, A.; Berliner, P. A mono-window algorithm for retrieving land surface temperature from Landsat TM data and its application to the Israel-Egypt border region. *Int. J. Remote Sens.* **2001**, *22*, 3719–3746. [CrossRef]

50. Liu, L.; Zhang, Y. Urban heat island analysis using the Landsat TM data and ASTER data: A case study in Hong Kong. *Remote Sens.* **2011**, *3*, 1535–1552. [CrossRef]

51. Rodriguez-galiano, V.; Pardo-Iguzquiza, E.; Sanchez-Castillo, M.; Chica-Olmo, M.; Chica-Rivas, M. Downscaling Landsat 7 ETM + thermal imagery using Land Surface Temperature and NDVI images. *Int. J. Appl. Earth Obs. Geoinform.* **2012**, *18*, 515–527. [CrossRef]

52. Abas, K.A. *Analysis of Climate and Drought Conditions in the Fedral Region of Kurdistan;* College of Basic Education, Salahaddin University: Erbil, Iraq, 2012.

53. Rogan, J.; Ziemer, M.; Martin, D.; Ratick, S.; Cuba, N.; DeLauer, V. The impact of tree cover loss on Land Surface Temperature: A case study of central Massachusetts using Landsat Thematic Mapper thermal data. *Appl. Geogr.* **2013**, *45*, 49–57. [CrossRef]

54. Essa, W.; Verbeiren, B.; van der Kwast, J.; Van de Voorde, T.; Batelaan, O. Evaluation of the DisTrad thermal sharpening methodology for urban areas. *Int. J. Appl. Earth Obs. Geoinform.* **2012**, *19*, 163–172. [CrossRef]

55. Adegoke, J.O.; Carleton, A.M. Relations between soil moisture and satellite vegetation indices in the US Corn Belt. *J. Hydrometeorol.* **2002**, *3*, 395–405. [CrossRef]

56. Hanamean, J.R., Jr.; Pielke, R.A., Sr.; Castro, C.L.; Ojima, D.S.; Reed, B.C.; Gao, Z. Vegetation greenness impacts on maximum and minimum temperatures in northeast Colorado. *Meteorol. Appl.* **2003**, *10*, 203–215. [CrossRef]

12

Hydroclimatic Characteristics of the 2012–2015 California Drought from an Operational Perspective

Minxue He *, Mitchel Russo and Michael Anderson

Division of Flood Management, California Department of Water Resources, 3310 El Camino Avenue, Sacramento, CA 95821, USA; Mitchel.Russo@water.ca.gov (M.R.); Michael.L.Anderson@water.ca.gov (M.A.)
* Correspondence: Kevin.He@water.ca.gov

Academic Editor: Christina Anagnostopoulou

Abstract: California experienced an extraordinary drought from 2012–2015 (which continues into 2016). This study, from an operational perspective, reviewed the development of this drought in a hydroclimatic framework and examined its characteristics at different temporal and spatial scales. Observed and reconstructed operational hydrologic indices and variables widely used in water resources planning and management at statewide and (hydrologic) regional scales were employed for this purpose. Parsimonious metrics typically applied in drought assessment and management practices including the drought monitor category, percent of average, and rank were utilized to facilitate the analysis. The results indicated that the drought was characterized by record low snowpack (statewide four-year accumulated deficit: 280%-of-average), exceptionally low April-July runoff (220%-of-average deficit), and significantly below average reservoir storage (93%-of-average deficit). During the period from 2012–2015, in general, water year 2015 stood out as the driest single year; 2014–2015 was the driest two-year period; and 2013–2015 tended to be the driest three-year period. Contrary to prior studies stating that the 2012–2015 drought was unprecedented, this study illustrated that based on eight out of 28 variables, the 2012–2015 drought was not without precedent in the record period. Spatially, on average, the South Coast Region, the Central Coast Region, the Tulare Region, and the San Joaquin Region generally had the most severe drought conditions. Overall, these findings are highly meaningful for water managers in terms of making better informed adaptive management plans.

Keywords: California 2012–2015 drought; hydroclimatic characteristics; operational perspective

1. Introduction

Drought is an economically and environmentally disruptive natural hazard typically characterized by deficits in water resources. Different from other natural hazards including floods, hurricanes, and earthquakes, drought is a gradual hazard with a slow onset but can last for months to years and cover extensively large areas [1–3]. Drought affects millions of people and causes widespread damages in the world every year [1]. In the United States alone, drought causes an annual loss of $6–8 billion on average [4,5]. The specific impacts of drought vary from region to region depending on regional resilience and coping capacities [6]. Advanced management practices are required to mitigate the adverse impacts of drought, which is particularly the case for dry areas including the state of California. Foremost in these practices is to understand the characteristics of drought events from an operational perspective. This understanding is critical in guiding drought response actions.

As a state with over 38 million people and a globally important economy, California is prone to drought with frequent drought events recorded. The most noticeably previous drought periods include 1929–1934, 1976–1977, 1987–1992, and 2007–2009. Currently, the state is in the fifth year of another prolonged drought originating in 2012. The first four years (2012–2015) of the drought stand out as the driest and warmest four consecutive years in terms of statewide precipitation and temperature, respectively. The drought had both well perceived and profound negative impacts on the economy, society, and environment of the state.

Large deficits in precipitation combined with higher than average temperatures caused progressive canopy water loss in California's forests and devastating wildfires in the mountainous areas. An estimate of 888 million large trees experienced measurable canopy water loss of which about 58 million trees had larger than 30% loss [7]. From 2011 to August 2015, it was estimated that about 27 million trees died in the forests [8]. If this drought continues, hundreds of millions of trees may suffer sufficient canopy water loss, resulting in death [7]. Forests undergoing significant canopy water loss are more vulnerable to wildfires. Two of the largest three California wildfires occurred within 2012–2015, including the Rush Fire (August 2012 in Lassen County) and Rim Fire (August 2013 in Tuolumne County), which collectively burned an area of about 2140 km^2 (0.53 million acres versus an area of 0.27 million acres burned by the largest fire in the record period) [9]. Two of the top 10 most damaging wildfires also occurred in the drought period, including the Valley Fire (September 2015 in Lake, Napa, and Sonoma counties) and the Butte Fire (September 2015 in Amador and Calaveras counties). Those two fires caused six fatalities and damaged 2876 structures (versus 25 fatalities and 2900 structures claimed by the most damaging fire in the record period) [10].

Furthermore, low river flows induced from shortfalls in precipitation and snowpack combined with high temperatures caused widespread water quality deterioration in streams and wetlands, leading to degraded habitat for native fishes and water birds and significantly reduced their populations [11]. Record low water allocation (5%) for State Water Project contractors was registered in calendar year 2014. Additionally, groundwater was largely over pumped during the drought to offset the deficits in surface water. This led to significant land subsidence, most notably in the Central Valley. Specifically, parts of the San Joaquin Region have been falling at a rate of five centimeters (two inches) per month [12], putting the integrity of the water supply and conveyance facilities in this region at greater than ever risk of damage.

The drought also had widespread impact on the economy of the state. On a single-year level, it was estimated that 17,100 jobs were lost and about 170 km^2 of farmland (428,000 acres) were fallowed in the 2014 drought, about 50% more severe than the 2009 drought, which was the last year of the previous modern drought (2007–2009) in the state [13]. The impacts of the 2015 drought were even more severe, with an estimation of 21,000 jobs lost and about 220 km^2 (54,000 acres) fallowed [14]. The monetary losses to all economic sectors were estimated at about $2.7 billion in 2015 versus $2.2 billion caused by the 2014 drought.

In light of the adverse impacts of the drought, water usage and management adaptions have been taken during the drought period across the state to mitigate those impacts. Water conservation efforts have been ongoing in urban areas, including the mandate of a 25% cut in urban water usage (California Governor's Executive Order B-29-15, https://www.gov.ca.gov/). Shifts toward perennial high revenue (per unit of water usage) crops have been in progress, leading to continued increases in the economic value of crops in spite of increasing acreages of field crop land fallowed during the drought [15]. The drought has also led to the enactment of the most comprehensive groundwater legislation, the Sustainable Groundwater Management Act (SGMA), in state history. The SGMA requires the most stressed basins to reach sustainability by 2040 [16].

Given the uniqueness of the 2012–2015 drought, a wealth of studies have been dedicated to exploring the cause and severity of the drought. The major cause of the drought has been identified as multi-year deficits in precipitation over the state linked to an abnormal high pressure ridge over the northeastern Pacific Ocean [17–19]. It was also reported that warming temperatures contributed to the evolvement of the drought [20–27].

Despite the high consensus on the cause of the drought, there was low consensus across the literature on the severity of the drought. Using paleoclimate reconstruction data (dating back to AD 800) for Central and Southern California, Griffin and Anchukaitis [28] found out that (1) according to the

June/July/August (June-August) Palmer Drought Severity Index (PDSI) [29], the 2012–2014 drought was the worst in the analysis period (800–2014) while 2014 was the driest year in at least the past 1200 years; (2) evaluating precipitation deficits, however, the conditions in 2014 and 2012–2014 were not unique in the analysis period. In contrast, using the same raw PDSI reconstructions but with different spatial average and bias correction procedures, Robeson [30] observed that based on June-August PDSI, the 2014 drought had a return period of 140–180 years, the 2012–2014 drought was almost a 10,000 year event, and the 2012–2015 drought was unprecedented with a nearly incalculable return period. In spite of having strikingly different findings, both studies shared a number of limitations including (1) the PDSI is determined from precipitation and temperature data. While both variables are indispensable in drought analysis, they need to be considered together with other variables including streamflow, snowpack, reservoir storage, and groundwater level; (2) the wet season of California is typically from November to April (rather than June-August) when the majority of the annual precipitation falls. The June-August PDSI may not be the most appropriate index in assessing the severity of the drought for California; (3) the results tend to be sensitive to the statistical methods employed. However, with significant theoretical implications in the research community, their practical meaning in terms of guiding real-world drought management operations seems to be circumscribed; (4) the study area is focused on Central and Southern California. Northern California is also important, particularly due to the fact that most of the state's annual precipitation and runoff occurs in the northern half of the state.

This study aims to present a comprehensive assessment of the 2012–2015 California Drought from an operational perspective. The study extends previous studies [28,30] in the context of (1) examining typical hydroclimatic variables and operational drought assessment metrics applied in routine drought management practices; and (2) covering all hydrologic regions across the state including the Northern half. The study addresses the following science questions, which are critical for drought managers in making adaptive plans: (1) how does the 2012–2015 drought evolve in a hydroclimatic framework; (2) what are the temporal characteristics of the drought in a historical context at one- to four-year levels; and (3) what is the spatial pattern of the drought in terms of which regions have the most severe conditions and by what means? The rest of the paper is organized as follows. Section 2 describes the study variables and metrics employed. Section 3 presents the results and findings. Discussion and conclusions are provided in Section 4.

2. Materials and Methods

2.1. Study Variables

Due to its Mediterranean-like climate, California receives the majority (85%) of its annual precipitation in the wet season (November–April) and very limited (3%) precipitation in the summer (June–August) (refer to Figure A1 in the Appendix A) when the demand for water is typically the highest [31]. The orographic effects of the Cascade Range in the northern part of the state and the Sierra Nevada make much of the winter precipitation fall as snowfall on the windward side of the mountains. Snowmelt in late spring and early summer contributes to runoff in major rivers and groundwater recharge. Precipitation also varies in amount among regions, with the North Coast area receiving about 1223 mm annually on average while the Southeast Desert area receives only 156 mm. While most precipitation falls in the northern half, most of the state's population and farmlands are located in the southern half. To counter the spatially and temporally unbalanced supply versus demand on water, the state has traditionally relied on water storage and transfer projects including the State Water Project (SWP) and the Central Valley Project (CVP) to redistribute water resources following preset protocols along with current and projected hydroclimatic conditions. These projections are typically designed based on historical precipitation, runoff, groundwater level, and reservoir storage information.

The California Department of Water Resources (CDWR) divides the state into 10 major hydrologic regions (Table 1; Figure 1a). Of these 10 regions, three regions in the Central Valley are of the primary interest from a water supply perspective: the Sacramento River Region (SAC), the San Joaquin

River Region (SJQ), and the Tulare Lake Region (TUL). CDWR maintains a network of precipitation, streamflow and reservoir level gauges, snow courses and pillows, and groundwater wells within these regions along with a range of cooperating federal, local, and private agencies. CDWR relies on this network of stations to monitor water conditions year-around in the state. CDWR conducts quality control of raw measurements from this observational network along with cooperating agencies. Based on quality-controlled data, CDWR calculates a number of hydroclimatic indices (refer to access links provided in Appendix A) and uses them in water resource planning and management practices. Those indices include three precipitation indices, three runoff indices, and two water supply indices [32]. The precipitation indices include the Northern Sierra 8-Station Precipitation Index (8SI), San Joaquin 5-Station Precipitation Index (5SI), and Tulare Basin 6-Station Precipitation Index (6SI) (Figure 1a; Table A1). These indices are calculated by averaging the precipitation recorded at corresponding amount of representative stations located in a specific region. The runoff indices include the SAC Four River Index (SAC4), SJQ Four River Index (SJQ4), and TUL Four River Index (TUL4). These four rivers in the SAC region are the Sacramento River above Bend Bridge (SBB), the Feather River (FTO), the Yuba River (YRS), and the American River (AMF); for the SJQ region, they are the Stanislaus River (SNS), the Tuolumne River (TLG), the Merced River (MRC), and the San Joaquin River inflow to Millerton Lake (SJF); four rivers in the TUL region include the Kings River (KGF), the Kaweah River (KWT), the Tule River (SCC), and the Kern River (KRI) (Figure 1c). These indices are determined by summing the unimpaired full natural runoff (FNF) for four rivers of each of these three regions. The water supply indices include the Sacramento Valley 40-30-30 Index (SAC WYI) and the San Joaquin Valley 60-20-20 Index (SJQ WSI). The former (latter) is determined by summing up 40% (60%) of the current year's April through July SAC4 (SJQ4) unimpaired runoff, 30% (20%) of the current year's October through March SAC4 (SJQ4) unimpaired runoff, and 30% (20%) of the previous year's index with a cap of 10 (4) million acre-feet.

Table 1. General geographic and hydrologic characteristics of the 10 hydrologic regions.

ID	Region Name	Area (km^2)	Annual Precipitation (10^9 m^3)	Annual Runoff (10^9 m^3)	Annual Runoff Ratio (%)
NC	North Coast	49,859	69.0	35.6	51.7
SAC	Sacramento River	69,750	64.6	27.6	42.7
NL	North Lahontan	15,672	7.4	2.3	31.7
SF	San Francisco Bay	11,535	6.8	1.5	21.8
SJQ	San Joaquin River	38,948	26.9	9.7	36.2
CC	Central Coast	28,995	15.2	3.1	20.3
TUL	Tulare Lake	43,604	17.1	4.1	23.7
SL	South Lahontan	68,434	11.5	1.6	14.0
SC	South Coast	27,968	13.3	1.5	11.1
CR	Colorado River	51,103	5.3	0.2	4.7

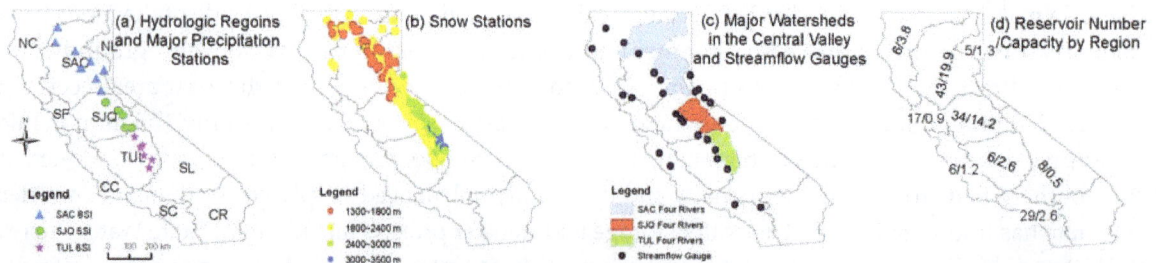

Figure 1. (**a**) Hydrologic regions and precipitation stations used in determining precipitation indices; (**b**) spatial distribution of 262 active snow stations within different elevation bands; (**c**) 12 major watersheds in the Central Valley; (**d**) number of reservoirs and the corresponding aggregated capacity (in 10^9 m^3) by hydrologic region.

In addition to those hydroclimatic indices, snowpack, reservoir storage, groundwater level, and reconstructed FNF data are also applied in water resource-related operations across the state. Winter snowfall mostly occurs in the high elevations of the three Central Valley regions. Most of the snow stations are therefore located in these regions (Figure 1b). Snowpack is traditionally deemed to peak around 1 April [33,34]. 1 April Snow Water Equivalent (A1 SWE) of the snowpack is typically used as the snow index to represent the annual peak snow condition. CDWR monitors the storage of 154 reservoirs with a total capacity of 47 billion m^3 across the state (Figure 1d). The end of the (water) year reservoir storage is often used as an index in water supply planning operations. CDWR monitors groundwater levels with cooperating agencies semiannually (Spring and Fall) in a large number of wells mostly located in the Central Valley. The groundwater level information is critical for groundwater management practices. On top of field observations which are often characterized with a short record period (less than a century), CDWR also uses reconstructed flow data (from tree ring chronologies) for major watersheds in assessing flow variability over centuries to millennia in support of drought management. Particularly, reconstructed annual FNF for major rivers in the SAC region and the SJQ region from 900–2010 are available [35]. In this study, the observed annual FNF of these two regions in the instrumental period are used to replace (up to 2010) or supplement (from 2011–2015) the reconstructed annual FNF to produce a hybrid annual FNF record. The hybrid dataset represents the best available flow data for these two regions in the context of record period (from 900 to 2015) and quality (using observed data in the instrumental period).

This study focuses on the aforementioned operational hydroclimatic indices and variables. Specifically, when analyzing the development of the drought, the focus is on precipitation, runoff, snow water equivalent, reservoir storage (September), temperature, and groundwater level data. When assessing the temporal characteristics of the drought in a historical context, the groundwater level data are not considered due to limited availability of long-term relevant data. The variables considered are tabulated in Table 2. When investigating the spatial characteristics, based on data availability, only precipitation, temperature, AJ FNF, annual FNF, and reservoir storage are considered. Reconstructed FNF data and groundwater level data are obtained from CDWR (http://www.water.ca.gov). Climate-divisional precipitation and temperature data are acquired from the National Centers for Environmental Information Climate Divisional Database (http://www.ncdc.noaa.gov) and converted to hydrologic-regional values using a simple (area-based) weighting method following our practical operations. Historical records of other data are obtained from the California Data Exchange Center (CDEC, http://www.cdec.water.ca.gov), an open-access CDWR data archive. It is worth noting that when examining the development and temporal characteristics of the drought, focus is placed on statewide conditions and the conditions over the Central Valley. When exploring the spatial characteristics of the drought, however, all ten hydrologic regions are considered. It should also be noted that detailed information on data sources and access links are provided in Appendix A.

Table 2. Record period length of study variables.

Variables	State	SAC	SJQ	TUL
Precipitation	1896–2015	1921–2015	1913–2015	1922–2015
Temperature	1896–2015	1896–2015	1896–2015	1896–2015
A1 SWE	1950–2015	1930–2015	1930–2015	1930–2015
AJ FNF	1941–2015	1906–2015	1901–2015	1931–2015
Annual FNF	1941–2015	1906–2015	1901–2015	1931–2015
WSI	-	1906–2015	1901–2015	-
Hybrid Annual FNF	-	900–2015	900–2015	-
Reservoir Storage (September)	1976–2015	1976–2015	1976–2015	1976–2015

2.2. Study Metrics

Drought conditions are often quantified by drought indices which normally measure the departure of interested variables from their normal conditions according to their corresponding historical distributions. They often serve as a link between the data, the public, and decision-makers to understand, assess, and act in response to drought events. Numerous drought indices have been developed for the purposes of drought monitoring, assessment, and prediction [36–38], ranging from parsimonious measures (e.g., percentiles or rank of interested hydro-meteorological variables) to more sophisticated statistical or physical model-based measures (e.g., the Standardized indices [39–41] and the Palmer Index [29]). Despite their popularity in the research community, the sophisticated metrics generally possess a number of limitations for operational usage. For example, those indicators often lack spatial and temporal consistency in a statistical sense. Particularly, a specific value (or drought severity level) of an index may have different occurrence probabilities at different locations and different times [42]. In addition, they are generally not comparable with each other and are disconnected from historic conditions and relative risk [42]. Those properties make it difficult for decision-makers to make sound decisions based on them. A recent study conducted a survey of 19 drought managers from western U.S. states. The study found out that 16 out of 18 states did not use their indicators in their drought plans since they found them "not at all useful", "never tested", "a guess", among others [43]. In light of this observation, this study employs parsimonious metrics including the percentiles-based drought categories (Table 3), percent of average, and historical rank, mainly because these metrics are the primary metrics applied in drought management practices across California.

Table 3. Drought monitor categories [1].

Drought Monitor Category	Description	Percentile
D0	Abnormally Dry	0.20–0.30
D1	Moderate Drought	0.10–0.20
D2	Severe Drought	0.05–0.10
D3	Extreme Drought	0.02–0.05
D4	Exceptional Drought	0.00–0.02

[1] Adapted from [44].

3. Results

3.1. Development of the 2012–2015 California Drought

Water year 2011 preceding the 2012–2015 drought has above average precipitation, A1 SWE, AJ FNF, and reservoir storage statewide and in the SAC, SJQ, and TUL regions (Figure 2). In contrast, water years 2012–2015 indicate significantly below average precipitation, A1 SWE, and AJ FNF on the annual scale. Reservoir storages by the end of the water year 2012 (September 2012), however, are slightly above average statewide as well as for SAC and SJQ regions. In general, an overall declining trend is evident in all four variables from 2011–2015 except for that statewide precipitation in 2015 is higher than the previous three years. Among four variables, A1 SWE has the largest deficit followed by the AJ FNF. In 2015, the statewide A1 SWE (AJ FNF/reservoir storage) is only 5% (23%/58%) of average. Among the three regions, the TUL region tends to be impacted the most with the lowest percent of average (POA) values in four variables in 2015. Overall, from 2012–2015, the progressively increasing severity of the drought is evident.

Four-year (2012–2015) accumulated deficiencies in those four variables presented in Figure 2 are determined (Table 4). Statewide, a shortfall of 117% of average (660 mm) in precipitation is observed from 2012–2015. This shortfall has a cascading effect, leading to shortfalls in snowpack (280% accumulated deficiency in A1 SWE), streamflow (220% deficiency in AJ FNF), and reservoir storage (93% deficiency). Similar deficits are also evident for the other three regions. In general, the TUL region has the largest deficiencies (in percent of average) for all four variables, indicating that the region has the most severe drought conditions in the Central Valley. This is in line with what Figure 2 illustrates. Among the four variables, snowpack indicates the most significant shortage on average, with a percent of average value consistently above 250% for all study regions. This is also consistent with what Figure 2 shows regarding the snowpack conditions. It is worth noting that for the SJQ region, the deficiency in AJ FNF (260% of average) is slightly higher than that of the A1 SWE (250% of average). This is due to the fact that snow is not the unique source for AJ FNF. Precipitation within the period from April to July is another contributor. The deficiency in AJ FNF partially stems from the lack of April-July precipitation over the SJQ region. It should also be highlighted that the deficiencies in snowpack are far more significant than precipitation. This is due to the fact that temperature is an important factor in snowfall and rainfall partition from precipitation as well as the accumulation of snowpack.

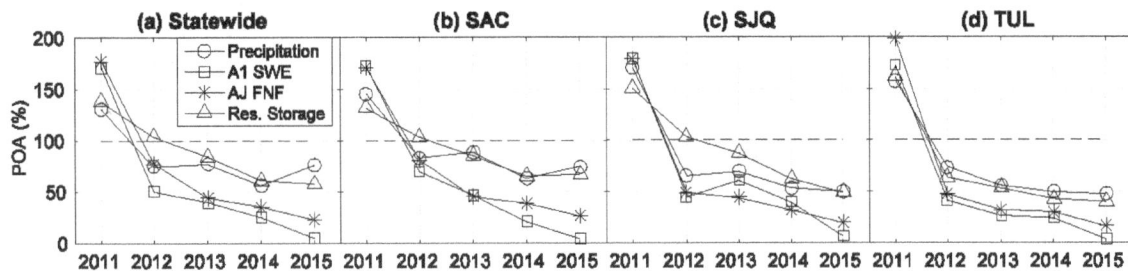

Figure 2. Percent of Average of (POA) of annual precipitation, A1 SWE, AJ FNF, and reservoir storage (September) from water year 2011–2015 for regions: (**a**) Statewide; (**b**) Sacramento (SAC); (**c**) San Joaquin (SJQ); (**d**) Tulare (TUL). X-axis indicates water years. The average value (applied in POA calculation) is determined in the entire record period as tabulated in Table 2.

Table 4. Four-year (2012–2015) accumulated deficiencies of study variables [1].

Region	Precipitation		A1 SWE		AJ FNF		Reservoir Storage	
	POA (%)	Amount (mm)	POA (%)	Amount (mm)	POA (%)	Amount (10^9 m^3)	POA (%)	Amount (10^9 m^3)
Statewide	117	660	280	2007	220	46.1	93	25.4
SAC	91	1168	260	1727	210	17.5	80	9.7
SJQ	164	1575	250	1981	260	12.3	99	7.8
TUL	176	1295	310	1549	280	6.8	203	1.7

[1] The average value (applied in POA calculation) is determined for the entire record period as tabulated in Table 2.

The departures of annual average temperature from the long-term (1896–2015) average values for the past five years are calculated (Figure 3a). In 2011, statewide temperature is slightly (0.05 °C) above the long-term average. For three Central Valley regions, it is slightly below the long-term average (−0.32 °C for SAC; −0.09 °C for SJQ and TUL). However, since 2011, annual average temperature for each region is consistently above the long-term mean. Statewide, increases (over long-term average) in annual average temperature from 2012–2015 are 0.83 °C, 1.12 °C, 1.71 °C, and 2.24 °C, respectively. A similar monotonically increasing trend in the departure is also observed for three Central Valley regions. In comparison, increases in the SAC region are the least significant. In the record period (1896–2015), year 2015 is the hottest year followed by 2014 statewide. High temperature along with shortfalls in precipitation leads to significantly lower than average A1 SWE (Figure 2; Table 4).

Groundwater provides about 40% of the urban and agricultural water demand across the state. Groundwater usages in regions SAC, SJQ, and TUL account for 30%, 38%, and 53% of respective total regional water use. During drought periods when surface water supplies are reduced, groundwater usage typically increases to mitigate the adverse impact of water shortage. This is the case for the 2012–2015 drought across the state (Figure 3b). Out of 2109 groundwater monitoring wells, in the five-year period from Fall 2010 to Fall 2015, about 68% (1436) of the wells experience declines in water level larger than 0.8 m (2.5 ft). Particularly, 37.8% (797) of the wells observe declines more than 3 m (10 ft) in magnitude. A small number of wells (5.9%), however, observe increases (greater than 0.8 m) in water level in this five-year period.

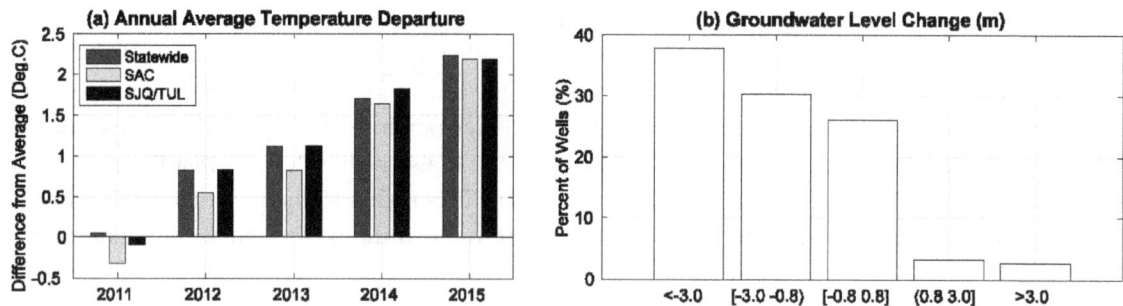

Figure 3. (a) Annual average temperature departures from long-term average values (°C) statewide and over three hydrologic regions in the Central Valley. X-axis represents water years. (b) Ground level change (m) of statewide (2109) wells between Fall 2010 and Fall 2015. X-axis denotes five categories of groundwater level changes. Negative (positive) numbers represent decrease (increase).

3.2. Temporal Characteristics of the Drought

To obtain a general understanding on the severity of the drought at different temporal scales, the 28 variables tabulated in Table 2 are ranked in their corresponding record periods according to their specific values at one- to four-year levels. For temperature, the highest value is ranked first, indicating the hottest condition. For other variables, the lowest value is ranked first, indicative of the driest condition. Either way, rank one generally indicates the most severe drought condition.

Drought monitor categories of statewide precipitation, 8SI, 5SI, and 6SI at one- to four-year levels are shown first (Figure 4). At the one-year scale, during the period from 2012–2015, on average year 2014 tends to be the driest year (D3 for 5SI, 6SI, and statewide; D2 for 8SI). Over the entire record period, however, precipitation conditions (statewide, 8SI, 5SI, and 6SI) of year 1977 and 1924 are both exceptional (D4). At the two-year level, 1976–1977 stands out as the driest two-year period, with statewide precipitation, 8SI, and 5SI deemed as exceptionally low (D4) and 6SI extremely low (D3). The period 2014–2015 is also dry with exceptionally low statewide precipitation, 5SI, and 6SI (D4) and severe conditions (D2) when looking at the 8SI. At the three-year scale, for statewide condition and 5SI, both 2012–2014 and 2013–2015 are exceptional (D4); for 8SI, both 1931–1933 and 1975–1977 are exceptional (D4), while 2012–2015 is categorized as severe (D2); for 6SI, the period 2013–2015 is exceptional (D4) along with 1959–1961. At the four-year level, for statewide conditions and the 5SI, both periods 2012–2015 and 1987–1990 are exceptional (D4). For 6SI, the period 2012–2015 is exceptional (D4) but not unprecedented (1959–1962 is also with D4 category). For 8SI, the period 2012–2015 is severe (D2); however, periods 1929–1932 and 1931–1934 are drier with exceptionally low amount of precipitation observed (D4). Overall, no single year within the period 2012–2015 is unprecedented in terms of statewide precipitation, 8SI, 5SI, and 6SI. At the two-year level, both 2013–2014 and 2014–2015 for 6SI are unparalleled before 2012. At the three-year scale, statewide precipitation and 5SI within 2012–2014 and 2013–2015 are unprecedented before 2012. Among the three precipitation indices, 8SI has relatively less severe conditions.

In contrast to the observation that statewide precipitation along with 8SI, 5SI, and 6SI are not all exceptional in different sub-periods within 2012–2015, AJ FNF in 2015 across the state and over three Central Valley regions are all in the exceptional (D4) category (Figure 5). It is the same case for 2014–2015, 2013–2015, and 2012–2015 at two-year, three-year, and four-year levels, respectively. Particularly, statewide AJ FNF of 2014–2015 and 2012–2015 is unparalleled at two-year and four-year levels, respectively. At the three-year level, statewide AJ FNF from 2013–2015 is unparalleled; for the SJQ and TUL regions, both 2012–2014 and 2013–2015 are unprecedented before 2012 in the record period. This contrast between precipitation and AJ FNF in the past four years (Figures 4 and 5) is due to the fact that AJ FNF largely depends on the snowmelt. Record low snowpack (Figure 1) reduces snowmelt contribution to AJ FNF. Additionally, record high temperature (Figure 3a) likely shifts the snowmelt timing earlier, contributing to less snowmelt from April–July.

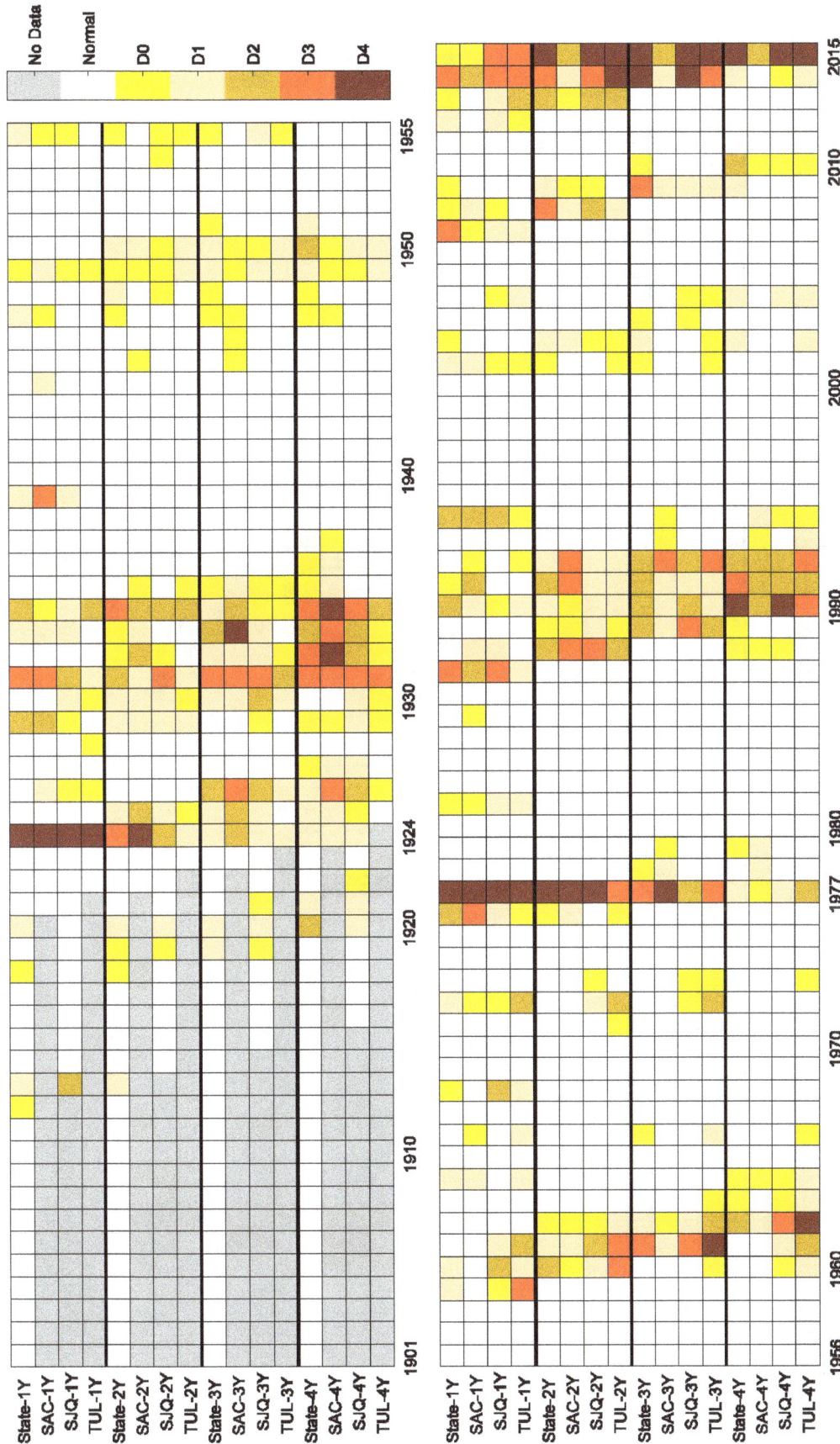

Figure 4. Drought monitor categories for statewide precipitation as well as major precipitation indices at different levels of periods (one- to four-year). At two-year, three-year, and four-year levels, the label of X-axis indicates the end year of the specific period. "Y" in the label of Y-axis stands for "Year".

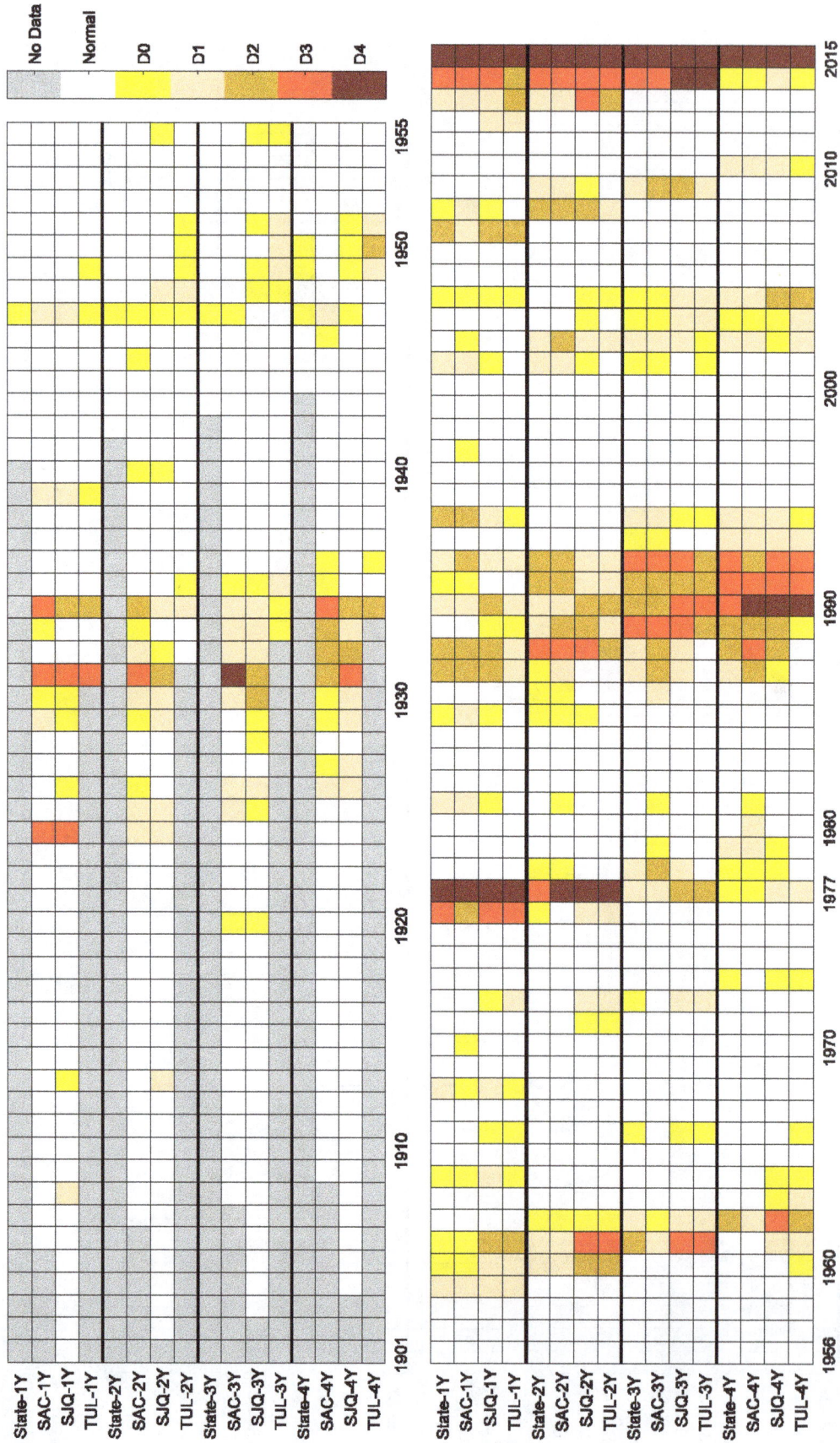

Figure 5. Drought monitor categories for statewide and central valley regional AJ FNF at different levels of periods (one- to four-year). At two-year, three-year, and four-year levels, the label of X-axis indicates the end year of the specific period. "Y" in the label of Y-axis stands for "Year".

In addition to drought monitor category, it is also a common practice in drought planning and management to compare the driest periods at different levels of periods. The ten driest periods in terms of statewide precipitation at one- to four-year levels are applied as an example (Figure 6). It is clear that year 1924 receives the lowest annual precipitation (Figure 6a) followed by 1977. At two-year, three-year, and four-year levels, the periods 1976–1977, 2012–2014, and 2012–2015 observe the lowest amount of precipitation, respectively (Figure 6b–d). This indicates that, at one-year and two-year levels, no sub-periods of 2012–2015 are unprecedented in terms of precipitation, whilst it is the case at three-year and four-year levels. Though not making the top of the list, water year 2014 is the third driest year (Figure 6a); the precipitation amounts of 2014–2015 and 2013–2014 rank the second the third lowest, respectively, at the two-year level. These observations are generally consistent with what is shown in Figure 4.

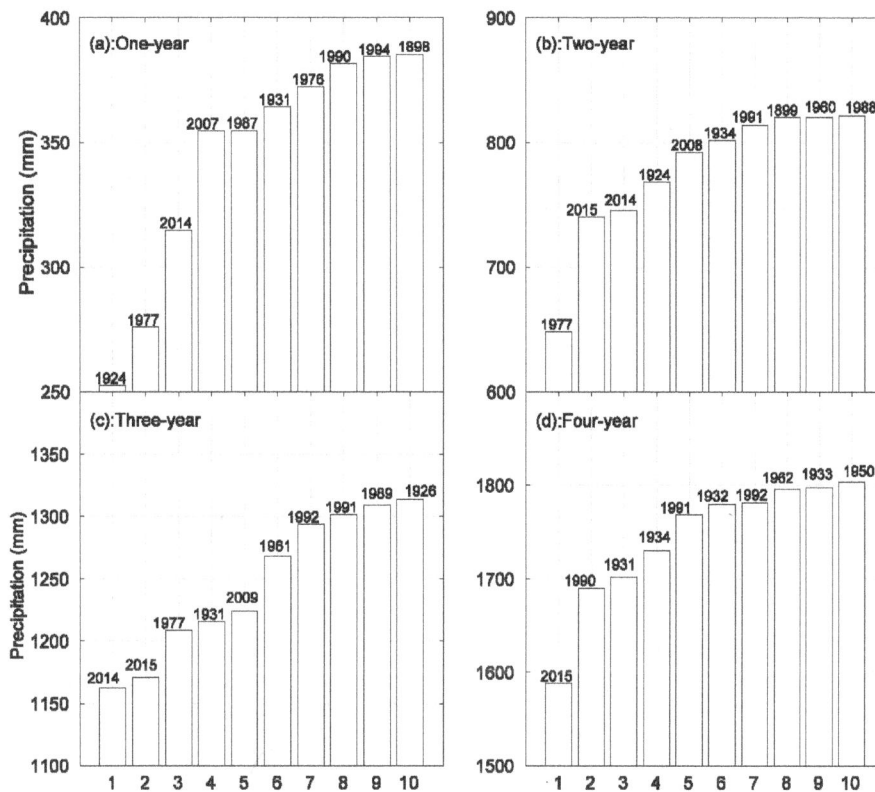

Figure 6. Ten driest (**a**) one-year; (**b**) two-year; (**c**) three-year; and (**d**) four-year periods in terms of statewide (accumulated over study period) precipitation. *X*-axis shows the ranks with rank 1 indicating the driest conditions. The numbers above bars represent the end year of the corresponding period.

Instrumental data are often limited by the length of the record period. Reconstructed paleoclimatic data supplement the instrumental data in terms of extending the analysis period, making it feasible to assess current drought events on a millennial scale. Based on reconstructed flow data (dating back to 900) and corresponding observed data (in the instrumental period), the SAC4 and SJQ4 annual FNFs at different levels of years within 2012–2015 are analyzed in a paleoclimatic context (Figure 7). Looking at a single year, only water year 2015 is a top 10 driest (rank sixth) year in terms of annual FNF for the SJQ four rivers. At the two-year level, the period 2014–2015 ranks third for the SJQ4 and fifth for the SAC4. On the three-year scale, average annual FNF of 2013–2015 for the SJQ4 and the SAC4 ranks first and seventh, respectively. At the four-year level, 2012–2015 FNF of the SJQ4 is also unparalleled, while it is the second lowest for SAC4. This shows that at three-year and four-year levels, the drought conditions (measured by annual FNF) in San Joaquin region are exceptional even within a prolonged millennial period (900–2015), while it is not the case for the Sacramento region. At one-year and two-year levels, neither of these two regions observes unprecedented annual FNFs.

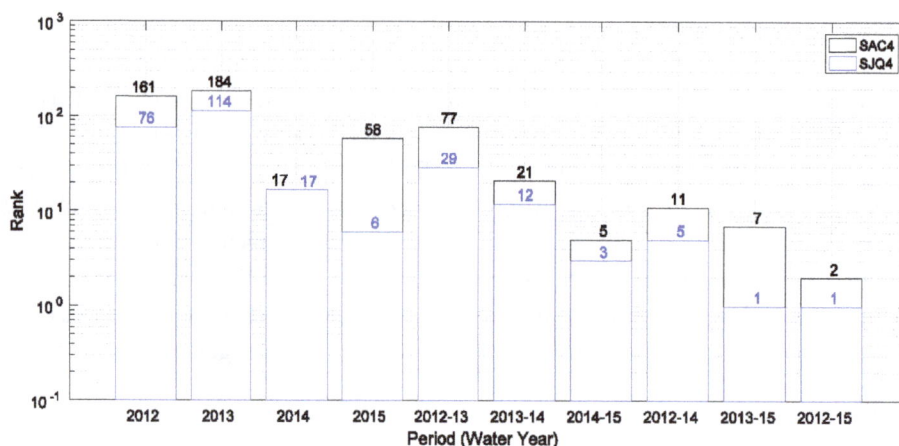

Figure 7. Rank of Sacramento Four River flow and San Joaquin Four River flow in a paleoclimatic context (from 900–2015) in different sub-periods within 2012–2015.

In addition to statewide precipitation and annual FNF for the SAC4 and the SJQ4, other variables (as tabulated in Table 2) are also ranked in their record periods. The corresponding periods at one- to four-year levels when these variables rank first are presented in this study (Table 5). For temperature, a period ranking first means the hottest (highest temperature) period. For other variables, a period ranking first indicates that the period is with the lowest value of the specific variable in the record period. Statewide, the driest year is 1924 (1977) in terms of precipitation (annual FNF and reservoir storage). However, water year 2015 has the highest temperature, lowest A1 SWE and AJ FNF. At the two-year level, the period 1976–1977 stands out as the driest in terms of precipitation, annual FNF, and reservoir storage. For the other three variables, the period 2014–2015 has the most severe conditions in the record period. Looking at the three-year level, the period 1990–1992 has the lowest annual FNF and reservoir storage and the period 2012–2014 is the driest (lowest precipitation). For the remaining three variables, the period 2013–2015 has the lowest values. At the four-year level, the period 2012–2015 tends to have the driest conditions in terms of all variables except for reservoir storage. The period 1989–1992 is the four-year period with the lowest reservoir storage. For the TUL region, the periods (one- to four-year levels) with the lowest reservoir storage all occur in the drought from 1988–1992. The driest single year is 1977 in terms of precipitation. For other variables, the driest conditions all occur within the 2012–2015 drought at one- to four-year levels. Except for variables WSI and hybrid FNF, which are specific to the SAC region and the SJQ region, the driest single years of these two regions for other variables are consistent with the statewide conditions. The driest two-year periods of these two regions are also consistent with the corresponding statewide conditions with one exception. The period 1976–1977 has the lowest AJ FNF for the SJQ region while it is 2014–2015 for the SAC region and statewide. At the three-year level, the conditions of the SAC region and the SJQ region are mostly similar to those of the statewide conditions with one exception for each region. The driest three-year (with lowest precipitation) period for the SAC region is 1975–1977 other than 2012–2014. The three-year period 2013–2015 (instead of 1990–1992) has the lowest annual FNF for the SJQ region. At the four-year level, the conditions over the SJQ region are in line with the corresponding statewide conditions. However, for the SAC region, the periods with the lowest precipitation and annual FNF are both 1931–1934 rather than 2012–2015. Focusing on the WSI, the driest one- to four-year periods for the SAC region differ from those of the SJQ region. The single year with the lowest WSI is 1977 (2015) for SAC (SJQ). The corresponding four-year periods are 2012–2015 and 1928–1931 for the SAC region and the SJQ region, respectively. In a paleoclimatic context, the one-year and two-year periods with the lowest annual FNF are 1580 and 1579–1580, respectively, for both the SAC region and the SJQ region. At three-year and four-year levels, as also indicated in Figure 7, the SJQ region observes the lowest annual FNF during the periods 2013–2015 and 2012–2015, respectively. For the SAC region, the corresponding periods are 1578–1580 and 1931–1934, respectively. In general, though differences exist across different regions, the period 2012–2015 and sub-periods within it are not unparalleled for all variables.

Table 5. Periods with hydroclimatic variables ranked first in the record period [1].

Variables	Statewide				TUL			
	One-Year	Two-Year	Three-Year	Four-Year	One-Year	Two-Year	Three-Year	Four-Year
Precipitation	1924	1976–1977	**2012–2014**	**2012–2015**	1977	**2014–2015**	**2013–2015**	**2012–2015**
Temperature	**2015**	**2014–2015**	**2013–2015**	**2012–2015**	**2015**	**2014–2015**	**2013–2015**	**2012–2015**
A1 SWE	**2015**	**2014–2015**	**2013–2015**	**2012–2015**	**2015**	**2014–2015**	**2013–2015**	**2012–2015**
AJ FNF	**2015**	**2014–2015**	**2013–2015**	**2012–2015**	**2015**	**2014–2015**	**2013–2015**	**2012–2015**
Annual FNF	1977	1976–1977	1990–1992	**2012–2015**	**2015**	**2014–2015**	**2013–2015**	**2012–2015**
Reservoir Storage	1977	1976–1977	1990–1992	1989–1992	1990	1989–1990	1988–1990	1989–1992

Variables	SAC				SJQ			
Precipitation	1924	1976–1977	1975–1977	1931–1934	1924	**2014–2015**	**2012–2014**	**2012–2015**
Temperature	**2015**	**2014–2015**	**2013–2015**	**2012–2015**	**2015**	**2014–2015**	**2013–2015**	**2012–2015**
A1 SWE	**2015**	**2014–2015**	**2013–2015**	**2012–2015**	**2015**	**2014–2015**	**2013–2015**	**2012–2015**
AJ FNF	**2015**	**2014–2015**	**2013–2015**	**2012–2015**	**2015**	1976–1977	**2013–2015**	**2012–2015**
Annual FNF	1977	1976–1977	1990–1992	1931–1934	1977	1976–1977	**2013–2015**	**2012–2015**
Hybrid FNF	1580	1579–1580	1578–1580	1931–1934	1580	1579–1580	**2013–2015**	**2012–2015**
WSI	1977	1930–1931	1922–1924	**2012–2015**	**2015**	1976–1977	**2012–2014**	1928–1931
Reservoir Storage	1977	1976–1977	1990–1992	1989–1992	1977	1976–1977	1990–1992	1989–1992

[1] The periods within 2012–2015 with variables ranked first are highlighted in bold.

To identify which sub-periods within 2012–2015 have the most severe conditions, the number of variables ranked first is counted for each one- to four-year sub-period (Figure 8). It is evident that at the one-year level, no variables in (water) year 2012, year 2013, or year 2014 rank first in the record period. However, in 2015, 14 out of 28 variables rank first (three statewide and for the SAC region; four for the SJQ region and the TUL region each). At the two-year level, 14 variables rank first in 2014–2015 (three statewide, the SAC region, and the SJQ region each; five for the TUL region) versus none in other two two-year sub-periods. Looking at the three-year level, the sub-period 2013–2015 tends to be the drier one with 16 variables ranked first (three statewide and for the SAC region; five for the SJQ region and the TUL region each) versus one in 2012–2014. At the four-year scale, 20 out of 28 variables rank first (five statewide and for the TUL region; four for the SAC region; six for the SJQ region). Overall, it is evident that, within the period 2012–2015, 2015, 2014–2015, and 2013–2015 have the most (more) significant drought conditions at one- to three-year levels, respectively. At the four-year level, 2012–2015 is not unprecedented for eight out of 28 variables considered.

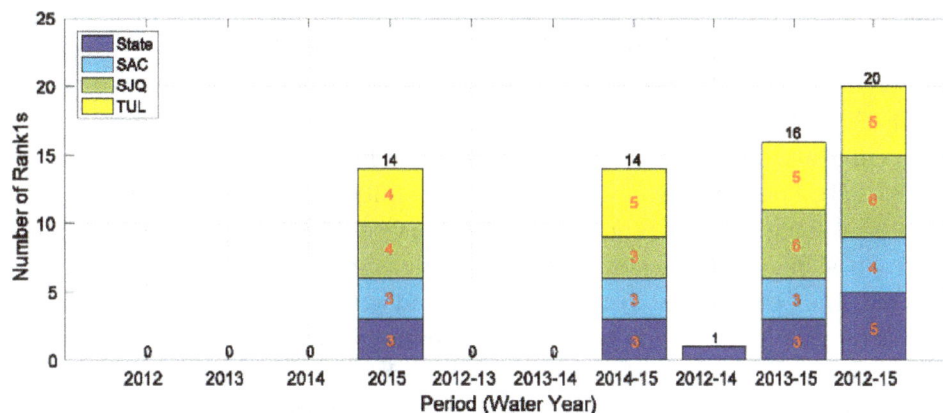

Figure 8. Number of variables (out of 28) ranked first within different periods from 2012–2015.

3.3. Spatial Characteristics of the Drought

To investigate the spatial characteristics of the 2012–2015 drought, precipitation, temperature, AJ and annual FNF, and reservoir storage (September) are compared to the average conditions at the (hydrologic) regional scale in the driest one- to four-year sub-periods within 2012–2015 (Figure 9). It should be highlighted that the Colorado Region depends on the Colorado River for the majority of its water resources. The full natural flow and reservoir storage data of this region are normally not collected by CDWR and thus are not presented in this study.

It is clear that all regions observe less than average (percent of average less than 100%) precipitation during the 2012–2015 drought (Figure 9a–d). Particularly, both the SJQ region and the TUL region have the lowest percent of average among all regions at the one-year level (Figure 9a). At two- to four-year levels, the South Coast Region tops the list with percent of average values consistently less than 60%, followed by the SJQ Region and the TUL region (Figure 9c,d). The Central Coast Region is the next driest region following those three regions in all four periods. In comparison, three northern regions (North Coast, Sacramento, and North Lahontan) have distinctly higher percent of average (mostly above 70%) in all four periods considered, indicating that these regions have relatively less significant shortage in precipitation during the drought.

It is also evident that all regions have higher than average temperature in the past four years (Figure 9e–h). Particularly, the temperatures of all regions in 2015 exceed their corresponding long-term averages by an amount over 2 °C consistently (Figure 9e). Among all regions, the South Coast Region observes the most significant increases (above the long-term average) in temperature at all four temporal levels investigated, followed by the Central Coast Region. In comparison, temperature increases in the Central Valley regions are relatively milder. For instance, in 2015, the temperature of the South Coast region is 2.6 °C higher than the long-term average, while it is 2.2 °C for three Central Valley regions (Figure 9e).

It is noticeable that all regions observe less than one third of their average AJ FNF in 2015 (Figure 9i). This is most likely caused by record high temperature (Figure 9e), record low statewide snowpack conditions (Figure 2), below average precipitation (Figure 9a), and increased evapotranspiration (induced from high temperature) in 2015. At one- to three-year levels, the Central Coast Region has the lowest percent of average of AJ FNF (Figure 9i–k). The South Coast Region has the second lowest percent of average. The specific percent of average values of these two regions are persistently below 10%. At the four-year level (2012–2015), however, the South Coast Region observes the lowest AJ FNF (16% of average), followed by the Central Coast Region (20% of average). In general, the TUL region and the SJQ region also have significantly lower than average AJ FNF. Their percent of average values rank fourth (TUL) and fifth (SJQ) lowest in 2015 (Figure 9i), respectively. The corresponding ranks are third and fourth, respectively, at other three temporal scales (Figure 9j–l). The two wettest regions (North Coast and Sacramento) have relatively higher percent of average values in AJ FNF. In comparison to AJ FNF, all regions normally have higher percent of average values in annual FNF (Figure 9m–p). The difference is the most remarkable in 2015 (Figure 9m) when the statewide percent of average of annual FNF is twice of that for AJ FNF (46% versus 23%). These two regions with the lowest percent of average in AJ FNF (Central Coast: 4%; South Coast: 9%) have much higher percent of average values for annual FNF (Central Coast: 28%; South Coast: 19%). It is evident that the South Coast Region has the lowest percent of average values consistently at four temporal scales considered. The Central Coast Region has the second lowest percent of average values at two- to four year levels (Figure 9n–p). The TUL region also observes significantly less than average annual FNF, ranking second at the one-year level (2015) and third at two- to four-year levels. Similar to the case for the AJ FNF, the two wettest regions (North Coast and Sacramento) have relatively higher percent of average values in annual FNF. At the four-year level (2012–2015), both regions observe 57% (versus 24% of the South Coast Region) of average annual FNF.

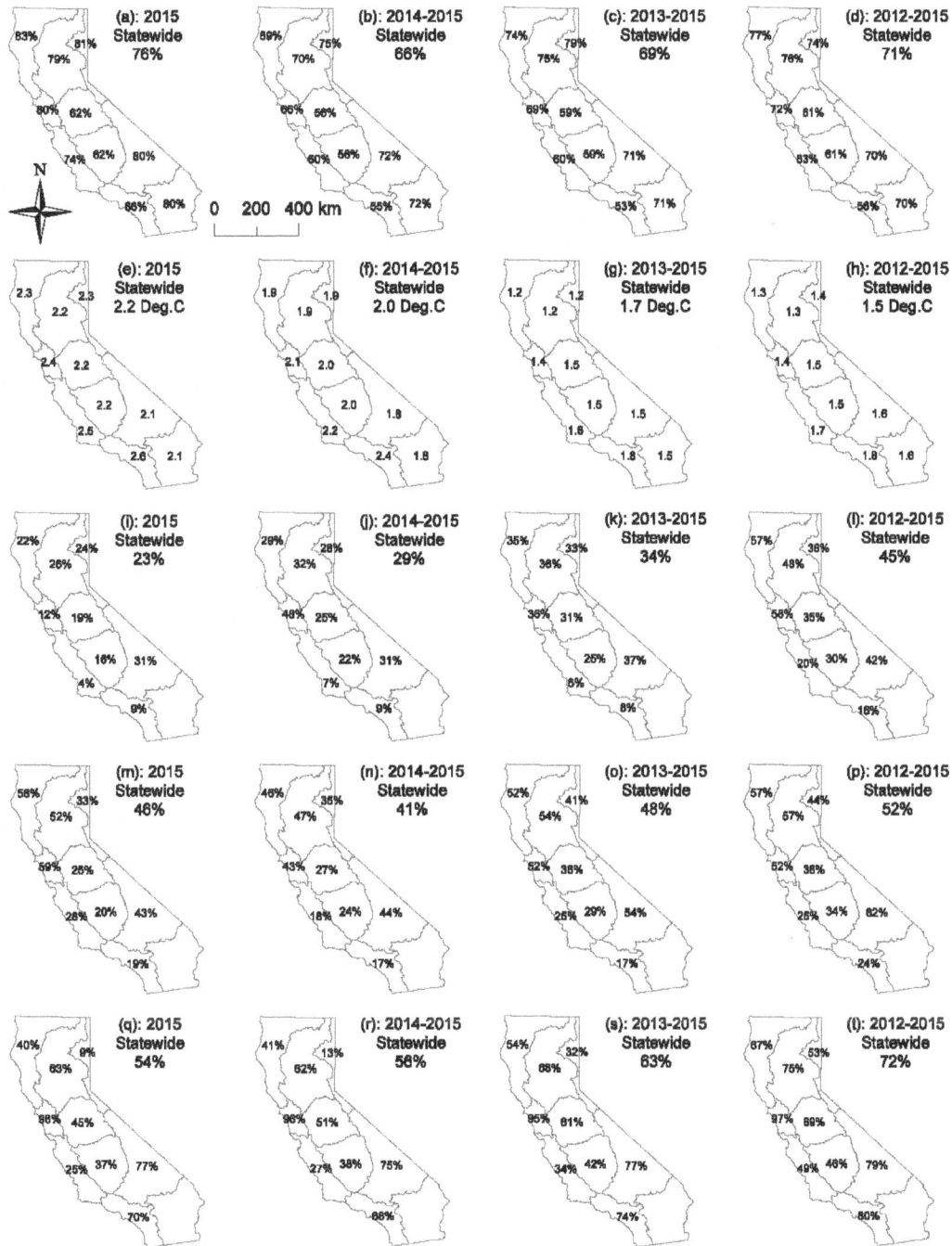

Figure 9. Precipitation, temperature, AJ FNF, annual FNF, and September reservoir storage (first to fifth row, respectively) of the driest one- to four-year (first to fourth column, respectively) periods within 2012–2015 by hydrologic region. For temperature (**e–h**), the numbers highlighted in each region denote the amount above average value (in °C); for other variables, the numbers represent the percent of average values (in %).

Looking at reservoir storage (September), the North Lahontan Region has the lowest percent of average values (the TUL region ranks third) at one- to three- year levels while the TUL region has the lowest percent of average (North Lahontan region ranks third) at the four-year level. The Central Coast region and the North Coast region rank second and fourth, respectively, at all four temporal levels. In contrast, the San Francisco region reservoir storages are nearly at the average (ranging from 95% to 98% of average) in all four periods investigated.

In general, large differences are observed across different regions and different temporal scales. Water year 2015 tends to be the hottest year for all regions with the lowest AJ FNF recorded for every region. During the four-year period (2012–2015), all regions also have higher than normal temperature along with significant shortages in precipitation, AJ FNF, annual FNF, and reservoir storage. On average, the shortage in AJ FNF is the most significant. Judging by precipitation, the South Coast Region is the driest during the 2012–2015 drought, followed by the San Joaquin Region and TUL region. Assessing by temperature, the South Coast Region is the hottest during the drought with the Central Coast Region ranked second. Looking at streamflow runoff, the South Coast Region, Central Coast Region, and TUL region observe the most deficiencies. Measuring by reservoir storage, the North Lahontan Region, Central Coast Region, and TUL region generally have the most severe conditions.

Statewide, the percent of average values of precipitation, AJ FNF, annual FNF, and reservoir storage at four-year level are generally the highest comparing to their corresponding counterparts at other three temporal scales. The above average temperature is also the lowest at the four-year level. Those observations indicate that drought severity conditions are relatively milder in a longer period. This is due to the fact that drought conditions vary year to year and the severity is normally averaged out when looking over a longer period.

4. Discussion and Conclusions

This study examined the development of the 2012–2015 California drought and assessed the temporal and spatial patterns of the drought in a hydroclimatic framework. Observed and reconstructed operational hydrologic indices and variables employed in water resources planning and management practices at statewide and (hydrologic) regional scales were applied for this purpose. Parsimonious metrics typically used in operational drought assessment and management actions including the drought monitor category, percent of average, and rank were utilized to facilitate the analysis. The drought was characterized by below average precipitation and well above average temperature in the drought period, record low snowpack (statewide accumulated deficit: 280%-of-average), exceptionally low April-July runoff (statewide accumulated deficit 220%-of-average), and significantly below average reservoir storage (statewide four-year accumulated deficit 93%-of-average). Within the period from 2012–2015, in general, water year 2015 stood out as the driest single year; 2014–2015 was the driest two-year period; 2013–2015 tended to be the driest three-year period. Contrary to some previous studies stating that the 2012–2015 drought was unprecedented [28,30], this study illustrated that measuring by a certain number of variables (8 out of 28), the 2012–2015 drought was not without precedent in the record period.

On the spatial scale, based on precipitation, temperature, and full natural flow data, the South Coast Region, the Central Coast Region, the Tulare Region, and the San Joaquin Region generally had the most severe drought conditions. However, the North Lahontan Region, the Central Coast Region, the Tulare Region, and the North Coast Region had the lowest reservoir storage in the record period in terms of percent of average. This difference most likely stemmed from the fact that reservoirs are largely regulated for different purposes including water supply, flood control, hydropower generation, and recreation, among others. Reservoir storage is thus a variable reflecting both natural and anthropogenic footprints, which is different from other variables including precipitation, temperature, and full natural flow.

This extraordinary drought had both well perceived and more profound adverse impacts on the economy, society, and environment of the state. It caused progressive canopy water loss in California's forests and devastating wildfires in the mountainous areas [7]. It also caused widespread water quality deterioration in streams and wetlands, leading to degraded habitat for native fishes and water birds and significantly reduced their populations [11]. The drought also had significant impact on the economy of the state particularly on the agriculture sector [13]. In light of these adverse impacts, water usage and management adaptions have been taken across the state to mitigate those impacts. Water conservation efforts have been ongoing, including the mandate of 25% cut in urban water usage and shifts toward perennial high revenue (per unit of water usage) crops in agriculture. The drought has also led to the enactment of the most comprehensive groundwater legislation, the Sustainable Groundwater Management Act (SGMA), in state history.

In a nutshell, this drought highlighted that California's water supply system and ecosystem are susceptible and vulnerable to extreme drought events, as is the case for many other regions facing water scarcity. Projected warming [45–47], intensified occurrence of hydroclimatic extremes [31,48–51], and population increase [52] in the near future along with aging water infrastructure [53] pose further challenges for water managers striving to maintain the balance between reliable water supplies and healthy ecosystems. This study, from a retrospective standpoint, provides meaningful information for water managers to make effective adaptive management plans (in the context of guiding when and where to focus the adaption and recovery efforts based on the temporal and spatial patterns of the drought identified). From a prospective perspective, this study lays foundation for further drought responses should drought events (similar to the 2012–2015 drought in terms of severity and extent) reoccur.

Acknowledgments: The authors would like to thank their colleagues Stephen Nemeth, Maury Roos, John King, David Parker, Matt Winston, and Boone Lek for their valuable help in data collection. The authors would also like to thank two anonymous reviewers for their valuable comments that helped improve the quality of this study. Any findings, opinions, and conclusions expressed in this paper are solely the authors' and do not reflect the views or opinions of their employer.

Author Contributions: The study was conceived by the authors together. Minxue He conducted the study and wrote the paper. Mitchel Russo and Michael Anderson provided critical discussions.

Conflicts of Interest: The authors declare no conflict of interest.

Appendix A

The Appendix A provides detailed information on the sources and access links of data applied in this study, including (1) two figures showing long-term monthly precipitating and temperature values of 10 hydrologic regions (Figure A1) and the location of precipitation, streamflow, and snow course stations (Figure A2), respectively; (2) two tables shows the precipitation gauge information (Table A1) and streamflow gauge information (Table A2), respectively; and (3) access links.

Table A1. Precipitation gauges applied in calculating precipitation indices.

Station ID	Station Name	Hydrologic Region	Latitude(°)	Longitude(°)	Elevation (m)
MSC	Mount Shasta City	Sacramento River	41.314	−122.317	1094
SHA	Shasta Dam	Sacramento River	40.718	−122.42	325
MNR	Mineral	Sacramento River	40.35	−121.6	1486
QRD	Quincy	Sacramento River	39.9366	−120.948	1042
BCM	Brush Creek	Sacramento River	39.694	−121.34	1085
SRR	Sierraville RS	Sacramento River	39.583	−120.367	1516
BYM	Blue Canyon	Sacramento River	39.283	−120.7	1609
PCF	Pacific House	Sacramento River	38.765	−120.5	1036
CVT	Calaveras Big Trees	San Joaquin River	38.283	−120.317	1431
HTH	Hetch Hetchy	San Joaquin River	37.95	−119.783	1180
YSV	Yosemite Headquarters	San Joaquin River	37.74	−119.583	1209
NFR	North Fork RS	San Joaquin River	37.233	−119.5	802
HNT	Huntington Lake	San Joaquin River	37.228	−119.221	2133
BAL	Balch PH	Tulare Lake	36.909	−119.089	524
GNF	Giant Forest	Tulare Lake	36.562	−118.765	2027
ASM	Ash Mt	Tulare Lake	36.483	−118.833	521
SGV	Springville	Tulare Lake	36.1631	−118.707	1240
PSC	Pascoes	Tulare Lake	35.967	−118.35	2789
ISB	Isabella Dam	Tulare Lake	35.646	−118.473	803

Table A2. Streamflow gauges applied in determining runoff indices and water supply indices.

Station ID	Station Name	Hydrologic Region	Latitude(°)	Longitude(°)	Elevation (m)
0	Russian R at Healdsburg	North Coast	38.61	−122.84	33
1	Eel River at Scotia	North Coast	40.49	−124.10	49
2	Trinity River at Lewiston Lake	North Coast	40.73	−122.79	570
3	Klamath River from Copco to Orleans	North Coast	41.30	−123.53	131
4	Napa River Near St Helena	San Francisco Bay	38.51	−122.46	59
5	Arroyo Seco Near Soledad	Central Coast	36.28	−121.32	122
6	Nacimiento below Nacimiento Dam	Central Coast	35.76	−120.85	182
7	Arroyo Seco Near Pasadena	South Coast	34.22	−118.18	426
8	Santa Ana River Near Mentone	South Coast	34.11	−117.10	594
9	Truckee River from Tahoe oo Farad	North Lahontan	39.43	−120.03	1574
10	East Fork Carson near Gardnerville	North Lahontan	38.85	−119.71	1519
11	West Fork Carson at Woodfords	North Lahontan	38.77	−119.83	1754
12	East Walker near Bridgeport	North Lahontan	38.33	−119.21	1951
13	West Walker below Little Walker	North Lahontan	38.52	−119.45	1682
14	Owens River below Long Valley Dam	South Lahontan	37.59	−118.71	629
15	Sacramento River above Bend Bridge	Sacramento River	40.29	−122.19	57
16	Feather River at Oroville	Sacramento River	39.52	−121.55	45
17	Yuba River near Smartville	Sacramento River	39.24	−121.27	85
18	American River at Folsom Lake	Sacramento River	38.68	−121.18	0
19	Stanislaus River at Goodwin Reservoir	San Joaquin River	37.85	−120.64	77
20	Tuolumne River at La Grange Reservoir	San Joaquin River	37.67	−120.44	52
21	Merced River below Merced Falls	San Joaquin River	37.52	−120.33	95
22	San Joaquin River below Millerton L	San Joaquin River	36.98	−119.72	90
23	Kings River below Pine Flat Res	Tulare Lake	36.83	−119.34	296
24	Kaweah River at Terminus Reservoir	Tulare Lake	36.41	−119.00	145
25	Tule River below Lake Success	Tulare Lake	36.06	−118.92	211
26	Kern River near Bakersfield	Tulare Lake	35.64	−118.48	742

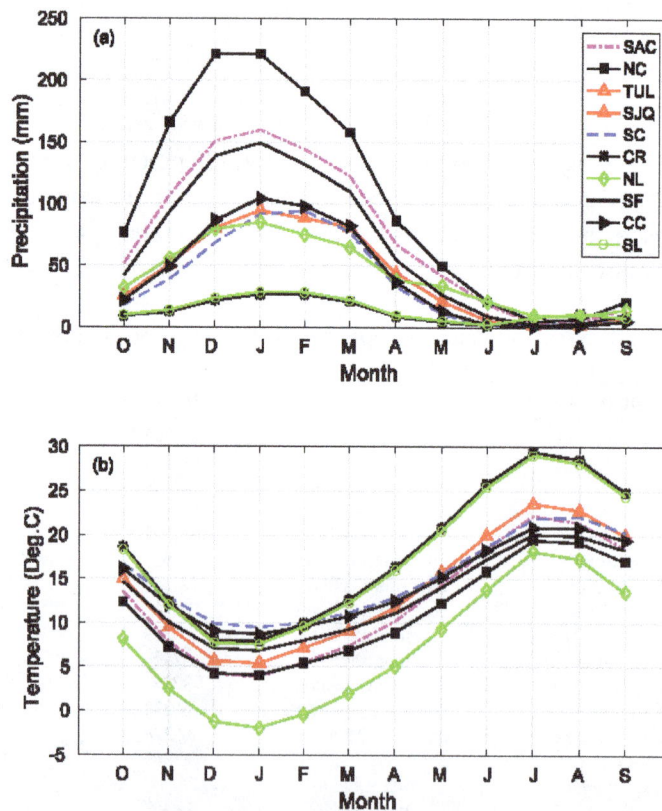

Figure A1. (a) Mean monthly precipitation; and (b) mean monthly average temperature of the hydrologic regions for October (O), November (N), December (D), January (J), February (F), March (M), April (A), May (M), June (J), July (J), August (A), and September (S).

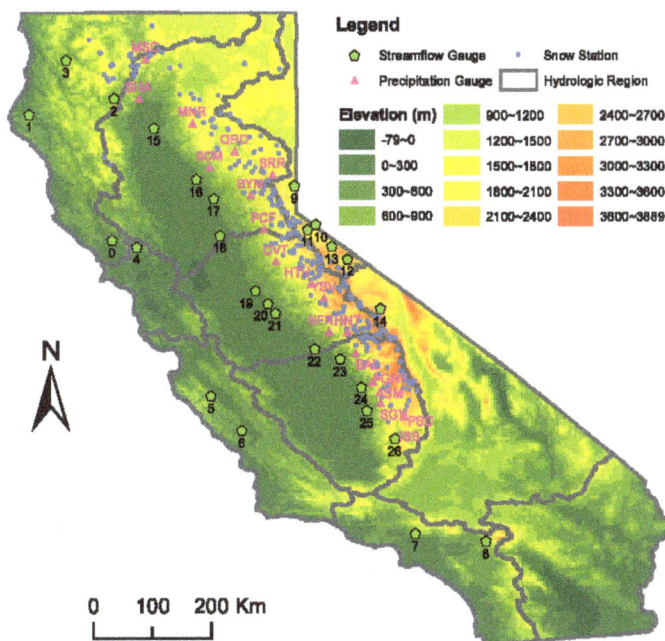

Figure A2. Location map of streamflow gauges (applied in determined runoff indices and water supply indices as well as regional runoff volume), precipitation gauges (applied in calculating precipitation indices), snow stations (applied in computing April 1st Snow Water Equivalent).

Access links:

(a). Daily 8-Station (Sacramento Region), 5-Station (San-Joaquin Region), and 6-Station (Tulare Lake Region) precipitation indices are available at:

http://cdec.water.ca.gov/cgi-progs/products/PLOT_ESI.pdf
http://cdec.water.ca.gov/cgi-progs/products/PLOT_FSI.pdf
http://cdec.water.ca.gov/cgi-progs/products/PLOT_TSI.pdf

(b). A list of streamflow gauges used in determined regional runoff for each hydrologic region is available at:
http://cdec4gov.water.ca.gov/cgi-progs/reports.cur?s=flowout.201509

The Water Supply Index for the Sacramento Valley and San Joaquin Valley is available at:
http://cdec4gov.water.ca.gov/cgi-progs/iodir/WSIHIST

The reconstructed flow data from tree ring for Sacramento four rivers and San Joaquin four rivers is available at:
http://www.water.ca.gov/waterconditions/docs/tree_ring_report_for_web.pdf

(c). A complete list of snow courses is available at:
http://cdec.water.ca.gov/misc/SnowCourses.html

(d). A complete list of major reservoirs is available at:
http://cdec4gov.water.ca.gov/cgi-progs/reservoirs/STORAGE

The end-of-year reservoir storage information for each hydrologic region is available at:
http://cdec.water.ca.gov/cgi-progs/reservoirs/STORAGEW.09

(e). Groundwater level data is available at:
http://www.water.ca.gov/groundwater/maps_and_reports/MAPS_CHANGE/DOTMAP_F2015-F2010.pdf

(f). Climate-divisional precipitation and temperature data is available at:
https://www.ncdc.noaa.gov/monitoring-references/maps/us-climate-divisions.php

References

1. Wilhite, D. Drought as a natural hazard: Concepts and definitions. In *Drought, A Global Assessment*; Routledge: New York, NY, USA, 2000; pp. 3–18.
2. Mishra, A.K.; Singh, V.P. A review of drought concepts. *J. Hydrol.* **2010**, *391*, 202–216. [CrossRef]
3. Dahal, P.; Shrestha, N.S.; Shrestha, M.L.; Krakauer, N.Y.; Panthi, J.; Pradhanang, S.M.; Jha, A.; Lakhankar, T. Drought risk assessment in Central Nepal: Temporal and spatial analysis. *Nat. Hazards* **2016**, *80*, 1913–1932. [CrossRef]
4. Wilhite, D.; Hayes, M.; Knutson, C.; Smith, K. Planning for drought: Moving from crisis to risk management. *J. Am. Water Resour. Assoc.* **2000**, *36*, 697–710. [CrossRef]
5. Federal Emergency Management Agency (FEMA). *National Mitigation Strategy: Partnerships for Building Safer Communities*; FEMA: Washington, DC, USA, 1995.
6. Dai, A. Drought under global warming: A review. *WIREs Clim. Chang.* **2011**, *2*, 45–65. [CrossRef]
7. Asner, G.P.; Brodrick, P.G.; Anderson, C.B.; Vaughn, N.; Knapp, D.E.; Martin, R.E. Progressive forest canopy water loss during the 2012–2015 California drought. *Proc. Natl. Acad. Sci. USA* **2016**, *113*, 249–255. [CrossRef] [PubMed]
8. USFS. 2015 Forest Health Protection Arial Detection Survey. Available online: www.fs.usda.gov/detail/r5/forest-grasslandhealth/ (accessed on 5 April 2016).
9. CALFire. Top 20 Largest California Wildfires. Available online: http://cdfdata.fire.ca.gov/incidents/incidents_statsevents/ (accessed on 5 April 2016).
10. CALFire. Top 20 most damaging california wildfires. Available online: http://cdfdata.fire.ca.gov/incidents/incidents_statsevents/ (accessed on 5 April 2016).
11. Hanak, E.; Mount, J.; Chappelle, C.; Lund, J.; Medellin-Azuara, J.; Moyle, P. What if California's drought continues? Available online: http://www.ppic.org/main/publication_quick.asp?i=1160/ (accessed on 1 April 2016).
12. Farr, T.G.; Jones, C.; Liu, Z. *Progress Report: Subsidence in the Central Valley, California*; NASA Jet Propulsion Laboratory (JPL): Pasadena, CA, USA, 2015.
13. Howitt, R.; Medellin-Azuara, J.; MacEwan, D.; Lund, J.; Sumner, D. *Economic Analysis of the 2014 Drought for California Agriculture*; Center for Watershed Sciences, University of California: Oakland, CA, USA, 2014.
14. Howitt, R.; MacEwan, D.; Medellin-Azuara, J.; Lund, J.; Sumner, D. *Economic Analysis of the 2015 Drought for California Agriculture*; Center for Watershed Sciences, University of California: Oakland, CA, USA, 2015.
15. Hanak, E.; Mount, J. Putting California's latest drought in context. *ARE Updat.* **2015**, *18*, 2–5.
16. CDWR. Sustainable Groundwater Management Act. Available online: http://groundwater.ca.gov/legislation.cfm/ (accessed on 5 April 2016).
17. Swain, D.L.; Tsiang, M.; Haugen, M.; Singh, D.; Charland, A.; Rajaratnam, B.; Diffenbaugh, N.S. The extraordinary California drought of 2013/2014: Character, context, and the role of climate change. *Bull. Am. Meteorol. Soc.* **2014**, *95*, 3–7.
18. Wang, S.Y.; Hipps, L.; Gillies, R.R.; Yoon, J.H. Probable causes of the abnormal ridge accompanying the 2013–2014 California drought: ENSO precursor and anthropogenic warming footprint. *Geophys. Res. Lett.* **2014**, *41*, 3220–3226. [CrossRef]
19. Seager, R.; Hoerling, M.; Schubert, S.; Wang, H.; Lyon, B.; Kumar, A.; Nakamura, J.; Henderson, N. Causes of the 2011–14 California drought. *J. Clim.* **2015**, *28*, 6997–7024. [CrossRef]
20. Diffenbaugh, N.S.; Swain, D.L.; Touma, D. Anthropogenic warming has increased drought risk in California. *Proc. Natl. Acad. Sci. USA* **2015**, *112*, 3931–3936. [CrossRef] [PubMed]
21. Mann, M.E.; Gleick, P.H. Climate change and California drought in the 21st century. *Proc. Natl. Acad. Sci. USA* **2015**, *112*, 3858–3859. [CrossRef] [PubMed]
22. AghaKouchak, A.; Cheng, L.; Mazdiyasni, O.; Farahmand, A. Global warming and changes in risk of concurrent climate extremes: Insights from the 2014 California drought. *Geophys. Res. Lett.* **2014**, *41*, 8847–8852. [CrossRef]

23. Shukla, S.; Safeeq, M.; AghaKouchak, A.; Guan, K.; Funk, C. Temperature impacts on the water year 2014 drought in California. *Geophys. Res. Lett.* **2015**, *42*, 4384–4393. [CrossRef]

24. Funk, C.; Hoell, A.; Stone, D. Examining the contribution of the observed global warming trend to the California droughts of 2012/13 and 2013/14. *Bull. Am. Meteorol. Soc.* **2014**, *95*, 11–15.

25. Mao, Y.; Nijssen, B.; Lettenmaier, D.P. Is climate change implicated in the 2013–2014 California drought? A hydrologic perspective. *Geophys. Res. Lett.* **2015**, *42*, 2805–2813. [CrossRef]

26. Williams, A.P.; Seager, R.; Abatzoglou, J.T.; Cook, B.I.; Smerdon, J.E.; Cook, E.R. Contribution of anthropogenic warming to California drought during 2012–2014. *Geophys. Res. Lett.* **2015**, *42*, 6819–6828. [CrossRef]

27. Richman, M.B.; Leslie, L.M. Uniqueness and causes of the California drought. *Procedia Comput. Sci.* **2015**, *61*, 428–435. [CrossRef]

28. Griffin, D.; Anchukaitis, K.J. How unusual is the 2012–2014 California drought? *Geophys. Res. Lett.* **2014**, *41*, 9017–9023. [CrossRef]

29. Palmer, W.C. *Meteorological Drought*; US Department of Commerce, Weather Bureau: Washington, DC, USA, 1965.

30. Robeson, S.M. Revisiting the recent California drought as an extreme value. *Geophys. Res. Lett.* **2015**, *42*, 6771–6779. [CrossRef]

31. He, M.; Gautam, M. Variability and trends in precipitation, temperature and drought indices in the state of California. *Hydrology* **2016**, *3*, 14. [CrossRef]

32. California Department of Water Resources (CDWR). *Bulletin 132–14: Management of the California State Water Project*; Sacramento, CA, USA, 2015.

33. Cayan, D.R. Interannual climate variability and snowpack in the Western United States. *J. Clim.* **1996**, *9*, 928–948. [CrossRef]

34. Serreze, M.C.; Clark, M.P.; Armstrong, R.L.; McGinnis, D.A.; Pulwarty, R.S. Characteristics of the Western United States snowpack from snowpack telemetry(SNOTEL) data. *Water Resour. Res.* **1999**, *35*, 2145–2160. [CrossRef]

35. Meko, D.M.; Woodhouse, C.A.; Touchan, R. Klamath/San Joaquin/Sacramento hydroclimatic reconstructions from tree rings. In *Draft Final Report to the California Department of Water Resources*; University of Arizona: Tucson, AZ, USA, 2014; p. 117.

36. Keyantash, J.; Dracup, J.A. The quantification of drought: An evaluation of drought indices. *Bull. Am. Meteorol. Soc.* **2002**, *83*, 1167–1180.

37. Heim, R.R. A review of twentieth-century drought indices used in the United States. *Bull. Am. Meteorol. Soc.* **2002**, *83*, 1149–1165.

38. Mishra, A.K.; Singh, V.P. Drought modeling—A review. *J. Hydrol.* **2011**, *403*, 157–175. [CrossRef]

39. McKee, T.B.; Doesken, N.J.; Kleist, J. The relationship of drought frequency and duration to time scales. In Proceedings of the 8th Conference on Applied Climatology, Anaheim, CA, USA, 17–22 January 1993.

40. Vicente-Serrano, S.M.; Begueria, S.; Lopez-Moreno, J.I. A multiscalar drought index sensitive to global warming: The standardized precipitation evapotranspiration index. *J. Clim.* **2010**, *23*, 1696–1718. [CrossRef]

41. Shukla, S.; Wood, A.W. Use of a standardized runoff index for characterizing hydrologic drought. *Geophys. Res. Lett.* **2008**. [CrossRef]

42. Steinemann, A.; Iacobellis, S.F.; Cayan, D.R. Developing and evaluating drought indicators for decision-making. *J. Hydrometeorol.* **2015**, *16*, 1793–1803. [CrossRef]

43. Steinemann, A. Drought information for improving preparedness in the Western States. *Bull. Am. Meteorol. Soc.* **2014**, *95*, 843–847. [CrossRef]

44. Svoboda, M. An introduction to the drought monitor. *Drought Netw. News* **2000**, *12*, 15–20.

45. Cayan, D.R.; Maurer, E.P.; Dettinger, M.D.; Tyree, M.; Hayhoe, K. Climate change scenarios for the California region. *Clim. Chang.* **2008**, *87*, 21–42. [CrossRef]

46. Dettinger, M.D. Projections and downscaling of 21st century temperatures, precipitation, radiative fluxes and winds for the southwestern US, with focus on Lake Tahoe. *Clim. Chang.* **2013**, *116*, 17–33. [CrossRef]

47. Scherer, M.; Diffenbaugh, N.S. Transient twenty-first century changes in daily-scale temperature extremes in the United States. *Clim. Dyn.* **2014**, *42*, 1383–1404. [CrossRef]

48. Berg, N.; Hall, A. Increased interannual precipitation extremes over California under climate change. *J. Clim.* **2015**, *28*, 6324–6334. [CrossRef]

49. Das, T.; Dettinger, M.D.; Cayan, D.R.; Hidalgo, H.G. Potential increase in floods in California's Sierra Nevada under future climate projections. *Clim. Chang.* **2011**, *109*, 71–94. [CrossRef]

50. Yoon, J.-H.; Wang, S.S.; Gillies, R.R.; Kravitz, B.; Hipps, L.; Rasch, P.J. Increasing water cycle extremes in California and in relation to ENSO cycle under global warming. *Nat. Commun.* **2015**. [CrossRef] [PubMed]

51. Wang, J.; Zhang, X. Downscaling and projection of winter extreme daily precipitation over North America. *J. Clim.* **2008**, *21*, 923–937. [CrossRef]

52. USCB. Interim Projections 2000–2030 Based on Census 2000. Available online: http://www.census.gov/population/projections/data/state/ (accessed on 5 April 2016).

53. Robinson, J.D.; Vahedifard, F. Weakening mechanisms imposed on California's levees under multiyear extreme drought. *Clim. Chang.* **2016**, *137*, 1–14. [CrossRef]

Permissions

All chapters in this book were first published in CLIMATE, by MDPI; hereby published with permission under the Creative Commons Attribution License or equivalent. Every chapter published in this book has been scrutinized by our experts. Their significance has been extensively debated. The topics covered herein carry significant findings which will fuel the growth of the discipline. They may even be implemented as practical applications or may be referred to as a beginning point for another development.

The contributors of this book come from diverse backgrounds, making this book a truly international effort. This book will bring forth new frontiers with its revolutionizing research information and detailed analysis of the nascent developments around the world.

We would like to thank all the contributing authors for lending their expertise to make the book truly unique. They have played a crucial role in the development of this book. Without their invaluable contributions this book wouldn't have been possible. They have made vital efforts to compile up to date information on the varied aspects of this subject to make this book a valuable addition to the collection of many professionals and students.

This book was conceptualized with the vision of imparting up-to-date information and advanced data in this field. To ensure the same, a matchless editorial board was set up. Every individual on the board went through rigorous rounds of assessment to prove their worth. After which they invested a large part of their time researching and compiling the most relevant data for our readers.

The editorial board has been involved in producing this book since its inception. They have spent rigorous hours researching and exploring the diverse topics which have resulted in the successful publishing of this book. They have passed on their knowledge of decades through this book. To expedite this challenging task, the publisher supported the team at every step. A small team of assistant editors was also appointed to further simplify the editing procedure and attain best results for the readers.

Apart from the editorial board, the designing team has also invested a significant amount of their time in understanding the subject and creating the most relevant covers. They scrutinized every image to scout for the most suitable representation of the subject and create an appropriate cover for the book.

The publishing team has been an ardent support to the editorial, designing and production team. Their endless efforts to recruit the best for this project, has resulted in the accomplishment of this book. They are a veteran in the field of academics and their pool of knowledge is as vast as their experience in printing. Their expertise and guidance has proved useful at every step. Their uncompromising quality standards have made this book an exceptional effort. Their encouragement from time to time has been an inspiration for everyone.

The publisher and the editorial board hope that this book will prove to be a valuable piece of knowledge for researchers, students, practitioners and scholars across the globe.

List of Contributors

Hamza Ouatiki and Abdelghani Boudhar
Faculté des Sciences et Techniques, Université Sultan Moulay Slimane, B.P. 523, Béni-Mellal 23030, Maroc

Yves Tramblay
IRD-HydroSciences, Montpellier 34090, France

Lionel Jarlan and Abdelghani Chehbouni
CESBIO (Université de Toulouse, CNRS, CNES, IRD), 18 Av. Edouard Belin BPI 280, Toulouse 31401 CEDEX 9, France; lionel

Tarik Benabdelouhab
Institut National de la Recherche Agronomique, B.P 415 R.P, Rabat 10000, Maroc

Lahoucine Hanich
Faculté des Sciences et Techniques, Université Cadi ayyad, B.P 549, Marrakech 40000, Maroc

M. Rachid El Meslouhi
Agence du bassin Hydraulique d'Oum Er Rabia, B.P 511, Béni Mellal 23000, Maroc

Shonam Sharma and Prasoon Kumar Singh
Department of Environmental Science and Engineering, Indian Institute of Technology (Indian School of Mines) Dhanbad, Jharkhand Pin-826004, India

Mélanie Trudel, Pierre-Louis Doucet-Généreux and Robert Leconte
Université de Sherbrooke, Department of Civil Engineering, 2500, boul de l'Université, Sherbrooke, QC J1K 2R1, Canada; pierre-louis

Udo Schickhoff and Jürgen Böhner
Center for Earth System Research and Sustainability, Institute of Geography, University of Hamburg, Bundesstraße 55, 20146 Hamburg, Germany

Ramchandra Karki
Center for Earth System Research and Sustainability, Institute of Geography, University of Hamburg, Bundesstraße 55, 20146 Hamburg, Germany
Department of Hydrology and Meteorology, Government of Nepal, 406 Naxal, Kathmandu, Nepal

Shabeh ul Hasson
Center for Earth System Research and Sustainability, Institute of Geography, University of Hamburg, Bundesstraße 55, 20146 Hamburg, Germany
Department of Space Sciences, Institute of Space Technology, Islamabad 44000, Pakistan

Thomas Scholten
Soil Science and Geomorphology, University of Tübingen, Department of Geosciences, Rümelinstrasse 19-23, 72070 Tübingen, Germany

Bernd Diekkrüger
Department of Geography, University of Bonn, 53115 Bonn, Germany

Mulugeta Dadi Belete
Department of Geography, University of Bonn, 53115 Bonn, Germany Institute for Technology and Water Resources Management in the Tropics and Subtropics, Cologne University of Applied Sciences, 50679 Köln (Deutz), Germany; Jackson.roehrig@fh-koeln.de
Institute of Technology, School ofWater Resources Engineering, Hawassa University, Hawassa P.O. Box 005, Ethiopia

Jackson Roehrig
Institute for Technology and Water Resources Management in the Tropics and Subtropics, Cologne University of Applied Sciences, 50679 Köln (Deutz), Germany

Pham Quy Giang and Le Thi Giang
Faculty of Land Management, Vietnam National University of Agriculture, Trau Quy, Gia Lam, Hanoi, Vietnam

Kosuke Toshiki
Faculty of Regional Innovation, University of Miyazaki. 1-1, Gakuenkibanadainishi, Miyazaki 8892192, Japan

Boris Bonn and Jürgen Kreuzwieser
Chair of Tree Physiology, Albert Ludwig University, Georges-Koehler-Allee 053, D-79110 Freiburg i.Br., Germany

Felicitas Sander and Rasoul Yousefpour
Chair of Forestry Economics and Forest Planning, Albert Ludwig University, Tennenbacher Str. 4, D-79106 Freiburg i. Br., Germany

Tommaso Baggio
Department of Land, Environment, Agriculture and Forestry, University of Padova, Agripolis, Viale dell'Università 16, I-35020 Legnaro (PD), Italy

Oladeinde Adewale
UMR LERFoB, AgroParisTech, INRA, 54000 Nancy, France

K C Gouda and Himesh Shivappa
CSIR Fourth Paradigm Institute, Wind Tunnel Road, Bengaluru, Karnataka 560037, India

Sanjeeb Kumar Sahoo and Payoshni Samantray
CSIR Fourth Paradigm Institute, Wind Tunnel Road, Bengaluru, Karnataka 560037, India
Visvesvaraya Technological University, Belagavi, Karnataka 590018, India

Ernest O. Asare and Leonard K. Amekudzi
Department of Physics, Kwame Nkrumah University of Science and Technology, Kumasi 00233, Ghana

Maikon Passos A. Alves, Rafael Brito Silveira and Alberto Elvino Franke
Applied Climatology Laboratory (LabClima), Federal University of Santa Catarina (UFSC), Trindade, Florianópolis 88040-900, Brazil

Rosandro Boligon Minuzzi
Agricultural Climatology Laboratory (Labclimagri), Federal University of Santa Catarina (UFSC), Admar Gonzaga Str., Itacorubi, Florianópolis 88034-000, Brazil

Gaylan Rasul Faqe Ibrahim
Geography Department, Faculty of Arts, Soran University, Soran 44008, Iraq
Tourism Department, Rawandz Private Technical Institute, Soran 44008, Iraq

Minxue He, Mitchel Russo and Michael Anderson
Division of Flood Management, California Department of Water Resources, 3310 El Camino Avenue, Sacramento, CA 95821, USA

Index